T0251251

NEXT GENERATION COMPUTING AND INFORMATION SYSTEMS

The proceedings of the 2nd International Conference on Next-Generation Computing and Information Systems (ICNGCIS'23) includes research papers in diverse emerging domains such as AI, blockchain, web 3.0, metaverse and edge computing apart from traditional domains such as distributed computing and networks and cybersecurity.

The proceedings include papers addressing currently relevant research issues such as smart contract security, interoperability in the metaverse, AI applications in healthcare, agriculture and related domains. The proceedings encompass findings with real-world implications for the growth and evolution of modern computing and information systems by addressing various challenges related to their design, deployment, operational aspects, performance and shortcomings.

The intended audience for the proceedings of ICNGCIS'23 include researchers from industry and academia, practitioners, students, technology enthusiasts and even general audience looking to explore interesting applications, use-cases and fundamental issues in modern computing and information systems.

Prof. Ankur Gupta is the Director at the Model Institute of Engineering and Technology, Jammu (India), besides being a Professor at the Dept. of Computer Science and Engineering. He has 25+ years of experience spanning industry and academia. Prior to joining MIET, he worked as a Team Leader at Hewlett-Packard, India at Bengaluru. He has 85 published research papers in international journals/conferences. Prof. Gupta holds B.E (Hons.) CS and MS degrees from BITS, Pilani and PhD from NIT, Hamirpur. He has 11 patents granted and 34 patents filed at the Indian Patents Office. He is the inventor of the Performance Insight 360, quality analytics framework for higher education which has received several accolades. He has received the DSFT-FIST Grant in 2012, first in the private sector in J&K and received competitive grants over Rs. 2.5 Crore from various funding agencies. He is a recipient of the AICTE Career Award, faculty awards from IBM, EMC and Rs. 2 crores in funding from Govt. agencies. He is also Senior member IEEE and ACM and founder of the International Journal of Next-Generation Computing. His research interests are in cloud computing, P2P networks, network management, artificial intelligence and metaverse.

PROCEEDINGS OF THE 2ND INTERNATIONAL CONFERENCE ON NEXT GENERATION COMPUTING AND INFORMATION SYSTEMS (ICNGCIS 2023), DECEMBER 18–19, 2023, JAMMU, J&K, INDIA

Next Generation Computing and Information Systems

Edited by

Ankur Gupta
Model Institute of Engineering and Technology, Jammu, India

CRC Press
Taylor & Francis Group
Boca Raton London New York Leiden

CRC Press is an imprint of the
Taylor & Francis Group, an **informa** business

A BALKEMA BOOK

First published 2025
by CRC Press/Balkema
4 Park Square, Milton Park, Abingdon, Oxon, OX14 4RN

and by CRC Press/Balkema
2385 NW Executive Center Drive, Suite 320, Boca Raton FL 33431

CRC Press/Balkema is an imprint of the Taylor & Francis Group, an informa business

© 2025 selection and editorial matter, Ankur Gupta; individual chapters, the contributors

The right of Ankur Gupta to be identified as the author of the editorial material, and of the authors for their individual chapters, has been asserted in accordance with sections 77 and 78 of the Copyright, Designs and Patents Act 1988.

All rights reserved. No part of this book may be reprinted or reproduced or utilised in any form or by any electronic, mechanical, or other means, now known or hereafter invented, including photocopying and recording, or in any information storage or retrieval system, without permission in writing from the publishers.

Although all care is taken to ensure integrity and the quality of this publication and the information herein, no responsibility is assumed by the publishers nor the author for any damage to the property or persons as a result of operation or use of this publication and/or the information contained herein.

British Library Cataloguing-in-Publication Data
A catalogue record for this book is available from the British Library

Library of Congress Cataloging-in-Publication Data
A catalog record has been requested for this book

ISBN: 978-1-032-73865-9 (hbk)
ISBN: 978-1-032-73866-6 (pbk)
ISBN: 978-1-003-46638-3 (ebk)

DOI: 10.1201/9781003466383

Typeset in Times New Roman
by MPS Limited, Chennai, India

Table of Contents

Preface

It gives us immense pleasure to bring to you the proceedings of the 2nd International Conference on Next Generation Computing and Information Systems (ICNGCIS 2023). The second edition of the conference is taking place at MIET, Jammu after a gap of six years. The first edition of ICNGCIS was a milestone as it was for the first time that a computing focussed conference was organized at Jammu. ICNGCIS 2023 is bigger, better and more comprehensive than before.

Papers were invited in different themes such as Artificial Intelligence, Machine Learning, Deep Learning, Internet of Things, Distributed Computing, Metaverse, Multimedia Processing, Networks, Information Security, etc. This year the conference proceedings are being published by CRC Press (Taylor and Francis) and will be submitted for potential indexing in Scopus. We received over 117 papers for the conference from diverse geographies and after a rigorous review process selected 38 papers for final inclusion in the conference proceedings with an acceptance rate of 32.4%. With a focus on advances in computing and information systems, we intended for the conference to be a small one affording the participants deeper and more relaxed engagement with their peers and experts.

The first keynote address was delivered by Prof. Sartaj Sahni, Distinguished Professor of Computer and Information Sciences and Engineering at the University of Florida, USA. The session stressed on the significance of the development of Fast Algorithms. Following this, second keynote address was delivered by Prof. Sitharama Iyengar. Prof. Iyengar serving currently as the Distinguished University Professor, Founding Director of the Discovery Lab and Director of the US Army-funded Center of Excellence in Digital Forensics at Florida International University, Miami, USA. The session emphasized on the rising era of Digital Forensics.

The third keynote address was delivered by Prof. Chakravarthy, Distinguished Professor of Computer Science and Engineering Department at The University of Texas at Arlington, USA. Prof. Chakravarthy focused on the importance of development of Multi-Layer Networks for modelling complex datasets. The fourth keynote address was delivered by Prof. Arif Sarwat, Eminent Scholar Chaired Professor in the Department of Electrical and Computer Engineering and the Director of the Solar Research Facility and Energy Power Sustainability group at Florida International University, Miami, USA. The session focused on the Impact of Artificial Intelligence on High-Penetration Renewable Physical Infrastructure.

The fifth keynote address was delivered by Prof. Bharat Bhargava, Professor of Computer Science at Purdue University, USA and Fellow of IEEE. The session focused on the systematic understanding of the threats imposed by collaborative attacks and formulate intelligent and effective defenses. The last keynote address was delivered by Prof. Ankur Gupta, Director, Model Institute of Engineering and Technology, Jammu. The session focused on building a safe and secure Metaverse.

The conference served as a platform for the exchange of cutting-edge research and ideas in the field of next-generation computing and information systems. Attendees had the opportunity to engage with and learn from the diverse perspectives shared by presenters, fostering collaboration and knowledge dissemination among scholars and professionals. The active involvement of expert chairs further enriched the discussions, reflecting the conference's commitment to advancing the discourse on emerging technologies and their implications on computing and information systems.

We are thankful to the authors, technical program committee members and Coordinators of this conference for their tremendous support in making this conference a successful event. We believe that future editions of the conference will continue to encourage research in the region and connect Jammu and Kashmir with the rest of the country and the world more effectively. The general chairs of the conference are deeply committed to this cause.

General Chairs
ICNGCIS 2023

Prof. Sartaj Sahni
Prof. Bharat Bhargava
Prof. Vijay Kumar
Prof. Ankur Gupta
Prof. Deepak Garg

Keynote address

Next Generation Computing and Information Systems – Gupta. (Ed.)
© 2025 The Author(s), ISBN 978-1-032-73865-9

Fast reinforcement learning

Sartaj Sahni
Professor, University of Florida

ABSTRACT: Reinforcement Learning (RL) is a highly effective approach for teaching intelligent agents to make decisions in challenging and ever-changing contexts. With the increasing use of RL in real-world situations, there is a rising demand for rapid reinforcement learning to improve effectiveness, decrease training duration, and facilitate immediate decision-making across many fields. Fast Reinforcement Learning can be achieved through various methods, one of which is Parallelization. This involves splitting RL computations across numerous processors or GPUs to take advantage of parallel processing capabilities, resulting in a large reduction in training time. Another method involves utilizing specialized hardware such as GPUs and TPUs that can enhance the speed of RL training by executing parallel computations more effectively than conventional CPUs. Developing and executing RL algorithms that are designed to be more efficient and have lower computational complexity leads to quicker learning. Cache memory is also essential in accelerating reinforcement learning by minimizing the access time for frequently utilized data. It enhances the performance of RL algorithms by storing and retrieving crucial information, such as state-action pairs and Q-values. Cache memory usage optimizes data access patterns, resulting in reduced latency and improved speed of learning rounds. The landscape of fast reinforcement learning will be shaped by overcoming challenges such as transfer learning and meta-learning, algorithmic complexity, scalability to handle data sets of complex environments, generalization, etc. will continue to shape the landscape of fast reinforcement learning.

BIO: Prof. Sartaj Sahni is a Distinguished Professor of Computer and Information Sciences and Engineering at the University of Florida. He is also a member of the European Academy of Sciences, a Fellow of IEEE, ACM, AAAS, and Minnesota Supercomputer Institute, and a Distinguished Alumnus of the Indian Institute of Technology, Kanpur. In 1997, he was awarded the IEEE Computer Society Taylor L. Booth Education Award for contributions to Computer Science and Engineering education in the areas of data structures, algorithms, and parallel algorithms", and in 2003, he was awarded the IEEE Computer Society W. Wallace McDowell Award for contributions to the theory of NP-hard and NP-complete problems. Prof. Sahni was awarded the 2003 ACM Karl Karlstrom Outstanding Educator Award for outstanding contributions to computing education. In 2016, Prof. Sahni was awarded the IEEE Technical Committee on Scalabale Computing (TCSC) Award for Excellence in Scalable Computing. Prof. Sahni has published over 420 research papers and written 15 texts. He has 15 US patents. His research publications and patents are on the design and analysis of efficient algorithms, parallel computing, interconnection networks, design automation, and medical algorithms. He is past Editor-in-Chief of ACM Computing Surveys and past co-Editor-in-Chief of the Journal of Parallel and Distributed Computing.

DOI: 10.1201/9781003466383-1

Next Generation Computing and Information Systems – Gupta. (Ed.)
© 2025 The Author(s), ISBN 978-1-032-73865-9

Global cooperative digital forensics in the era of artificial intelligence

S. Sitharama Iyengar
Professor, Florida International University, Miami

ABSTRACT: As Artificial Intelligence (AI) continues to reshape industries and societies, a critical examination of the ethical dimensions surrounding its development becomes imperative. This talk delves into the multifaceted landscape of ethical considerations within AI, aiming to guide developers, researchers, and industry stakeholders towards global responsible innovation. The first part of the discussion explores key ethical challenges, including bias in algorithms, data privacy concerns, and the societal impact of AI technologies. Drawing on real-world examples and case studies, the talk illuminates the complexities of navigating these ethical quandaries and underscores the importance of a holistic, inclusive approach to AI development. In the second part, the talk proposes practical strategies and frameworks for integrating ethical considerations into the AI lifecycle for Global development. Emphasizing the significance of interdisciplinary collaboration, transparency, and user involvement, the discussion provides actionable insights for cultivating a culture of responsibility in the AI community. Attendees will gain valuable perspectives on fostering ethical awareness, aligning AI initiatives with societal values, and contributing to the evolution of a more ethically sound and sustainable AI landscape.

BIO: Prof. Iyengar is currently the Distinguished University Professor, Founding Director of the Discovery Lab and Director of the US Army-funded Center of Excellence in Digital Forensics at Florida International University, Miami. He is also the Distinguished Chaired Professor at National Forensics Sciences University, Gandhinagar, India. Prof. Iyengar is a Member of the European Academy of Sciences, a Life Fellow of the Institute of Electrical and Electronics Engineers (IEEE), a Fellow of the Association of Computing Machinery (ACM), a Fellow of the American Association for the Advancement of Science (AAAS), a Fellow of the Society for Design and Process Science (SDPS), and a Fellow of the American Institute for Medical and Biological Engineering (AIMBE). He was recently bestowed with the 2023 Karnataka Rajyotsava Award (Karnataka State's 2nd Highest Civilian Award) on 1st November 2023. In addition to this he has been awarded the Lifetime Achievement Award for his contribution to the field of Digital Forensics and IEEE Cybermatics Congress "Test of Time Award" for his work on creating Brooks-Iyengar Algorithm and its impact in advancing modern computing. Prof. Iyengar in collaboration with HBCUs were awarded a $2.25 M funding for setting up a Digital Forensics Center of Excellence over a period of 5 years (2021-2026). He has published more than 600 research papers, has authored/co-authored and edited 32 books. His vision, expertise, and dedication have made him an asset in the academic community at large.

Next Generation Computing and Information Systems – Gupta. (Ed.)
© 2025 The Author(s), ISBN 978-1-032-73865-9

Why multilayer networks are needed for complex data analysis

Upendranatha Sharma Chakravarthy
Professor, The University of Texas at Arlington

ABSTRACT: We are on the cusp of holistically analyzing a variety of data being collected in every walk of life in diverse ways. For this, current analytics and science are being extended (Big Data Analytics/Science) along with new approaches to benefit humanity at large in the best possible way. These warrants developing and/or using new approaches, technological, scientific, and systems, in addition to building upon and integrating with the ones that have been developed so far. In the first part of the presentation, we introduce multilayer networks (or MLNs) and illustrate their elegance for modeling complex data sets from well-known applications (IMDb and DBLP). Flexibility of analysis comes from this approach to modeling. We will show a dashboard that we have developed for MLN analysis. In second part of this talk, we discuss situation analysis on videos using a novel approach. We extract contents and apply stream processing techniques. This approach has the potential to process videos in real-time and raise alerts for situations, such as Nancy Pelosi's house break-in and Las Vegas shooting.

BIO: Prof. Chakravarthy is Professor of Computer Science and Engineering Department at The University of Texas at Arlington. He is an ACM Distinguished Scientist and an ACM Distinguished Speaker. He is also an IEEE Senior Member and was instrumental in establishing the Information Technology Laboratory at University of Texas at Arlington. Prof. Chakravarthy has also established the NSF funded, Distributed and Parallel Computing Cluster at UT Arlington. His current research includes big data analysis using multi-layered networks, stream data processing for disparate domains (e.g., video analysis), scaling graph mining algorithms for analyzing very large social and other networks, active and real-time databases, distributed and heterogeneous databases, query optimization (single, multiple, logic-based, and graph), and multi-media databases. He has published over 200 papers/book chapters in refereed international journals and conference proceedings. He has supervised 15 PhD theses and 100+ MS thesis. He has published over 200 papers and book chapters in refereed international journals and conference proceedings. He is listed in Who's Who Among South Asian Americans and Who's Who Among America's Teachers. Prior to joining UTA, he was with the University of Florida, Gainesville for 10 years. Prior to that, he worked as a Computer Scientist at the Computer Corporation of America for 3 years and as a Member, Technical Staff at Xerox Advanced Information Technology, Cambridge for a year. Prof Chakravarthy received his B.E. degree in Electrical Engineering from the Indian Institute of Science Bangalore and M.Tech from IIT Bombay, India. He worked at the Tata Institute of Fundamental Research Bombay, for a few years and received his M.S. and Ph.D degrees from the University of Maryland in College park.

DOI: 10.1201/9781003466383-3

Next Generation Computing and Information Systems – Gupta. (Ed.)
© 2025 The Author(s), ISBN 978-1-032-73865-9

Impact of artificial intelligence on high-penetration renewable physical infrastructure

Arif Sarwat

Professor, Florida International University, Miami

ABSTRACT: The efficient incorporation of substantial quantities of renewable energy into the power grid is crucial for ensuring a sustainable future and promoting environmentally-friendly smart cities. The current methods for integrating solar power rely on imprecise weather forecasting and inadequate load distribution, resulting in an overall grid performance that is both unreliable and inefficient. Therefore, it is necessary to adopt a more comprehensive perspective to address these issues. Artificial Intelligence (AI) is one such technology that is applied in high-penetration renewable physical infrastructure. AI-powered smart grids provide the real-time monitoring and control of renewable energy sources, such as solar and wind, to ensure the most efficient use of energy. Smart algorithms optimize battery lifespan, enhancing efficiency and lowering maintenance expenses. AI-powered predictive maintenance models are utilized to identify possible problems in physical infrastructure components prior to system failures. Smart buildings utilize AI to control heating, ventilation, and air conditioning (HVAC) systems by analyzing occupancy trends, weather forecasts, and user preferences. Although AI has mostly good effects on high-penetration renewable physical infrastructure, it also presents problems like as data security concerns, ethical considerations, and the requirement for experienced people to handle and advance these intricate systems. Tackling these obstacles is essential to fully harness the capabilities of AI in enhancing the implementation and effectiveness of widespread renewable energy solutions.

BIO: Prof. Arif Sarwat is an Eminent Scholar Chaired Professor in the Department of Electrical and Computer Engineering, the Director of the FPL-FIU Solar Research Facility and Energy Power Sustainability (EPS) group at FIU. He is also the Principal Investigator for the state-of-the-art grid-connected 3MW/9MWh AI-based Renewable (AIR) Microgrid long-term project funded by FPL. His research interests include smart grids, electric vehicles, high penetration renewable systems, battery management systems, grid resiliency, large-scale data analysis, artificial intelligence, smart city infrastructure, and cybersecurity. He has published more than 200 peer-reviewed articles and multiple patents. He currently has multiple funded projects; funded by the National Science Foundation (NSF), industry, and the Department of Energy (DOE). This list includes the NSF CAREER Award. Previously, Prof. Sarwat worked at Siemens for more than nine years, winning three recognition awards. Prof. Sarwat received the Faculty Award for Excellence in Research & Creative Activities in 2016, College of Engineering & Computing Worlds Ahead Performance in 2016, and FIU TOP Scholar Award in 2015 and 2019. He has been the chair of the IEEE Miami Section VT and Communication since 2012. He is an associate editor of the journals ACM Computing Surveys and IEEE IAS.

DOI: 10.1201/9781003466383-4

Next Generation Computing and Information Systems – Gupta. (Ed.)
© 2025 The Author(s), ISBN 978-1-032-73865-9

Collaborative attacks and defense

Bharat Bhargava
Professor, Purdue University

ABSTRACT: Ad Hoc Networks, especially Flying Ad Hoc Networks (FANETs), are becoming increasingly crucial in a variety of civil and military applications, including wildfire monitoring and suppression, search and rescue operations in hazardous scenarios, and transport of supplies or personnel to remote locations. Going beyond protecting networks from single forms of attacks, the ongoing research work addresses collaborative attacks (CA) within these networks. A collaborative attack occurs when attackers synchronize their malicious actions against a target network. The collaboration can occur simultaneously, where multiple attackers attempt to compromise the system at the same time or split in time where one attacker gathers the information about the network and subsequently another attacker executes the actual exploit. There is a need to develop a systematic understanding of the threats imposed by collaborative attacks and formulate intelligent and effective defenses. In future, Machine Learning algorithms can be used to analyze large amounts of data to detect advanced threats and reduce false positives/negatives.

BIO: Bharat Bhargava is a Professor of Computer Science at Purdue University and Fellow of IEEE and IETE. He led a Northrup Grumman sponsored consortium on Real Applications of Machine Learning (REALM) with MIT, CMU, and Stanford. He contributed to the Department of Defense on The Science of Artificial Intelligence and Learning for Open-world Novelty (SAIL-ON) project. He works with Sandia Corporation to maintain mission capabilities of the US Space Enterprise, Jet Propulsion Lab to predict attacks on space systems and Ford Corporation on software defined networking for V2V communication. Prof. Bhargava is a distinguished alumni of Indian Institute of Science and DAV college. He is also a recipient of eight best paper awards in various international computer science conferences. He has received the IEEE Technical Achievement Award and has also been awarded the charter Gold Core Member distinction by the IEEE Computer Society for his distinguished service. He is major thesis advisor of the very first African American woman to receive her Ph. D in the history of Computer Science department at Purdue in May 2019. He has served on the IEEE Computer Society on Technical Achievement Award and Fellow committees. Prof. Bhargava is the founder of the IEEE Symposium on Reliable and Distributed Systems, IEEE conference on Digital Library, and the ACM Conference on Information and Knowledge Management.

DOI: 10.1201/9781003466383-5

Next Generation Computing and Information Systems – Gupta. (Ed.)
© 2025 The Author(s), ISBN 978-1-032-73865-9

Building a safe and secure metaverse

Ankur Gupta

Professor, Model Institute of Engineering and Technology, Jammu

ABSTRACT: The metaverse is an exciting domain representing an amalgamation of advances in computing and communications along with immersive technologies (AR/VR/ XR). It is therefore intuitive for large corporations, research organizations and businesses cutting across industry verticals to adopt a metaverse-centric approach and explore diverse use-cases catering to business/individual needs. However, the transition from traditional applications to metaverse-native applications is not expected to be seamless. Privacy, safety and security concerns exist in the early versions of the metaverse with several flaws endangering individual user safety being highlighted. Plus, there is increased scrutiny on the ill-effects of social media, gaming and immersive environments by Governments and regulatory agencies. Thus, it is increasingly clear that the final iteration of the metaverse will need to address the concerns of individual users while navigating a complex legal and regulatory landscape. This keynote examines the different aspects and challenges which the future metaverse will need to address. A set of "first principles" are put forth towards building an inclusive, safe and secure metaverse.

BIO: Prof. Ankur Gupta is the Director at the Model Institute of Engineering and Technology (MIET), Jammu, India, besides being a Senior Professor at the Department of Computer Science and Engineering. He has 25 years of experience spanning industry and academia. Prior to joining MIET, he worked as a Team Leader at Hewlett-Packard, India Software Operations, Bangalore. He has 5 patents granted, 35 patents pending at the Indian Patents Office and over 90 published research papers in international journals/conferences. Prof. Gupta holds B.E (Hons.) CS and MS degrees from BITS, Pilani and PhD from NIT, Hamirpur. He is the inventor of the Performance Insight 360, quality analytics framework for higher education which has received several accolades. His main areas of interest include peer-to-peer networks, network management, software engineering, cloud computing and higher technical education. He is a recipient of the AICTE Career Award, faculty awards from IBM and EMC. He has received the DSFT-FIST Grant in 2012, first in the private sector in J&K and received competitive grants over Rs. 2.5 Crore from various funding agencies. He is the founder of the International Journal of Next-Generation Computing, Senior member IEEE, ACM and a life-member of the Computer Society of India.

DOI: 10.1201/9781003466383-6

Regular papers

Next Generation Computing and Information Systems – Gupta. (Ed.)
© 2025 The Author(s), ISBN 978-1-032-73865-9

Distributed-computing based versatile healthcare services framework for diagnostic markers

Venkata Chunduri
Senior Software Developer, Department of Mathematics & Computer Science, Indiana State University, USA

Bhargavi Posinasetty
Masters in Public Health, The University of Southern Mississippi, Hattiesburg, MS, US

Mukesh Soni
Dr. D. Y. Patil Vidyapeeth, Pune, Dr. D. Y. Patil School of Science & Technology, Tathawade, Pune, India

Haewon Byeon
Department of Digital Anti-Aging Healthcare, Inje University, Gimhae, Republic of Korea

ABSTRACT: Recently, there has been a boom in imaginative medical services technologies that have enhanced the delivery of medical care data. Human-to-human and gadget-to-gadget networks are crucial in people's lives and function 24 hours a day, seven days a week. Mobile-Cloud-Computing becomes a crucial tool in this regard. Given the tremendous improvements in the Internet of Things, such a combination has become a major worry. This study developed a framework that included two essential application components: flexible usage and a server request. The server application records patient data while a Counterfeit Neural Network (CNN) module distinguishes between the two stroke subtypes. Similarly, as a framework for stroke patients, our model ensures accessibility, security, and adaptability by utilising the Stroke dataset for CNN computation and the Multilayer Perceptron Algorithm (MLP), which has been completed without precedent for working with large amounts of data in this extension.

1 INTRODUCTION

In recent years, data innovation has been increasingly employed in the medical profession. Similarly, since it is received from a range of sources in a short amount of time, information has risen in bulk and complexity. As a result, this becomes a duty for dealing with huge information difficulties, given that this sort of information is stored in many configurations and is created quickly (Campbell 2009). One of the most important issues to address in the domain of medical services applications and administrations is giving more helpful assistance and a medical care environment. Accessible and visible elements of Electronic Health Records (EHRs), biological information base, and general well-being have been enhanced. The focus was oriented that developing distributed calculating to deliver registration facilities will be advantageous with regards to adaptability, along with possessing the capability to be one of the prospective open doors required after through desperate instructional foundations. Written a similar assessment that focuses on intelligent analyses of the promise as well as issues with regards to wearables (wearable-technology) with regards to clinical training along with clinical service provision (Campbell 2010). A new convolutional brain network for motion recognition with particular direction messages from humans was recommended. The study showed how brain organizations might be linked to enhanced realism as an emerging area (Yang *et al.* 2018). We want to provide a successful and beneficial Stroke-Diagnosis-Healthcare-System considering stroke convalescents that would offer stroke convalescents to manage their infection prevention plus follow up, ensure so individuals may carry it to themselves throughout their everyday routines (Karan *et al.* 2012).

2 STRATEGIES

2.1 *Medical nuances*

Approx 1250 persons were observed. Clinical stroke subtypes were found in the patients. The informative collection in this independent study covers 1250 ischemic stroke subtypes, including cryptogenic (560) and cardioembolic (690) (Karlik and Olgac 2011). This work looks at those who have been diagnosed with ischemic stroke (Figure 1). Since they are assessed with the use of 'NIHSS' (National-Institutes-of-Health-Stroke-Scale), ischemic strokes on the equator's dominant side result in greater utilitarian impairments than those on the other side of the planet (Kim *et al.* 2009). The variables about the patients are used to determine whether stroke sub-kind (cardioembolic/cryptogenic) the convalescents in a Stroke-Diagnosis-Healthcare-System have a place.

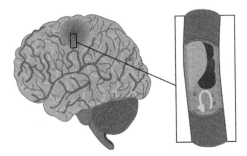

Figure 1. An illustration of an ischemic stroke.

2.2 *Techniques*

The argument depends on concepts related to utility registration. The information is available on the "cloud," where it may be used by stroke patients, master frameworks, and clinical staff in addition. MySQL is used in cloud computing environments together with PHP programming to store data (Li *et al.* 2013). As illustrated in the Table 1 Age breakdown of stroke sufferers. MySQL is a trustworthy database management architecture for server systems that are open source. Because they use the open-source model, Java developments have several significant strengths. As a result, Android is regarded as the most well-known and adaptable operating system in the world (Lian *et al.* 2014).

Table 1. Age breakdown of stroke sufferers.

Stroke/oldness	1–25	25–35	36–45	46–65	66–85	86–100
Cryptogenic	24	14	55	230	200	18
Cardioembolic	20	6	22	196	394	60

Our method is dependent on the following means:

(1) Frameworks for demographic information, clinical history, test results from research facilities, treatment information, and prescription information are to be used.

(2) These details will be utilized to differentiate between different stroke subtypes for patient medical services, and Android-based mobile phones will be used collect the necessary data.

(3) Cloud-based education for stroke patients about their condition and way of life will be transmitted or generated, providing a CNN application programming point of contact (API). For informative collection size, energy size, and emphasis size, the programming interface exactness rate is tested (Tatta *et al.* 2021).

(4) The process for obtaining, managing, and deciding depending upon that patient's information via smartphone having Android-OS has been assured.

2.2.1 *Use of CNN in the setup of the framework*

Clinical data frameworks have expanded to include increasingly diverse clinical and wellness contexts. However, this pattern has emerged with all of its substantial complications and problems, including difficulties in obtaining crucial information for preferred emotionally supporting networks (Sanober *et al.* 2021). Old-fashioned on-automatic information investigation shave turned out to be ineffective, then fresh approaches, likely to be the usage of CNNs aimed at an effective mobile dependent examination, have been important to identify ailments. The back proliferation calculation is shown together with its important steps as follows:

Step. 1 Beginning: Weights are established, and predispositions are modified under small amounts of really arbitrary characteristics.

Step. 2 Demonstration of information as well as preferred yields. An information path has been demonstrated like $y(1), y(2), y(3) \ldots \ldots y(M)$ also alludes to the ideal reaction as $i(1), i(2), i(3) \ldots \ldots i(M)$, each pair in turn, where M is the number of prepared designs.

Step.3 Calculating real outcomes the calculation of the result signals may be done using Equation.1.

$$z_j = \vartheta \left(\sum_{k=1}^{M_{N-1}} x_{kl}^{(N-1)} y_l^{(N-1)} + p_k^{(N-1)} \right), \quad k = 1, \quad \ldots \ldots \quad M_{N-1} \quad (1)$$

Step.4 Predispositions p_k and x_{kl} weights adaptation, for this reason, it is necessary to use the following conditions.

$$\Delta p_k^{(p-1)}(m) = \gamma . \delta_k^{(p-1)}(m) \quad (2)$$

$$\Delta x_{kl}^{(p-1)}(m) = \gamma . y_l(m) . \delta_k^{(p-1)}(m) \quad (3)$$

Here p has been a layer, r is a number of result hubs in the brain organization, N is the yield layer, and is the capacity to initiate, and $y_l(m)$ = the result of hub l at emphasis m. Their net contributions employing a scalar-to-scalar capacity known as the enactment capability may very well be the central component of a brain network topology. The subsequent modules are considered as well as executed for a training technique of this review: for identifying the 2-stroke sub-kinds in the **MLP** prototype, this has shown in Figure 2. For the diagnosis of stroke subtypes, there are 22 sources of information and 2 findings.

2.2.2 *Flexible frameworks for cloud-based medical services*

With regard to numerous perspectives, such as efficiency and programmability, versatile controls the expense of the fundamentals of brilliant and competent mists that are simple to learn. As a result, we have chosen to use the cloud, a free smartphone OS built on Linux that has been created by the Open Handset Alliance and Google. Due to the distributed computing administration's great handling power, information inquiry is very swift (Zhang *et al.* 2021). The Stroke Healthcare System is compatible with any Android phone using an Android-OS 5.0 (Lollipop)else a later version, the iOS operating system, or both. As shown, foundation types from 5G Networks or higher echelon organizations have been used.

3 RESULTS OF EXPLORATION

This section of the evaluation is concerned with the creation and execution cycle of the stroke diagnosis healthcare system then the hypothesis and computation are elaborated. The findings of the characterization of the Stroke Subtypes using CNN are covered as an outcome (Maadeed *et al.* 2019).

3.1 *Hypothesis/computation*

It is impossible for stroke patients to have a meaningful conversation with their physicians when they lack adequate info regarding their ailment as well as their current state of healthiness (Rida *et al.* 2019). The in-question partnerships are shown in the major cycle in Figure 2.

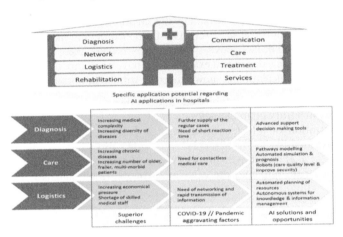

Figure 2. Stroke-Diagnosis-Healthcare-System collaboration stream to fix the issues that develops.

Patients with strokes need tools that are easy to utilize as well as conveniently accessible to them during their daily activities. This will benefit from brilliant and skilled mists that are simple to grasp in terms of programmability and efficiency (Khan *et al.* 2022). The following major level requirements shown in Figure 3 should make the Stroke Use Case-Model clear. Using Enterprise Architect, it has been possible to get the Case-Model for the healthcare system utilization for stroke diagnosis (Mehmood *et al.* 2022).

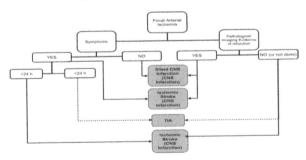

Figure 3. Utilization of the Healthcare System for stroke diagnostic as a scenario.

In the Stroke-Diagnosis-Healthcare-System Use Case-Model, a crucial performer, needs, focused situation, elective situation, along with key requirements are explained as follows. The Stroke-Diagnosis-Healthcare-System Use Case-Model's main needs, focus scenario, essential entertainer, and requirements (Ullah *et al.* 2019).

3.2 Interaction between the Stroke-Diagnosis-Healthcare-System's development as well as implementation

The client (convalescent) mobile application and the application of the server, which contains two major segments (modules): the CNN-module along with the Server-module, make up the Stroke-Diagnosis-Healthcare-System constructed for such study. Applications of the client, as well as applications of servers, rely upon the CNN calculations for the CNN-module (Rasheed *et al.* 2023). The shown concept verification by MLP computation was used in this work. It would also be beneficial to keep in mind that this approach may be used to study illnesses other than stroke (Gupta *et al.* 2011).

Stage 1: Computation of the CNN model on a Virtual Dedicated Server
The following portion of this research has provided further detail on the framework's operating system. The testbed is made up of the following components. Thorough explanation of the methods for the technique anticipated for the stroke diagnosis healthcare system.

(a)   a mobile device running Android working foundation provides UI.
(b)   The Stroke Diagnosis Healthcare System offers a solution that includes medical care facilities backed by a cloud-dependent system and communication that run on Windows with stroke sufferers via concept verification.
(c)   The cloud medical services platform provides robust system management, registration, and a resource for real-world data analysis and health demonstration.

Android is the main controller of the operating system for mobile devices, as should be seen from Figure 4. Java is used for driver development, connection point implementation, and the CNN computation implant. Both the local and remote customers are furnished with a mobile phone with enrolment so they may manage their stroke information (Sharma *et al.* 2022). Following the administration of the stroke testing, the information is put in storage within the cloud after that itis used within the consumer interface. That may be used as a helpful independent direction for stroke subtype therapy (cardioembolic, cryptogenic) (Panchal *et al.* 2023).

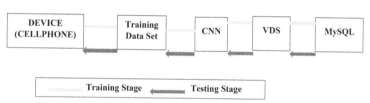

Figure 4.   Verification of the concept plan using a representation of the suggested stroke diagnosis healthcare system.

Stage 2: The customer (client)/stroke patient's mobile application
One enrolled (joined) individual may connect to the processor using their smartphone app. A customer who is signed in may completely insert the accompanying data as the new test results:

– Date of Birth info has been delivered from the processor. The person's birth date, which the server recorded as part of the framework during the recruitment interaction, is used to calculate age.
– During the recruitment interaction, orientation data is also included as a contribution to the framework.

A few credits for the server module are as follows:

– If the logged-in client does not take any action, the system will log the meeting out as a result. Similar arrangements may be made for this duration.
– Whenever necessary, any of the server module's bounds may be designed.
– The associations' port amounts may also be planned.

– Any new patient or customer might also include in the framework, along with their information must be kept (Soni *et al.* 2022).

Stage 3: Utilizing as well as keeping knowledge of data for CNN-Train
The information is prepared using the CNN module. Following the preparation, it is time to use the module that will result from information on a patient who has been enrolled. In the CNN module, MPL calculations are used. For the framework, arrangement features like learning rate worth may be specified.

Stage 4: Subtype identification and progression in light of the server-client
There are 19 patient ascribes of information in the stroke diagnosis healthcare system. On the server, the CNN interaction is used to apply two results for different subtypes of stroke. The android application finally displays the result.

3.3 *Characterization of the stroke subtypes' aftereffects using CNN*

This research is the first of its kind in the field of literature that uses MLP calculations to validate a notion. The overall goal and motivation behind this framework are to increase availability, organisation, accuracy rate, and quality of health correspondence while providing financial benefits by reducing medical costs (Bhatt and Sharma 2023; Liu *et al.* 2022). In the Stroke Diagnosis Healthcare System, the preparation stage focuses on identifying the energy, information size, and emphasis number attributes that provide the most notable precision degree to classify stroke sub-kind (Table 2).

Table 2. For the preparation stage, MLP computation in the Stroke-Diagnosis-Healthcare-System exactness rate according to (i) emphasis number, (ii) energy, (iii) informative gathering size.

(i)		(ii)		(iii)	
Exactness rate (%)	Emphasis number	Exactness rate (%)	Energy	Exactness rate (%)	Informative gathering size
82	220	88	0.6	75	35
85	450	89	0.7	78	65
88	650	90	0.75	79	95
89	850	88	0.9	82	125
90	1050	88	1.0	89	350

As demonstrated in Table 2 the most notable precision rate result for the MLP calculation into the Stroke-Diagnosis-Healthcare-System is achieved as 90% (training rate 0.002, stored layer neuron number is 12), for the number of emphases = 1200, energy = 0.7, and informational collection size = 320 qualities. The MLP computation to the Stroke-Diagnosis-Healthcare-System was created by being installed within the cloud.

4 RESULTS

The topic of theory prototype confirmation considering the Stroke-Diagnosis-Healthcare-System is interestingly addressed in this work. Additionally, MLP calculations (Figure 5) have been suggested for a critical and challenging medical situation, i.e., to provide the differentiating evidence of the many stroke subtypes. The continuing evaluation makes usage of a VDS(Virtual-Dedicated-Server) with 5-VCPU and 32-GB of RAM to address the problem.
The following process might be used to determine the server limit:

$$Users = \frac{(P \times Q \times R \times S)}{52}$$

Where is the CPU rate stacking that must be viewed, Q represents the CPU speed lnes represented as GHz, R represents the mainboard front-side transport speediness represented as MHz, and S represents CPU cores quantity.

A typical framework configuration recognized the server RAM including32-MB for every synchronous patient and an additional4-GB for the Organisation-Operating-System (OOS) house maintenance. If required, Server settings could be updated.

$$RAM = (Users \times 0.032) + 4$$

Here RAM isa memory amount represented as gigabytes; Consumers are selected by the first condition; 0.032GB = 32MB for each concurrent client; and 4GB is the operating system for the company. Utilizing these calculations:

$$Users = \frac{(1 \times 3.2 \times 1700 \times 4)}{52}$$

$$Users = 418.46$$

$$RAM = (418.46 \times 0.032) + 4$$

$$RAM = 17.390GB$$

Regarding that specific topic, this needs for demonstrating how the Stroke-Diagnosis-Healthcare-System gives individuals in command of their sickness. Our framework has a built-in backup. Our server is part of a server farm with an approximate of 100% uptime guarantee and TIER certification. As a result, the patients do not lose consumer faith and really wish to continue receiving the framework's support, which suggests that the framework can provide care for many patients at once.

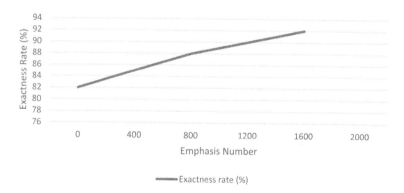

Figure 5. A realistic depiction of MLP calculation in the Stroke-Diagnosis-Healthcare-System.

5 CONCLUSION

The main aim of this paper is to provide a novel approach for analysing the features of primary stroke subtypes in relation to mobile phone use using a cloud infrastructure. Analysis of the Stroke in Health-restructures been a cardioembolic, cryptogenic selection tool, which gives information on 1242 stroke patients. Additionally, the Stroke-Analysis-Healthcare Arrangement has to be enhanced like, this makes the situation possible in characterisation and determine several issues at once under the umbrella of the Analysis-Healthcare Arrangement.

REFERENCES

Bhatt, M.W. and Sharma, S., 2023. An object recognition-based neuroscience engineering: A study for future implementations. *Electrica, 23*(2).

Gupta, A., Kapoor, L. and Wattal, M., 2011. C2C (cloud-to-cloud): An ecosystem of cloud service providers for dynamic resource provisioning. In *Advances in Computing and Communications: First International Conference, ACC 2011, Kochi, India, July 22–24, 2011. Proceedings, Part I 1* (pp. 501–510). Springer Berlin Heidelberg.

Karan, O., Bayraktar, C., Gümüşkaya, H. and Karlık, B., 2012. Diagnosing diabetes using neural networks on small mobile devices. *Expert Systems with Applications, 39*(1): 54–60.

Karlik, B. and Olgac, A.V., 2011. Performance analysis of various activation functions in generalized MLP architectures of neural networks. *International Journal of Artificial Intelligence and Expert Systems, 1*(4): 111–122.

Khan, M.A., Abbas, S., Raza, A., Khan, F. and Whangbo, T., 2022. Emotion based signal enhancement through multisensory integration using machine learning. *Computers, Materials & Continua, 71*(3).

Kim, D.K., Yoo, S.K., Park, I.C., Choa, M., Bae, K.Y., Kim, Y.D. and Heo, J.H., 2009. A mobile tele-medicine system for remote consultation in cases of acute stroke. *Journal of telemedicine and telecare, 15*(2): 102–107.

Li, M., Yu, S., Zheng, Y., Ren, K. and Lou, W., 2012. Scalable and secure sharing of personal health records in cloud computing using attribute-based encryption. *IEEE transactions on parallel and distributed systems, 24*(1): 131–143.

Lian, J.W., Yen, D.C. and Wang, Y.T., 2014. An exploratory study to understand the critical factors affecting the decision to adopt cloud computing in Taiwan hospital. *International Journal of Information Management, 34*(1): 28–36.

Liu, Q., Zhang, W., Bhatt, M.W. and Kumar, A., 2022. Seismic nonlinear vibration control algorithm for high-rise buildings. *Nonlinear Engineering, 10*(1): 574–582.

Maadeed, S.A., Jiang, X., Rida, I. and Bouridane, A., 2019. Palmprint identification using sparse and dense hybrid representation. *Multimedia Tools and Applications, 78*: 5665–5679.

Mehmood, S., Ahmad, I., Khan, M.A., Khan, F. and Whangbo, T., 2022. Sentiment analysis in social media for competitive environment using content analysis. *Computers, Materials & Continua, 71*(3).

Panchal, A.V., Patel, S.C., Bagyalakshmi, K., Kumar, P., Khan, I.R. and Soni, M., 2023. Image-based plant diseases detection using deep learning. *Materials Today: Proceedings, 80*: 3500–3506.

Rasheed, Z., Ma, Y.K., Ullah, I., Ghadi, Y.Y., Khan, M.Z., Khan, M.A., Abdusalomov, A., Alqahtani, F. and Shehata, A.M., 2023. Brain tumor classification from MRI using image enhancement and convolutional neural network techniques. *Brain Sciences, 13*(9): 1320.

Ratta, P., Kaur, A., Sharma, S., Shabaz, M. and Dhiman, G., 2021. Application of blockchain and internet of things in healthcare and medical sector: applications, challenges, and future perspectives. *Journal of Food Quality, 2021*: 1–20.

Rida, I., Herault, R., Marcialis, G.L. and Gasso, G., 2019. Palmprint recognition with an efficient data driven ensemble classifier. *Pattern Recognition Letters, 126*: 21–30.

Sanober, S., Alam, I., Pande, S., Arslan, F., Rane, K.P., Singh, B.K., Khamparia, A. and Shabaz, M., 2021. An enhanced secure deep learning algorithm for fraud detection in wireless communication. *Wireless Communications and Mobile Computing, 2021*: 1–14.

Sharma, V., Gupta, A., Hasan, N.U., Shabaz, M. and Ofori, I., 2022. Blockchain in secure healthcare systems: state of the art, limitations, and future directions. *Security and Communication Networks, 2022*.

Soni, M., Gomathi, S., Kumar, P., Churi, P.P., Mohammed, M.A. and Salman, A.O., 2022. Hybridizing convolutional neural network for classification of lung diseases. *International Journal of Swarm Intelligence Research (IJSIR), 13*(2): 1–15.

Sultan, N. and Van De Bunt-Kokhuis, S., 2012. Organisational culture and cloud computing: coping with a disruptive innovation. *Technology Analysis & Strategic Management, 24*(2): 167–179.

Sultan, N., 2010, November. Cloud computing and SMEs: A match made in the recession. In *ISBE (The Institute for Small Business and Entrepreneurship) Conference* (pp. 2–4).

Ullah, I., Shen, Y., Su, X., Esposito, C. and Choi, C., 2019. A localization based on unscented Kalman filter and particle filter localization algorithms. *IEEE Access, 8*: 2233–2246.

Yang, P., Karambakhsh, A., Bin, S. and Li, P., 2018. Deep gesture interaction for augmented anatomy learning. *International Journal of Information Management*.

Zhang, X., Rane, K.P., Kakaravada, I. and Shabaz, M., 2021. Research on vibration monitoring and fault diagnosis of rotating machinery based on internet of things technology. *Nonlinear Engineering, 10*(1): 245–254.

Next Generation Computing and Information Systems – Gupta. (Ed.)
© 2025 The Author(s), ISBN 978-1-032-73865-9

Supervised context-aware Latent Dirichlet allocation-based drug recommendation model

Gopaldas H. Waghmare
Mechanical Engineering Department YCCE Nagpur, Maharashtra, India

Bhargavi Posinasetty
Masters in Public Health, The University of Southern Mississippi, Hattiesburg, MS, USA

Mohammad Shabaz
Model Institute of Engineering and Technology, Jammu, J&K, India

Saima Ahmed Rahin
United International University, Dhaka, Bangladesh

Abhishek Choudhary
Department of Political Science, University of Delhi, India

Seena K.
St. Marys College Thrissur, India

ABSTRACT: The simplicity of electronic medical records and the ambiguity of patient symptom descriptions make the diagnosis model vulnerable to the interference of high-incidence diseases and joint symptoms. At the same time, much contextual information other than the description of the condition, such as the patient's gender, age and additional personal information, the diagnosis and treatment process, examination results and the local weather, temperature difference and other external information, etc., are also helpful in refining the diagnosis of the patient. Based on effectively integrating multi-source heterogeneous context information, it provides an interpretable basis for disease diagnosis and drug recommendation. Then a contextual topic model Medicine-LDA based on the LDA model, was designed to solve the problem of contextual information combination explosion. The effectiveness and robustness of the method are proved by the comparative experiments based on the electronic medical record data set of a large tertiary hospital.

1 INTRODUCTION

The demand for high-quality medical services is on the rise, and the strain on the medical system is growing as a result of population expansion and the acceleration of the ageing process. According to the national medical and health statistics released by the National Health Commission, from January to November 2018, there were 3.23 billion people in hospitals across the country, a year-on-year increase of 5.3%. Correspondingly, as of 2019, the number of doctors per 1,000 population was 2.592, which is still a big gap compared with developed countries, which also leads to a heavy burden on medical staff and a massive hole in medical resources. On this basis, scholars in related fields have carried out a series of studies and achieved good results in competent guidance, medical image analysis and chronic disease follow-up (Al-Saffar and Baiee 2022; Murad *et al.* 2022; Wang *et al.* 2013). Firstly, the promotion of electronic medical records is relatively late, resulting in fewer reliable medical data sources and other problems such as less data accumulation and

DOI: 10.1201/9781003466383-8

inability to correlate the data of various departments. Secondly, the drug recommendation problem is more sensitive to the data quality, and mainly reflected in three aspects:

1) Most patients have only one visit record, and it is impossible to use the conventional personalized recommendation technology to model.
2) The distribution of diseases is seriously uneven, common diseases occupy most medical records, while niche groups and rare diseases are seriously sparse, and it is difficult to ensure the quality of their recommendations.
3) There is a high proportion of overlap between the symptoms of the disease. Taking this outbreak of new coronary pneumonia as an example, its early signs have many similarities with other pneumonia and even common flu, and even experienced doctors are at risk of misjudgment.

However, these data also bring new difficulties: comprehensively considering this contextual information to achieve effective diagnosis and treatment. To solve the above problems, this paper proposes a drug recommendation method named as medicine-LDA (MLDA), based on multi-source contextual co-awareness. This paper modeled the process of diagnosing a patient by simulating a doctor in reality. Inspired by this, this paper model this process through four steps:

1) Extract the patient's chief complaint text and check abnormality in the data, use the bag-of-words model to synthesize it into a document, and use this document to represent the patient, which is called the disease document;
2) Use the topic model to model the disease document, assuming that the paper has a potential topic, the possible topic represents the patient's disease, and has the corresponding drug distribution and word distribution;
3) Integrate all the contextual information corresponding to the patient into one document, called the contextual document, and use the LDA model to model it to obtain the patient's contextual topic distribution;
4) Assuming that the context topic will impact the patient's disease and the medicines suitable for the patient, a unified probabilistic model framework is used to integrate the drug topically and the context topic to comprehensively model the context information and disease and drug information.

2 RELATED WORK

This section introduces the related work of this paper from three aspects: a drug recommendation system, a context-aware recommendation system and a topic model.

2.1 *Drug recommendation system*

Presently, the research on drug recommendation is not very extensive, and most drug recommendations mainly use the interaction records between users and drugs to make recommendations based on collaborative filtering technology. For example, (Zitouni *et al.* 2020) proposed a drug recommendation algorithm based on user similarity and trust to solve the problem of low recommendation accuracy of collaborative filtering. The method reduces the time complexity by clustering drugs, introduces a standard.

2.2 *Context-aware recommender system*

Traditional recommendation systems, such as content-based recommendation and collaborative filtering, ignore that users have different preference behaviors in different situations (Dwivedi and Rawat 2016). Therefore, researchers have proposed "context-aware" data mining technology to improve the recommendation effect. Since the same user often has different preferences in different contexts, the context-aware recommendation system can give more relevant recommendation results according to the user's specific situation.

A topic model is a statistical model for clustering implicit semantic structure in the corpus, which is widely used in semantic analysis and text mining in natural language processing. Latent Dirichlet allocation (LDA) is the most common topic model, which assumes that the document selects a potential topic with a certain probability and then selects a word from the topic with a certain chance to generate the entire corpus (Bose and Sohi 2006; Kluckner *et al.* 2013; Yu 2021). LDA can be extended in various ways by introducing external label information.

The essence of the topic obtained by the LDA model is to learn the co-occurrence relationship between words implicitly. When counting short texts, such methods often do not work well because the words in the short texts do not co-occur frequently enough. Some works have studied this problem.

2.4 *Data preprocessing*

First, the problem is described by simulating a doctor's diagnosing of a patient in reality. In the actual process of diagnosis and treatment, doctors often record the patient's chief complaint in the patient's medical text, as well as the doctor's physical examination of the patient (Jawarneh *et al.* 2023). The basis of segmentation includes the following:

1) Common sense of life, such as gender is divided into male and female, the season is divided into spring, summer, autumn and winter;
2) Data characteristics, such as the types of anesthesia are divided into general anesthesia, local anesthesia and no anesthesia;
3) External knowledge, such as age segmentation according to standard age segmentation methods, insurance status segmentation according to insurance amount and conditions;
4) For temperature difference, temperature and other situations without reference knowledge, based on the common sense of life and data distribution, the data distribution in different segments is relatively uniform, and the segmentation method is close to people's cognition of life.

2.5 *Formal description*

In this sub-section, we formally describe the problem studied in this paper as follows: In the context-aware drug recommendation task, the training set contains the disease document $V_i = \{v_{i,1}, v_{i,2}, v_{i,p}\}$, where v_i, are the words obtained from the segmentation of the medical text or a single abnormal item. Correspondingly, we have context documents $U_i = \{u_{i,1}, u_{i,2},..,u_{i,q}\}$, where U_i, is a single context word. In addition, we have a set of medicines $E_i = \{e_{i,1}, e_{i,2},..,e_{i,s}\}$, where e_i, is one of all the medicines prescribed by the doctor for the patient (Ahmed *et al.* 2022; Iqbal *et al.* 2022; Khan *et al.* 2020).

3 CONTEXT-AWARE DRUG RECOMMENDATION METHOD

3.1 *Model overview*

The general steps of the model proposed in this paper are:

1) The model needs to process the context information. Although the value of context information has been segmented in the data preprocessing part, if it is assumed that there are K kinds of context information, and each context has N value segments, there will be NK combinations, resulting in the model (Rida *et al.* 2015).
2) This paper uses a probabilistic framework to jointly model context information, disease documents, and patient medicines. The idea of modeling comes from the diagnosis and treatment process in actual medical places; that is, the doctor first determines the disease that the patient may suffer from according to the patient's symptoms, physical

examination, and laboratory reports, and then gives the appropriate medicine for the disease according to the patient's corresponding situational information.

3.2 *Contextual information modeling methods*

To solve the problem of the explosion of patient contextual information combination, it is necessary to try to reduce the dimensionality of the contextual information (Rida *et al.* 2018). Clustering is a standard method to solve such problems, which can be roughly divided into two types: hard clustering and soft clustering. Among them, the hard clustering classifies the data exactly. Finally, after several iterations converge, each dish is assigned a fixed context topic, so the topic distribution of each context document can be calculated using Equation (1).

$$Q(a|e) = \frac{\eta_{e,a} + \alpha}{\sum_{i=1}^{k} \eta_{e,a_i} + L\alpha} \tag{1}$$

Where e and a represents the number of words in the context document d assigned to the topic a, α is the Dirichlet prior parameter, and L is the number of context topics.

3.3 *Context-aware drug recommendation method*

After the soft clustering of the context information is completed, the comprehensive modeling of all the data can be carried out. The modeling method proposed in this paper is based on two assumptions about the diagnosis and treatment process:

1) Context topics are soft clustering of context information, and each context topic may affect the suitable medicines and the prone diseases of patients.
2) The data of each patient corresponds to a latent variable, which represents the disease of the patient, called the disease theme.

Figure 1 shows the probability map representation of the model, and the symbol descriptions.

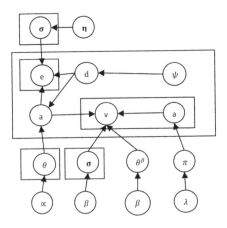

Figure 1. Probability graph representation of the MLDA model.

3.4 *Sampling and inference algorithms*

MLDA needs to infer the hidden variables z, c, and y. However, since the exact inference algorithm is unsolvable, this paper uses the collapsed Gibbs sampling (CGS) to sample the hidden variables and obtains the parameters θ, φ, f; then, the parameters obtained by training are used to recommend drugs to new patients.

First, random initialization is performed on all the data in the training set: for the nth data, the context topic latent variable d_n and disease topic latent variable a_n are randomly assigned, and for the mth disease document word in the nth data, the corresponding identifier is randomly assigned [15]. The conditional probability distribution of the latent variables z and c is

$$q(a_n = l, d_n = g \,|\psi_n, v.e.a.z.\alpha, \beta, \eta)\alpha$$

$$\psi_n^f \times \frac{m_f^k + \alpha}{\sum_{j=1}^{k}(m_f^j + \alpha)} \times \prod_{m=1}^{M_{n,m}.z_{n,m=1}} \frac{m_k^{v_{n,m}} + \beta}{\sum_{v=1}^{W}(m_k^v + \beta)} \times \prod_{e=1}^{N_{n,d}} \frac{m_{f,k}^{e_{n,e}} + \eta}{\sum_{e=1}^{E}(m_{f,k}^e + \eta)} \tag{2}$$

Among them, ψ_n^f is the probability q (f|ψ_n) of sampling f from ψ_n, m_f^k is the number of times that the word v is assigned to the topic k, m_f^k, k, n is the drug d is set as the drug topic k, and the number of times the situation topic c, both counts above exclude document n. For each word in the disease document, the conditional probability distribution of the corresponding identification variable y is

$$q(z_{n,m} = 0 \,|v, e, a, z, \alpha, \beta, \eta)\alpha$$

$$\times \frac{m_{(n,m)}^{z=0} + \lambda}{\sum_{z=0}^{1}(m_{n,m}^z + \lambda)} \times \frac{m_{z=0,(n,m)}^{x_{n,m}} + \beta}{\sum_{x=0}^{Z}(m_{y=0,n,m}^x + \beta)}, \tag{3}$$

Among them, $m_{n,m}^z$ is the number of words corresponding to the label y, m_y^x=0,(m,n) is the number of occurrences of w in the word corresponding to the label 0, $m_{a_{n,m},n,m}^x$ is the number of times the phrase $v_{n,m}$ is assigned the topic $a_{n,m}$, and the above counts exclude the mth word in the nth document (Gupta and Awasthi 2010; Gupta *et al.* 2022; Khan *et al.* 2020).

$$\theta_{f,k} = \frac{m_f^k + \alpha}{\sum_{j=1}^{M}\left(m_f^j + \alpha\right)} \tag{4}$$

Among them, m_f^j is the number of times the document with the situation topic f corresponding to the disease topic j.

$\varphi_{k,w}$ represents the probability that a certain disease topic generates a certain word:

$$\varphi_{k,w} = \frac{m_f^k + \beta}{\sum_{j=1}^{M}\left(m_f^j + \beta\right)} \tag{5}$$

Where $\varphi_{k,w}$ is the number of occurrences of word w in the document corresponding to disease topic k.

3.5 *Drug recommendation*

Suppose a set of test data sets are given, in which the corresponding drugs of patients are unknown. Then the conditional probability of the patient corresponding to any drug d can be obtained by formula (10).

$$q(e|X, \psi) = \sum_{f=1}^{F}\sum_{l=1}^{L} \sigma_{f,l,e} \times \left(\varphi_f \theta_{f,l} \prod_{v} \left(\pi_0 \varnothing_v^C + \pi_1 \varnothing_{l,e}\right)\right) \tag{6}$$

29

4 EXPERIMENT AND RESULT ANALYSIS

This section describes in detail the datasets, metrics, benchmark methods, parameter settings and analysis of experimental results used in the experiments in this paper.

4.1 Experimental dataset

The data used in this experiment comes from the electronic medical record database of the inpatient department of a large tertiary hospital from 2015 to 2018 and has been desensitized. The data set was cleaned and reduplicated, and 158,556 complete electronic medical records were obtained, each of which included patients' complete diagnosis and treatment process records from discharge to admission. Logarithmic form of drug frequency statistics is shown in Table 1.

Table 1. Logarithmic form of drug frequency statistics.

Serial	Medicine ID
0	1400
200	1200
400	1000
600	800
800	600
1000	400
1200	200

4.2 Experimental results

Table 2 shows the performance of different methods on the dataset, and each experimental result is obtained by 5-fold cross-validation. The subject word prior in the LLDA model is set to 0.01; the number of topics in Tag-LDA is set to 50; the subject word prior is set to 0.01, the vector dimension in CF is set to 30, and the number of similar documents is set to 40.

Table 2. Performance comparison of different methods.

Metric	Freq	LLDA	MLDT	MLP	NTM	MLDA	MLDA-pure	MLDA-z	MLDA-c
NDCG@20	0.5116	0.3113	0.7206	0.7283	0.7333	0.707	0.7144	0.7155	0.8155
NDCG@30	0.5492	0.3432	0.779	0.789	0.7961	0.7679	0.7757	0.7771	0.8771
NDCG@40	0.5779	0.3631	0.8125	0.825	0.8334	0.8057	0.8135	0.8149	0.9149
P@20	0.4137	0.2524	0.6084	0.611	0.6182	0.5925	0.6022	0.6103	0.7103
P@30	0.3484	0.2252	0.5121	0.5147	0.5215	0.5006	0.5079	0.5156	0.6156
P@40	0.3075	0.2032	0.4469	0.4481	0.4545	0.4369	0.4426	0.4486	0.5486
R@20	0.4374	0.2661	0.667	0.6777	0.6829	0.6539	0.6616	0.6624	0.7624
R@30	0.4962	0.3213	0.7635	0.7788	0.7878	0.756	0.7647	0.7654	0.8654
R@40	0.5521	0.3587	0.8239	0.8444	0.8559	0.8259	0.8341	0.8354	0.9354
F@20	0.3978	0.2432	0.5933	0.5982	0.605	0.5797	0.5893	0.591	0.691
F@30	0.3837	0.2398	0.571	0.576	0.584	0.5595	0.568	0.5723	0.6723
F@40	0.3663	0.2276	0.5374	0.5412	0.5494	0.5268	0.5341	0.539	0.639

In the following, the parameter sensitivity analysis of the proposed model is carried out. Figure 2 shows the effect of the number of contextual topics on the model effect, where K = 400, adjusting different F values (Limbasiya et al. 2018; Soni and Singh 2022). Figure 3 with Table 3 shows the impact of the number of disease subjects on the model performance where F = 10 and change different K values. When the parameters change within an extensive range, the model effect is only 0.01. It fluctuates in the field, which confirms the excellent performance of the proposed method (Bhatt and Sharma 2022; Liu *et al.* 2022).

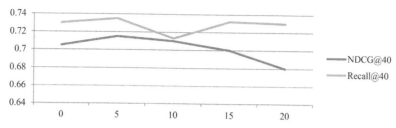

Figure 2.   Influence of context topic number.

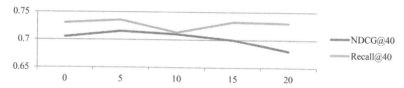

Figure 3.   Influence of the number of situational themes and disease themes on the model effect.

Table 3.   Influence of disease topic number.

Serial	NDCG@40	Recall@40
0	0.705	0.73
5	0.715	0.735
10	0.71	0.713
15	0.7	0.732
20	0.68	0.73

5   CONCLUSION

In this paper, a context-aware-based drug recommendation model, Medicine-LDA, is proposed that utilized a general probabilistic framework to describe the diagnosis and treatment process in real-life scenarios. The model inferred the possible diseases of patients based on patient documents composed of laboratory abnormalities and medical texts and proposed the use of disease topics to describe the probability of patients suffering from different diseases. Finally, through experiments on a data set from a large tertiary hospital, it is shown that the model proposed in this paper outperforms the commonly used traditional machine learning models, topic models and collaborative filtering models in accuracy. The analysis of the results can obtain a series of valuable information, which confirms the excellent performance of the model and its value in practical application.

REFERENCES

Ahmed, T.I., Bhola, J., Shabaz, M., Singla, J., Rakhra, M., More, S. and Samori, I.A., 2022. Fuzzy logic-based systems for the diagnosis of chronic kidney disease. *BioMed Research International, 2022.*
Al-Saffar, M. and Baiee, W., 2022, March. Survey on implicit feedbacks extraction based on yelp dataset using collaborative filtering. In *2022 Muthanna International Conference on Engineering Science and Technology (MICEST)* (pp. 125–130). IEEE.

Bhatt, M.W. and Sharma, S., 2023. An object recognition-based neuroscience engineering: A study for future implementations. *Electrica*, 23(2).

Bose, S. and Sohi, S., 2006, June. Enabling seamless contextual collaborations for mobile enterprises. In *15th IEEE International Workshops on Enabling Technologies: Infrastructure for Collaborative Enterprises (WETICE'06)* (pp. 119–124). IEEE.

Dwivedi, S.K. and Rawat, B., 2016, March. A review on improving recommendation quality by using relevant contextual information. In *2016 3rd International Conference on Computing for Sustainable Global Development (INDIACom)* (pp. 244–248). IEEE.

Gupta, A. and Awasthi, L.K., 2010. Toward a quality-of-service framework for peer-to-peer applications. *International Journal of Distributed Systems and Technologies (IJDST)*, 1(3): 1–23.

Gupta, A., Tripathi, M., Muhuri, S., Singal, G. and Kumar, N., 2022. A secure and lightweight anonymous mutual authentication scheme for wearable devices in Medical Internet of Things. *Journal of Information Security and Applications*, 68: 103259.

Iqbal, M.W., Naqvi, M.R., Khan, M.A., Khan, F. and Whangbo, T., 2022. Mobile devices interface adaptivity using Ontologies. *Computers, Materials & Continua*, 71(3).

Jawarneh, M., Arias-Gonzáles, J.L., Gandhmal, D.P., Malik, R.Q., Rane, K.P., Omarov, B., Mahapatra, C. and Shabaz, M., 2023. Influence of grey wolf optimization feature selection on gradient boosting machine learning techniques for accurate detection of liver tumor. *SN Applied Sciences*, 5(7): 178.

Khan, F., Khan, A.W., Khan, S., Qasim, I. and Habib, A., 2020. A secure core-assisted multicast routing protocol in mobile ad-hoc network. *Journal of Internet Technology*, 21(2): 375–383.

Khan, W., Wang, H., Anwar, M.S., Ayaz, M., Ahmad, S. and Ullah, I., 2019. A multi-layer cluster based energy efficient routing scheme for UWSNs. *IEEE Access*, 7: 77398–77410.

Kluckner, P.M., Buchner, R., Weiss, A. and Tscheligi, M., 2013, May. Collaborative reporting tools: An analysis of maintainace activites in a semiconductor factory. In *2013 International Conference on Collaboration Technologies and Systems (CTS)* (pp. 508–515). IEEE.

Limbasiya, T., Soni, M. and Mishra, S.K., 2018. Advanced formal authentication protocol using smart cards for network applicants. *Computers & Electrical Engineering*, 66: 50–63.

Liu, Q., Zhang, W., Bhatt, M.W. and Kumar, A., 2022. Seismic nonlinear vibration control algorithm for high-rise buildings. *Nonlinear Engineering*, 10(1): 574–582.

Murad, D.F., Hassan, R., Wijanarko, B.D., Leandros, R. and Murad, S.A., 2022, May. Evaluation of hybrid collaborative filtering approach with context-sensitive recommendation system. In *2022 7th International Conference on Business and Industrial Research (ICBIR)* (pp. 7–12). IEEE.

Rida, I., Al-Maadeed, S., Mahmood, A., Bouridane, A. and Bakshi, S., 2018. Palmprint identification using an ensemble of sparse representations. *IEEE Access*, 6: 3241–3248.

Rida, I., Jiang, X. and Marcialis, G.L., 2015. Human body part selection by group lasso of motion for model-free gait recognition. *IEEE Signal Processing Letters*, 23(1): 154–158.

Sadiq, M.T., Yu, X., Yuan, Z., Zeming, F., Rehman, A.U., Ullah, I., Li, G. and Xiao, G., 2019. Motor imagery EEG signals decoding by multivariate empirical wavelet transform-based framework for robust brain–computer interfaces. *IEEE access*, 7: 171431–171451.

Soni, M. and Singh, D.K., 2022. LAKA: Lightweight authentication and key agreement protocol for internet of things based wireless body area network. *Wireless personal communications*, 127(2): 1067–1084.

Wang, J., Li, H. and Zhao, H., 2013, September. The contextual group recommendation. In *2013 5th International Conference on Intelligent Networking and Collaborative Systems* (pp. 127–131). IEEE.

Yu, Y., 2021, December. Research on intelligent recommendation of learning resources based on collaborative filtering algorithm. In *2021 International Symposium on Advances in Informatics, Electronics and Education (ISAIEE)* (pp. 326–329). IEEE.

Zitouni, H., Bouchelik, K., Saidi, R. and Chekkai, N., 2020, November. Personalized Menu: A new contextual collaborative recommender system. In *2020 International Conference on Advanced Aspects of Software Engineering (ICAASE)* (pp. 1–6). IEEE.

Next Generation Computing and Information Systems – Gupta. (Ed.)
© 2025 The Author(s), ISBN 978-1-032-73865-9

Combination kernel support vector machine based digital twin model for prediction of dyslexia in distributed environment

H. Kareemullah
Electronics and Instrumentation Engineering, B.S.A Crescent Institute of Science and Technology, Chennai, Tamil Nadu, India

Aadam Quraishi
Interventional Treatment Institute Houston Texas USA, McAllen, Texas, USA

G.C. Prashant
Software Developer at Paycom, USA University, Texas Tech University, Lubbock State, Texas, USA

Mohammad Shabaz
Model Institute of Engineering and Technology, Jammu, J&K, India

Archana Kollu
Pimpri Chinchwad College of Engineering and Research, Ravet-Pune, India

Mukta Sandhu
Skill Department of Computer Science and Engineering, Shri Vishwakarma Skill University, Palwal, Haryana, India

ABSTRACT: Dyslexia is a learning disability that is widespread and marked with a persistent problem with word identification and spelling. It has an impact on a person's capacity to decipher letters and words accurately and fluently. Dyslexia has emerged as one of the most widespread learning disabilities, though the medical experts have not yet come out explaining its core causes. In this research, the prediction and classification of dyslexia is effectively carried out with the help of brain images which acts as input to machine learning techniques towards the prediction of dyslexia. The proposed dyslexic prediction model uses a Combination digital twin Kernel-based Support Vector Machine (KSVM) optimized by Whales algorithm. The combination kernel SVM shows better accuracy and less computational time compared to single kernel of SVM in a distributed environment. The proposed combinational kernel is less complex than deep learning models.

1 INTRODUCTION

Dyslexia represents a type of syndrome associated to reading impairment. Dyslexia does not deal with the cognitive development of children and other learning aspects of a child, like visual acuity and it is just a reading disability. This inability to read can turn into developmental dyslexia in many children. The effects of developmental dyslexia in children are impaired phonological processing and poor word identification or reading (Sahoo *et al.* 2016). Five to seven percent of children worldwide are facing difficulties because of developmental dyslexia. Studies like MRI brain imaging when conducted on children at an early stage, to observe reading skills can prove the theories related to the cause of dyslexia by accurately localizing neurobiological defects (Belle *et al.* 2015). Dyslexics perform poorly in digit span aptitude tests. They also show low capabilities in word arrangement and have less rapid naming capabilities (Mohamad *et al.* 2013). Dyslexia may represent a multifaceted ailment, and hence several constraints including the phonological factors can trigger the issues encountered while reading. Surface dyslexia is reading trouble that comprises of poor unpredictable word reading and discretion of non-words can be typical. This sort of reading trouble is found in an assortment of dialects such as English, Hebrew, Italian and Spanish. Both surface and phonological dyslexia have been shown to persist into adulthood (Frid and Breznitz 2012). Visual Dyslexia is perusing trouble because of either optical visual issue (physical causes) or visual preparing issues (psychological/neurological causes).

2 RELATED WORK

Sahoo et al. predicted the future health of a patient. Inputs for this research are radiology images, physiological data, 3D imaging, and genomic sequencing (Prabha and Bhargavi 2020). The obtained inputs are further sent to the Map-Reduce-based correlation analysis technique. The results from the correlation analysis are sent to the future health prediction model. The FHCP algorithm predicts the status of future health by analyzing current health parameters and status. The accuracy level is 98% (Appadurai and Bhargavi 2021). Using electroencephalogram (EEG) recorded channels, an algorithm is built to identify participants as control or dyslexic based on their reading skills. Ensemble of Support Vector Machine (SVM) is used for classification. Electroencephalogram-based analysis was used to predict dyslexia (EEG). SVC classifier model is used to extract features and categorize dyslexics. The pros and cons of the EEG approach were recognized, as well as optimization techniques for better dyslexia prediction (Prabhi et al. 2019).

3 PROBLEM DEFINITION

In the document, the effective prediction of dyslexia is carried out through MRI brain images. The research also focuses on utilization of appropriate ML techniques towards identification of dyslexia. The dyslexic prediction model proposed is implemented using combinational Kernel-based Support Vector Machine (KSVM). A classification model is built to extricate dyslexics from controls by combining Image Processing techniques and AI and ML algorithms.

- Accuracy has a vital role while classifying dyslexics and the existing methods in literature have low prediction accuracy.
- The computational complexity of the existing prediction system is high.
- Existing prediction algorithms such as fuzzy, neural network, etc required high processing time for a prediction.

4 PROPOSED METHODOLOGY

It is high time the due emphasis is afforded for the precise and prompt identification of dyslexia which causes untold suffering to the learning skills of school-going children. Our approach is to build a classification model to distinguish dyslexics from non-dyslexics (Prabha, Bhargavi and Harish 2019). The assessment of the amount of GMV is accomplished by using the RGB values in the image (Jothi Prabha and Bhargavi 2019). As a result, the white lesions and black lesion in the image is removed, thereby paving the way to extricate Grey matter Volume (GMV), the estimation of the white lesion in the image is carried out, and then results acts as input to the classification phase. Here we have designed a classification model with the help of the combinational kernel SVM technique (Dhiman *et al.* 2021). The classified images further undergo optimization by adopting Whale optimization technique.

4.1 Preprocessing

A median filter is used to smoothen the images in our proposed study. It can handle pulse noise in higher density when combined with a medium filter. The details of the images can be preserved while using this filter.

4.2 Feature extraction

The median filter approach is used to accomplish preprocessing, after which the pre-processed pictures are sent to the feature extraction step (Soni *et al.* 2021). The **MRI** brain image is used to extract three separate parameters, including Grey Matter Volume and cortical thickness.

4.3 Combination kernel for support vector machine

The maximum margin classifier works fine when there is a linear boundary (i.e) a hyperplane separating both the classes. It does not work when there are no solutions when M>0, for the optimization problem. To address non-linear data, the primary feature space is changed into a higher – dimensional feature space. The original feature space of the kernel is illustrated employing the following Equation (1).

$$K(U, V) = \varphi(U)^T \varphi(V) \tag{1}$$

Given below are the relations for the several kernels.
Linear Kernel:

$$linear_k(U, V) = u^T v + c \tag{2}$$

Where u, v characterize the dot product of linear kernel and constant c.
Quadratic Kernel:

$$quad_k(U, V) = 1 - \frac{\|u - v\|^2}{\|u - v\|^2 + c} \tag{3}$$

Here u, v signify the vectors in the original input space
Polynomial Kernel:

$$poly_k(U, V) = (\lambda u^T v + c)^\theta, \ \lambda > 0 \tag{4}$$

Sigmoid Kernel:

$$sig_k(U, V) = \tanh(\lambda u^T v + c), \ \lambda > 0 \tag{5}$$

4.4 Whale optimization

Whale Optimization (WO) Algorithm is adopted for the weight optimization. The final score gained y using the WO is dependent on optimization score value (Gupta and Awasthi 2008).

The whale optimization technique encompasses three distinct stages like the act of encircling prey, bubble-net attacking procedure, and hunt for the victim/prey.

4.4.1 Encircling the prey/victim
The Hump-back whales are competent for discriminating the position of the victim and surrounds them. For the vague location of the most favored is its search area, the current finest contender elucidation is the objective victim or is in its vicinity of the most favored in the WO technique (Shamra et al. 2022).

4.4.2 *Bubble-net attacking method (exploitation phase)*

Following the arithmetical illustration of the bubble-net activities of the humpback whales, two improved techniques are envisaged which are furnished below.

4.4.3 *Shrinking encircling mechanism*

Eq. (3) takes care of reducing the value of an effectively. Specifically, the variation array of V is also reduced through a. By means of stipulating the arbitrary values for A that lies between -1 and 1. The updated location of a search criteria shall be distinctly located by identifying the best among locations with that of the currently updated location

5 RESULT AND DISCUSSION

The proposed system is implemented in apache spark (Ullah *et al.* 2019) that has a system configuration as a i5 core processor, 8GB RAM, Windows 10 OS.

5.1 *Performance evaluation*

5.1.1 *Memory usage*

Java Programming with Cloud Sim program elegantly executes the efficient organization of the memory for utilization. Latest objects are produced and arranged in the stack. Memory utilization of the new-fangled technique is effectively estimated in Binary digits (bits) (Figure 1).

5.1.2 *Running time*

Based on the memory attributes, the runtime in the Java Programming with Cloud Sim program is firmly derived. The running time taken is illustrated to the absolute exclusion of the coefficients. The time interval is effectively estimated in milliseconds. Table 1 shows research memory measures were taken based on testing. The optimization time is calculated based on a testing % is 60, 70, 80, and 90. The optimization time for a testing % 60 obtains 6348ms to complete a process (Shah *et al.* 2023).

Table 1. Proposed research memory measures were taken based on testing %.

Training %	Memory
60	1877527
70	1966548
80	2236584
90	2684484

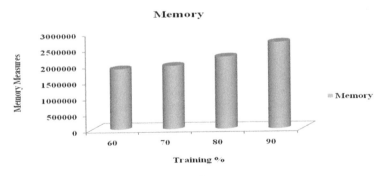

Figure 1. Graphical representation of our proposed memory measures.

We can attain a memory measures value for our proposed prediction of dyslexia it will vary from the training % 60, 70, 80, and 90. In our proposed technique has utilized a small amount of memory is required to complete a process in the testing % 60 has taken a 1877527 ms is required to complete a process. Figure 2 shows Graphical representation of proposed Accuracy measures.

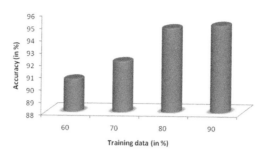

Figure 2.   Graphical representation of our proposed accuracy measures.

The efficiency of our proposed research here we have explained the accuracy measures of this study it will be varying from the testing % 60, 70, 80, and 90. Figure 3 shows the Graphical representation of proposed precision measures.

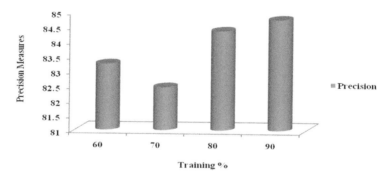

Figure 3.   Graphical representation of our proposed precision measures.

We can attain a precision measures value for our proposed prediction of dyslexia it will vary from the training % 60, 70, 80, and 90. In our proposed technique has utilized a least amount of memory is needed to end a process in the testing 60% has taken an 83.24% is required to complete a procedure. Figure 4 shows Graphical representation of proposed study recall measures. Table 2. shows Recall measures of proposed study taken based on training %.

Table 2.   Recall measures of our proposed study taken based on training %.

Training %	Recall
60	73.28
70	72.61
80	74.16
90	75.22

The proposed combinational kernel-based dyslexia detection approach is evaluated using dyslexia detectors based on SVM Kernel and its variants. The proposed model outperforms the ML and DL-based dyslexia detection models. The proposed model had a high accuracy of 90.36 percent.

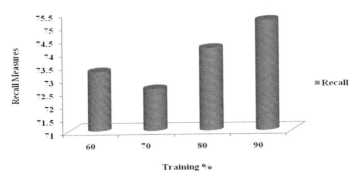

Figure 4. Graphical representation of our proposed study recall measures.

5.2 *Comparative analysis*

In the related work section, some of the existing methods are reviewed for the research of our work by considering the limitations of such state-of-art works and to tackle the problems. Figure 5 shows the Accuracy comparison of the proposed models.

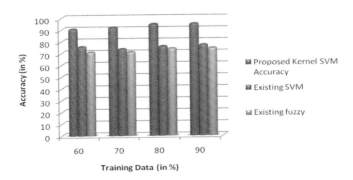

Figure 5. Accuracy comparison.

The proposed combinational kernel SVM performs better when compared to existing SVM and Fuzzy models. By using this comparison result, we can evaluate the performance of our proposed work whether it has very good classification efficiency or not (Khan *et al.* 2022). Table 3 shows the comparative study of the proposed model with the existing models for classification of dyslexia from different types of data such as fMRI images, MRI images, psychological assessments, Handwriting, Eye Movements, and Electroencephalogram signals (Hosseinzadeh *et al.* 2022).

Table 3. Comparative study of the proposed model with other existing models for detection of dyslexia (Aski *et al.* 2023; Bhatt and Sharma 2023; Rahkra et al. 2021; Rida *et al.* 2016; Tharewal *et al.* 2022; Vyas *et al.* 2022; Yao *et al.* 2021).

Type of Data	Features	Model Proposed	Accuracy %
Electroencephalogram Signals	Maximum peak amplitude, Spectral Flatness and positive region	Ensemble of Support Vector Classifiers	86.3%

(continued)

38

Table 3. Continued

Type of Data	Features	Model Proposed	Accuracy %
Reading and writing assessments	Assessment score of URAWSS test	Decision tree and Random Forest	Decision tree and Random forest performed better than traditional rule based approaches
Eye Movements while reading 12 different texts	Age, Font–type, Mean fixations, number of fixations, sum of fixations, no of visits, mean of visits, sum of visits	SVM-classifier for binary class	80.18 %
Reading or writing assessments conducted by teacher and Parent	Assessment score of reading and writing tests	Fuzzy Unordered Rules Induction Algorithm	Fuzzy model was reported to perform better than traditional rule–based approaches
Reading and writing performance	Three different range of scores were obtained	Artificial neural networks	75%
EEG signals	Features (Beta band power) extracted by discrete wavelet transform with Daubechies of order 2	Enhanced K-Nearest Neighbor classifier	Optimal electrode positions in the evaluation of dyslexic is suggested as 8
Handwritten texts, pictures, and audio	Features extracted from pictures and audio files.	Random Forest	90%

6 CONCLUSION

In this research, a combinational kernel SVM of linear and quadratic kernel uses brain images for the prediction of dyslexia in distributed environment. Apache Spark MlLib is implemented which is a distributed approach that provides support for big data and overcomes big data processing and execution concerns. As a future advancement, the proposed model can be further improved to predict dyslexia better by considering a greater number of input parameters. In the future, multimodal quantitative MRI techniques can be used for better characterization of brain tissue and fine structure of the developmental curves of developmental dyslexia. The ability and the performance of the existing system shall be improved.

REFERENCES

Appadurai, J.P. and Bhargavi, R., 2021. Eye movement feature set and predictive model for dyslexia: Feature set and predictive model for dyslexia. *International Journal of Cognitive Informatics and Natural Intelligence (IJCINI)*, *15*(4): 1–22

Aski, V.J., Dhaka, V.S., Parashar, A. and Rida, I., 2023. Internet of Things in healthcare: A survey on protocol standards, enabling technologies, WBAN architectures and open issues. *Physical Communication*: 102103.

Belle, A., Thiagarajan, R., Soroushmehr, S.M., Navidi, F., Beard, D.A. and Najarian, K., 2015. Big data analytics in healthcare. *BioMed research international*, *2015*.

Bhatt, M.W. and Sharma, S., 2023. An object recognition-based neuroscience engineering: A study for future implementations. *Electrica*, *23*(2).

Dhiman, G., Soni, M., Pandey, H.M., Slowik, A. and Kaur, H., 2021. A novel hybrid hypervolume indicator and reference vector adaptation strategies based evolutionary algorithm for many-objective optimization. *Engineering with Computers*, *37*: 3017–3035.

Frid, A. and Breznitz, Z., 2012, November. An SVM based algorithm for analysis and discrimination of dyslexic readers from regular readers using ERPs. In *2012 IEEE 27th Convention of Electrical and Electronics Engineers in Israel* (pp. 1–4). IEEE.

Gupta, A. and Awasthi, L.K., 2008, July. Secure Thyself: Securing Individual Peers in Collaborative Peer-to-Peer Environments. In *GCA* (pp. 140–146).

Hosseinzadeh, M., Tanveer, J., Masoud Rahmani, A., Yousefpoor, E., Sadegh Yousefpoor, M., Khan, F. and Haider, A., 2022. A Cluster-Tree-Based Secure Routing Protocol Using Dragonfly Algorithm (DA) in the Internet of Things (IoT) for Smart *Agriculture. Mathematics, 11*(1): 80.

Jothi Prabha, A. and Bhargavi, R., 2019. Prediction of dyslexia using machine learning—a research travelogue. In *Proceedings of the Third International Conference on Microelectronics, Computing and Communication Systems: MCCS 2018* (pp. 23–34). Springer Singapore.

Khan, A.W., Yaseen, G., Khan, M.I. and Khan, F., 2022. AHP-Based Prioritization Framework for Software Outsourcing Human Resource Success Factors in Global Software Development. *Evolving Software Processes: Trends and Future Directions*: 151–173.

Mohamad, S., Mansor, W. and Lee, K.Y., 2013, August. Review of neurological techniques of diagnosing dyslexia in children. In *2013 IEEE 3rd international conference on system engineering and technology* (pp. 389–393). IEEE.

Prabha, A.J. and Bhargavi, R., 2020. Predictive model for dyslexia from fixations and saccadic eye movement events. *Computer Methods and Programs in Biomedicine, 195*: 105538.

Prabha, J.A., Bhargavi, R. and Harish, B., 2019. Predictive model for dyslexia from eye fixation events. *Int. J. Eng. Adv. Technol, 9*(1S3): 20.

Prabha, J.A., Bhargavi, R. and Ragala, R., 2019. Prediction of dyslexia using support vector machine in distributed environment. *Int. J. Eng. Technol, 20*: 2795–2799.

Rakhra, M., Singh, R., Lohani, T.K. and Shabaz, M., 2021. Metaheuristic and machine learning-based smart engine for renting and sharing of agriculture equipment. *Mathematical Problems in Engineering, 2021*: 1–13.

Rida, I., Boubchir, L., Al-Maadeed, N., Al-Maadeed, S. and Bouridane, A., 2016, June. Robust model-free gait recognition by statistical dependency feature selection and globality-locality preserving projections. In *2016 39th International Conference on Telecommunications and Signal Processing (TSP)* (pp. 652–655). IEEE.

Sahoo, P.K., Mohapatra, S.K. and Wu, S.L., 2016. Analyzing healthcare big data with prediction for future health condition. *IEEE Access, 4*: 9786–9799.

Shah, A., Ali, B., Wahab, F., Ullah, I., Amesho, K.T. and Shafiq, M., 2023. Entropy-based grid approach for handling outliers: a case study to environmental monitoring data. *Environmental Science and Pollution Research*: 1–20.

Sharma, M., Gupta, A. and Singh, J., 2022. Resource discovery in inter-cloud environment: a review. *International Journal of Advanced Intelligence Paradigms, 23*(1–2): 129–145.

Soni, M., Dhiman, G., Rajput, B.S., Patel, R. and Tejra, N.K., 2021. Energy-effective and secure data transfer scheme for mobile nodes in smart city applications. *Wireless Personal Communications*: 1–21.

Tharewal, S., Ashfaque, M.W., Banu, S.S., Uma, P., Hassen, S.M. and Shabaz, M., 2022. Intrusion detection system for industrial Internet of Things based on deep reinforcement learning. *Wireless Communications and Mobile Computing, 2022*: 1–8.

Ullah, I., Liu, Y., Su, X. and Kim, P., 2019. Efficient and accurate target localization in underwater environment. *IEEE Access, 7*: 101415–101426.

Vyas, S., Shabaz, M., Pandit, P., Parvathy, L.R. and Ofori, I., 2022. Integration of artificial intelligence and blockchain technology in healthcare and agriculture. *Journal of Food Quality, 2022*.

Yao, Q., Shabaz, M., Lohani, T.K., Wasim Bhatt, M., Panesar, G.S. and Singh, R.K., 2021. 3D modelling and visualization for vision-based vibration signal processing and measurement. *Journal of Intelligent Systems, 30*(1): 541–553.

Next Generation Computing and Information Systems – Gupta. (Ed.)
© 2025 The Author(s), ISBN 978-1-032-73865-9

Time series analysis of vegetation change using remote sensing, GIS and FB prophet

Pushpendra Kushwaha, Azra Nazir, Rishita Bansal & Faisal Rasheed Lone
School of Computer Science & Engineering, VIT Bhopal University, Madhya Pradesh, India

ABSTRACT: Forecasting techniques are crucial for any industry or field that benefits from anticipating future events or trends, allowing for better planning, resource allocation, and decision-making. Modern agriculture necessitates efficient land use and the reconversion of wastelands to productive terrains. This study seeks to harness forecasting techniques tailored for time series data to detect potential wastelands ripe for conversion into Greenlands, aiming to boost agricultural yield. The time series analysis of NDVI (Normalized Difference Vegetation Index) assesses vegetation cover changes over time, identifies trends and patterns, and maps the areas with potential restoration. The time series data is processed and analysed using advanced techniques such as season decomposition and statistical modelling to identify the significant changes in vegetation behaviour, specifically focusing on incorporating the FBprophet model to elucidate the temporal patterns and underlying factors influencing vegetation dynamics. In this study, the model is employed on 9 years (2013–2022) data of a village in Katoria Bihar (India) that was restored from a wasteland into Greenland. The study's finding showcases the FBprophet model's effectiveness in analysing NDVI time series data and understanding vegetation dynamics. By providing more accurate trend decomposition and robust forecasting capabilities, the FBprophet model offers a sustainable approach to land utilisation, benefitting both local communities and the broader environment.

1   INTRODUCTION

Forecasting techniques have long been pivotal in diverse sectors, from economics to meteorology. Their application to agriculture, especially in the domain of land utilisation, remains relatively nascent but exceedingly promising. Time series data, with its chronologically ordered information, provides a rich source for analysis (Mahalakshmi *et al.* 2016). This research endeavours to predict where wastelands exist and can be rejuvenated by applying forecasting techniques to such data. Land use practices, deforestation, and climate change have led to the degradation of the local ecosystem and the depletion of critical ecosystem services. In response to this critical situation, this study seeks to find solutions utilising advanced remote sensing technology and the application of time series analysis of the Normalized Difference Vegetation Index (NDVI) to monitor and restore vegetation dynamics for better agricultural yield.

The core approach of this study involves leveraging the capabilities of Google Earth Engine (GEE), a powerful cloud-based platform specifically designed to analyse vast volumes of satellite imagery and geospatial data. The significance of utilising NDVI times series analysis lies in its ability to reveal long-term trends and changes in vegetation cover, which is crucial for understanding the dynamics of land transformation and recovery. The study seeks to discern vegetation growth and decline patterns by monitoring these trends, allowing for informed decision-making in implementing targeted restoration efforts. The village of Katoria was chosen as a case study for data collection. This village is situated in a

DOI: 10.1201/9781003466383-10

region vulnerable to land degradation and desertification. These variations make this area ideal for gathering time series data on soil health and vegetation cover.

Remote sensing and GIS have been used to describe the different features of the study area. Landsat 8, officially known as the Landsat Data Continuity Mission (LDCM), is a satellite operated by the United States Geological Survey (USGS) and the National Aeronautics and Space Administration (NASA). Landsat carries two main instruments: an Operational Land Imager (OLI) and a Thermal Infrared Sensor (TIRS). The OLI Sensor captures data in nine spectral bands, including visible, near-infrared, and short-wave infrared. The OLI sensor has improved signal-to-noise performance, radiometric resolution, and spatial resolution compared to its predecessor. The TIRS sensor measures the Earth's thermal infrared radiation, providing temperature data for monitoring surface temperature and heat emissions. Still, the main focus is on OLI sensors for finding the NDVI parameters.

Traditional forecasting models, like ARIMA, need extensive parameter adjustment and expertise in analytics and statistics (Newbold 1983). FBProphet is an open-source library created by Facebook that is simple to integrate into an automated production system and requires very little domain knowledge (Montgomery et al. 2015). A multifaceted approach has been employed to enhance NDVI prediction for vegetation cover. This method augments prediction accuracy and ensures the model's adaptability to dynamic environmental conditions. By implementing cloud removal algorithms on satellite imagery, the hindrance caused by cloud cover is mitigated, thereby increasing the frequency of available training data.

Additionally, a classification algorithm to generate a vegetation-only mask allows for a more precise calculation of NDVI by isolating vegetated areas. To further elevate the model's performance, the image source is visited with the shortest revisit times, enabling the capture of rapid changes in NDVI, which is especially crucial for real-time monitoring. This approach extends beyond routine predictions; it encompasses identifying and integrating significant events such as natural disasters, droughts, or forest fires, which can trigger substantial NDVI variations. By incorporating prediction errors as a regressor within the FBProphet model, a feedback loop continually refines forecasts and bolsters accuracy. This holistic and data-driven strategy ensures the comprehensive improvement of NDVI predictions for accommodating the ever-changing landscape and environmental influences.

2   RELATED WORK

Numerous studies have been conducted for NDVI analysis across various problem areas utilising Time Series Analysis. A multi-resolution analysis (MRA) based on the wavelet transform(WT) has been implemented in (Martínez and Gilabert 2009a) to study NDVI time series. (Waylen et al. 2014) provides an in-depth survey for finding vegetation and deforestation of the earth from image processing. (Martínez and Gilabert 2009b) deals with remote sensing and normalised difference vegetation index with various techniques. The authors extensively cover theoretical and empirical contributions to image processing in the current decade. Also, NDVI generation using various deep learning techniques has been examined in (Zaidi et al. 2017). A study consisting of the NDVI forecasting model, which is based on the combination of Time Series Decomposition (TSD), Long Short-Term Memory (LSTM) and Convolutional Neural Network, was proposed in (Gao et al. 2023). Two forecasting models of temperature and precipitation based on their historical information and four NDVI forecasting models based on temperature, rain, and their historical information were established to confirm the performance of the TSD-CNN-LSTM model and investigate the response of NDVI to climatic factors. When the findings of the correlation analysis are combined, it can be concluded that temperature has the greatest significant influence on NDVI changes, followed by precipitation, which has the most minor influence. (Khamchiangta and Dhakal 2020) conducted a comprehensive comparative analysis of different forecasting models for NDVI, including traditional time series methods, deep learning approaches, and FBprophet. Their work provided valuable insights into the strengths and limitations of each model. The results indicated that FBprophet consistently outperformed other methods in capturing the seasonality and trend patterns inherent in NDVI data. (Kumar Jha and Pande 2021) explored the application of machine learning models, including FBprophet, for NDVI forecasting. Their study demonstrated the potential of FBprophet in capturing the seasonal and long-term trends in NDVI time series, which is crucial for assessing vegetation health and land-use management. They highlighted the model's ability to handle missing data points and outliers, making it suitable for real-world applications.

3 METHODOLOGY

FBProphet model breaks down any time series data into components related to trends, seasonality, events, or holidays. It can be expressed using equation 1, where T(t) is the logistic or piecewise linear growth curve for trend component modelling. Cycles in the time series (daily, weekly, monthly, and quarterly) are denoted by the symbol S(t). H(t) is the impact of unplanned occurrences or holidays, a noisy phrase impossible to model mathematically.

$$Y(t) = T(t) + S(t) + H(t) + \epsilon \tag{1}$$

It is important to emphasise that FBProphet considers forecasting an endeavour of curve fitting rather than the typical time-based interdependencies. This new viewpoint is essential to FBProphet's ability to produce precise predictions with low requirements for domain-specific knowledge. The time series data is divided into trend, seasonality, and residual as part of the decomposition process. The trend is the time series' long-term trend or direction. Seasonality is the recurring and periodic patterns in a time series that happen daily, weekly, or annually. In residual, the prophet allows the inclusion of holidays and special events as additional components that can impact the time series. It's crucial to emphasise the importance of choosing the right type (Logistic Growth or Piecewise Linear) based on the characteristics of the data. For example, the Logistic Growth Model should be selected if the NDVI data exhibits signs of saturating growth. On the other hand, if the NDVI data shows linear features with past growth or shrinkage tendencies, the default Piecewise Linear Model is preferable. This model can accommodate linear trends, making it suitable for various business and environmental data. The mathematical representation of the model is given by equation 2. Here, C is the carrying capacity, representing the maximum limit a saturating growth model can reach. It should be set to reflect the data's capacity for growth. The growth rate parameter that influences the rate at which the data approaches the carrying capacity is given by k. It should match the data's actual growth rate. Also, m is the model's offset parameter for a vertical shift. It can be adjusted to align the model with the data.

$$g(t) = \frac{C}{1 + e^{-k(t-m)}} \tag{2}$$

The piecewise linear model is fit using statistical equation 3, where c is the trend change point, which can be fine-tuned per the requirement.

$$y = \begin{cases} \beta_0 + \beta_1 x & x \leq c \\ \beta_0 - \beta_2 c + (\beta_1 + \beta_2)x & x > c \end{cases} \tag{3}$$

4 EXPERIMENTAL SETUP AND DATASET

The study area in this paper is located in the outer regions of Bihar. The village of Katoria is near the Banka district. The village was barren, which made it unsuitable for cultivation, and hence, it had little vegetation in 2015. However, with determination, community participation, and afforestation projects, the green cover was reclaimed in the 2021–2022. The variation observed over these years is a perfect fit for quantitative forecasting techniques needed by AI/ML models. The dataset used in this study is the standard Landsat-8 Collection 1 Tier 1 8-Day TOA Reflectance Composite product[12]. The Landsat data products are organised into different collections. Collection 1 represents the first systematic processing of Landsat 8 data, following specific standards to ensure consistency and accuracy. The USGS provides Landsat data in different processing tiers. Tier 1 products are top-of-atmosphere (TOA) reflectance, which means the data represents the raw measurements of the amount of light reflected by the Earth's surface before any atmospheric correction. The data is composited over 8 days. Instead of providing single images, the USGS creates a composite image by combining data from multiple cloud-free scenes acquired during an 8-day window. Top-of-atmosphere (TOA) Reflectance is the amount of sunlight reflected by the Earth's surface, measured at the top of the atmosphere. This metric is valuable for comparing surface reflectance values across locations and dates. This 8-day composite is particularly useful for monitoring changes in land cover and vegetation over time, as it minimises the impact of cloud cover and atmospheric effects that can obscure the Earth's surface in individual images.

The image extraction was done using LANDSAT/LC08/C01/T1_SR with cloud coverage per cent at 20, and the images were extracted from 2013 to 2023. This typically means that the images were obtained from the satellite on the days when cloud coverage was less than 20% using the bands B4 and B5. Figure 1 compares the image of the village from 2013, when it was barren, to 2022 when part of the land was restored. From every image, the NDVI value was extracted and interpolated monthly to retrieve more consistent data around the period.

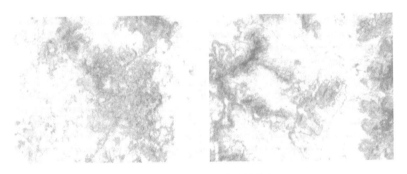

Figure 1.   Satellite Image of Katoria in 2013(left) vs. 2022(right).

The different features seen in the 3-band satellite image of the Katoria region are extracted using the NDVI. Based on the distinctive reflectance patterns of green vegetation, vegetation indices enable us to define the distribution of soil and vegetation. The NDVI is a straightforward numerical indicator that may be used to evaluate remote sensing data from a remote platform and determine whether or not the target or object being examined has live, green vegetation. The NDVI is determined using equation 4, where NIR stands for near-infrared reflectance, and RED is for visible red reflectance.

$$RNDVI = (NIR - RED)/(NIR + RED) \qquad where \ 0 < NDVI < 1$$
$$GNDVI = (NIR - GREEN)/(NIR + GREEN) \quad where \ 0 < NDVI < 1 \qquad (4)$$

Rock, sand, or snow-covered deserts correspond to very low NDVI values (0.1 and lower). High numbers denote temperate and tropical rainforests (0.6 to 0.8), whereas moderate values imply shrub and grassland (0.2 to 0.3). Water bodies are represented with negative NDVI values, while bare land is represented by NDVI values that are closest to 0 (Montgomery *et al.* 2015).

44

5 PERFORMANCE ANALYSIS AND VISUALISATION

Model performance is evaluated by comparing predicted values with test data. Comparison is made in terms of Mean Squared Error (MSE), Mean Absolute Error (MAE), and Root Mean Squared Error (RMSE). The average of these metrics is taken and named 'average_score' to get a sense of the overall fit of the prediction with test data. Figure 2 shows monthly interpolated data. The monthly analysis of the data shows that due to higher cloud cover and rain, less data is available for July, Figure 3. Also, September is the crop's flowering time, and yellow flowers can decrease the NDVI values. June is the hottest month in the region and has the worst NDVI, implying that the region is unsuitable for summer crops. All the winter months, starting from October to January, had the highest NDVI index, which means the region is suitable for the winter crops. The plants start to form grains during the month of October. Therefore, a significant peak in the NDVI index is seen. It is the period when farmers need to ensure that fields are kept flooded and that there is enough sunlight for the grains to mature. The NDVI index increased in the post-COVID year, which depicted better crop growth after COVID-19, and there could be multiple reasons for this change in the index.

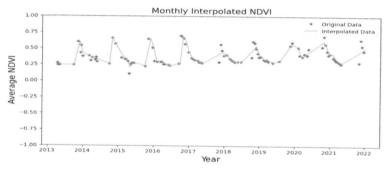

Figure 2.   Average NDVI vs Monthly interpolated data over the years.

The lowest NDVI index was in the year 2015, which has the most outliers in the data Figure 3. According to the Indian Meteorological Department (IMD), Bihar witnessed rainfall deficiency till the middle of 2015, declaring a drought as the rainfall touched 28% lower than the average. The horizontal line in the box represents the median NDVI, the lower and upper box edges represent the first and third quartiles, respectively, and the whiskers represent the data minimum and maximum. The whiskers show a few outliers in the data, with some NDVI values falling below 0.2 or above 0.7. However, these outliers are relatively rare.

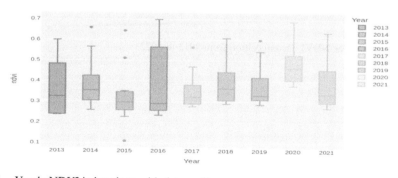

Figure 3.   Yearly NDVI index along with data outliers.

The health of the crops is visualized in Figure 4. It can be concluded that in all these years, most of the crops lay in the unhealthy region, and only a few crops lie in the moderately healthy region. The crops in the healthy region are significantly less as these crops don't grow over the entire year. There are specific months that had higher NDVI Index. NDVI analysis can direct where to investigate the further or broader damage assessment if the factors influencing plant health are known. The maximum NDVI was recorded as 0.69, and the minimum was recorded as 0.11. The NDVI, between 0.27 and 0.359, depicts moderately healthy plants.

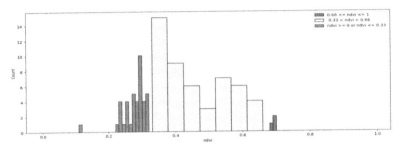

Figure 4.   NDVI index depicting health of cultivation.

The trend, seasonality and residual of the data from 2013 to 2022 are shown in Figure 5. For this study, the modelling exercise is limited to a foundational model structure. However, the potential for augmentation is substantial, encompassing the incorporation of FBProphet's potent features, such as accommodating saturating growth, detecting shifts in trends, and capturing the impacts of unique events.

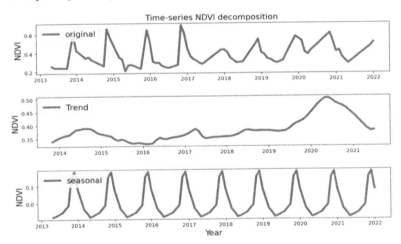

Figure 5.   Time-series NDVI decomposition from 2013 to 2022.

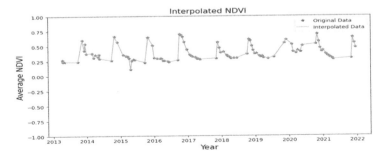

Figure 6.   Up sampling of data for daily frequency.

46

FBProphet is harnessed to model the prediction of the NDVI for the collected data. As opposed to time-based dependence, FBProphet considers forecasting to be a curve-fitting problem. The date column is upsampled to daily frequency by interpolating it daily to increase the data size to get more accurate predictions, and the NDVI values have been linearly interpolated to fill in any missing values, Figure 6.

The training data contains all the NDVI values from the beginning of the time series up to the cutoff date, while the testing data contains all the NDVI values from the cutoff date to the end of the time series. The plot in Figure 7 shows the NDVI values over time, with the training data in green and the testing data in blue. The y-axis ranges from −1 to 1, and the x-axis shows the years. The plot demonstrates how the data has been split into training and testing sets based on the cutoff date of 2020-01-01.

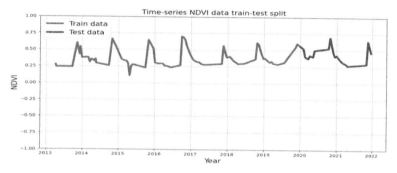

Figure 7.   Train and test data split.

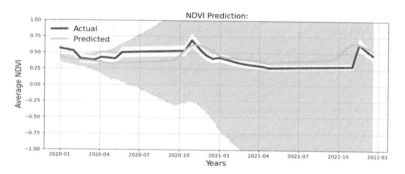

Figure 8.   Forecasting the results on original data.

The FB Prophet model is applied to NDVI data with daily seasonality set to false and frequency for the prediction set to daily. It can be seen from the results that actual NDVI data is infrequent, whereas the model trains based on daily observation. More frequent data is expected to improve the prediction significantly. The MSE of the trained model is **0.0137,** the MAE is **0.100,** the RMSE is **0.11,** and the average score is **0.077.** The shaded region in Figure 8 shows how uncertain the prediction is coming timeline. The wider the shaded region, the more uncertainty there is in the forecast, and the narrow part relatively offers certainty in the prediction. Our approach yields valuable results and can generate a list of potential wastelands targeted for agricultural enhancement. Finding special events that trigger a change in NDVI data, such as natural disasters, drought or forest fire, can be included in the FB Prophet Model for much more robust prediction. The prediction error can be identified and added as a regressor to enhance the model further.

6 CONCLUSION

In the nexus of modern technology and traditional agricultural practices lies the promise of increased yields and more sustainable land use. A comprehensive study is presented to transform wastelands into Greenlands by integrating forecasting techniques with time series analysis. The research monitors and restores vegetation dynamics by leveraging Google Earth Engine and the FBprophet model, contributing to ecological balance and climate change mitigation. The study highlights the urgency of addressing land degradation and desertification in regions like Katoria, emphasising the role of NDVI as an indicator of vegetation health. Notably, the analysis reveals insights into the impact of climate factors and special events, such as the COVID-19 pandemic, on NDVI trends. Integrating the FBprophet model proves promising in predicting NDVI values and accommodating event-based variations, turning barren lands into blooming fields. However, future improvements could include incorporating additional variables and a deeper understanding of local factors influencing vegetation dynamics.

REFERENCES

Cohen, W.B., Healey, S.P., Yang, Z., Zhu, Z. and Gorelick, N. 2020. Diversity of algorithm and spectral band inputs improves landsat monitoring of forest disturbance. *Remote Sensing 2020, Vol. 12, Page 1673* 12(10), p. 1673. Available at: https://www.mdpi.com/2072-4292/12/10/1673/htm.

Gandhi, G.M., Parthiban, S., Thummalu, N. and Christy, A. 2015. Ndvi: Vegetation change detection using remote sensing and Gis – A case study of Vellore district. *Procedia Computer Science* 57, pp. 1199–1210. doi: 10.1016/J.PROCS.2015.07.415.

Gao, P., Du, W., Lei, Q., Li, J., Zhang, S. and Li, N. 2023. NDVI forecasting model based on the combination of time series decomposition and CNN – LSTM. *Water Resources Management* 37(4), pp. 1481–1497. Available at: https://link.springer.com/article/10.1007/s11269-022-03419-3.

Khamchiangta, D. and Dhakal, S. 2020. Time series analysis of land use and land cover changes related to urban heat island intensity: Case of Bangkok metropolitan area in Thailand. *Journal of Urban Management* 9(4), pp. 383–395. doi: 10.1016/J.JUM.2020.09.001.

Kumar Jha, B. and Pande, S. 2021. Time series forecasting model for supermarket sales using FB-Prophet. *Proceedings – 5th International Conference on Computing Methodologies and Communication, ICCMC 2021*, pp. 547–554. doi: 10.1109/ICCMC51019.2021.9418033.

Mahalakshmi, G., Sridevi, S. and Rajaram, S. 2016. A survey on forecasting of time series data. *2016 International Conference on Computing Technologies and Intelligent Data Engineering, ICCTIDE 2016*. doi: 10.1109/ICCTIDE.2016.7725358.

Martínez, B. and Gilabert, M.A. 2009a. Vegetation dynamics from NDVI time series analysis using the wavelet transform. *Remote Sensing of Environment* 113(9), pp. 1823–1842. doi: 10.1016/J.RSE.2009.04.016.

Martínez, B. and Gilabert, M.A. 2009b. Vegetation dynamics from NDVI time series analysis using the wavelet transform. *Remote Sensing of Environment* 113(9), pp. 1823–1842. doi: 10.1016/J.RSE.2009.04.016.

Montgomery, D.C., Kulahci, Murat. and Jennings, C.L. 2015. Introduction to time series analysis and forecasting.

NDVI, *Mapping a Function over a Collection, Quality Mosaicking | Google Earth Engine | Google for Developers*. Available at: https://developers.google.com/earth-engine/tutorials/tutorial_api_06.

Newbold, P. 1983. ARIMA model building and the time series analysis approach to forecasting. *Journal of Forecasting* 2(1), pp. 23–35. Available at: https://onlinelibrary.wiley.com/doi/full/10.1002/for.3980020104.

Waylen, P., Southworth, J., Gibbes, C. and Tsai, H. 2014. Time series analysis of land cover change: Developing statistical tools to determine significance of land cover changes in persistence analyses. *Remote Sensing 2014*, 6(5), pp. 4473–4497. Available at: https://www.mdpi.com/2072-4292/6/5/4473/htm.

Zaidi, S.M., Akbari, A., Samah, A.A., Kong, N.S., Isabella, J. and Gisen, A. 2017. Landsat-5 Time Series Analysis for Land Use/Land Cover Change Detection Using NDVI and Semi-Supervised Classification Techniques. *umpir.ump.edu.mySM Zaidi, A Akbari, A Abu Samah, NS Kong, A Gisen, J IsabellaPolish Journal of Environmental Studies,2017•umpir.ump.edu.my* 26(6),pp. 2833–2840. Available at: http://umpir.ump.edu.my/19825/1/Pol.J.Environ.Stud.Vol.26.No.6.2833-2840.pdf.

Next Generation Computing and Information Systems – Gupta. (Ed.)
© 2025 The Author(s), ISBN 978-1-032-73865-9

Advances in smart farming for precision agriculture: Green-IoT and machine learning as a solution

Monia Digra & Preeti Rajput
Model Institute of Engineering and Technology, Jammu, J&K, India

Pooja Sharma
School of Computer Science and Engineering, Shri Mata Vaishno Devi University, Katra, J&K, India

ABSTRACT: Agricultural development is crucial to ending poverty and boosting economic growth in emerging nations. The global population is expected to reach 9.7 billion by 2050. Smart and precise agriculture must enhance agricultural productivity without compromising quality, nutrition, or natural resources. IoT and AI/ML have significant uses in agriculture due to internet technologies. Although IoT has many benefits, the enormous number of networked devices may raise carbon emissions and e-waste due to energy consumption. Inspired by the goal of sustainable smart agriculture, this article addresses major results and research gaps in IoT-based smart agriculture and provides solutions for energy-saving Green-IoT (G-IoT) applications in agriculture. We start with G-IoT's evolution and smart agriculture applications. We also examine how green architecture may efficiently apply AI/ML in smart agriculture. We propose an AI/ML G-IoT framework for smart agriculture with green sensor cloud and green AI. This unique G-IoT paradigm provides sustainable and energy-efficient agricultural architecture. We investigate statistical and experimental data and emphasise the limitations of adopting G-IoT and AI/ML architecture in agriculture. In conclusion, G-IoT and AI/ML architecture can enable sustainable, energy-efficient, smart precision agriculture.

1 INTRODUCTION

Agriculture plays an important role in the economic growth of any country. In 2014, one-third of the Gross Domestic Product (GDP) was accounted for through agriculture worldwide (The World Bank 2015). For example, India ranks second after the United States of America in agricultural output as per the Indian economic survey of 2018. It contributes 17–18% of India's GDP and employs 50% of the workforce (Ministry of Statistics and Program Implementation | Government Of India). Traditional agriculture is the major and dominant activity of India's economy, like in other developing countries. Every year, millions of food consumers have been added globally. Agricultural production is going to be affected by climate change and global warming. However, all developing nations like India confront multiple risks and uncertainty associated with agricultural output. Unplanned irrigation, poor management of crops and pesticides, and spreading of pathogens and weeds cause production deficits.

However, IoT devices and applications produce carbon emissions, consume massive energy, and embrace e-waste and toxic pollutants. These must also be green to address global warming and climate change. The solution to this problem is Green-IoT (G-IoT) in smart agriculture. G-IoT focuses on a green design by using low energy consumption devices, low data transmission energy, and carbon-free materials and promotes reusability and minimal greenhouse gas emissions to counter climate change (Arshad *et al.* 2017).

It has been seen in the past year from Google web search data as shown in Figure 1 that the popularity of keywords "Smart Agriculture", "IoT", and "Green Technology" is increasing.

DOI: 10.1201/9781003466383-11

Figure 1. Google trends identify the keywords "Smart Agriculture", "IoT", and "Green Technology" for the last 2 year.

2 RELATED WORK

2.1 *Green-IoT*

The exponential growth of the Internet of Things (IoT) has brought about a revolutionary shift in connectivity, revolutionizing various industries and our daily lives by seamlessly connecting devices, sensors, and data. Although IoT has brought about remarkable convenience and efficiency, there are growing concerns regarding its impact on the environment. There has been a significant increase in interest regarding the combination of IoT and sustainability due to concerns about energy consumption, resource utilization, and waste generation. This has resulted in the emergence of "Green IoT".

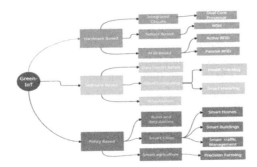

Figure 2. Taxonomy of Green-IoT.

This study paper seeks to conduct an extensive investigation on Green IoT, examining its fundamental principles, practical applications, obstacles, and possible remedies. Our study aims to explore the convergence of IoT and environmental sustainability, with a focus on how Green IoT can accelerate the transition towards a more environmentally responsible future.

2.2 *Smart agricultural G-IoT applications.*

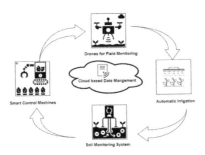

Figure 3. Applications of G-IoT in agriculture.

We examine smart agriculture applications and a G-IoT concept. Figure 3 shows an integrated model of smart agriculture. Green IoT has numerous applications in the various data hierarchies in smart agriculture, including data filtering or cleansing, data processing, data classification, data contextualization, and data condensation (Maddikunta *et al.* 2021; Ruan *et al.* 2019). The model above indicates that UAVs can monitor big amounts of land better than humans. These can readily identify nutrient deficiencies, pests, diseases, weeds, and temperature to determine plant stress. An SP2F framework provides security limitation, scalability, traceability, and fault tolerance (Kumar *et al.* 2021). Renewable energy sources, such as wind and solar energy, can improve UAV battery life (Adão *et al.* 2017; Maddikunta *et al.* 2021). Smart irrigation and agricultural management systems can be controlled by G-IoT.

Farmers may monitor their farm activities from anywhere using a user-friendly software interface to control energy-efficient and cost-effective autonomous irrigation systems. Intelligent grids, prognostics, and renewable energy systems can use G-IoT (Ku *et al.* 2016). Solar-powered smart irrigation systems with temperature, moisture, and humidity sensors solve the problem. Autonomous smart control robots like Agbots solve crop management

Table 1. Research findings and gaps in IoT based smart agriculture with G-IoT as a solution.

Ref.	Year	Focus area	Research findings and gaps with G-IoT as a solution.
(Farooq *et al.* 2020)	2020	Precision Agriculture	The author described the various IoT sensors, protocols, and networks in agriculture using IoT but lacks in providing the energy-efficient protocols, green network technology, green disposal, and efficient communication with less carbon emission.
(Khanna & Kaur 2019)	2019	Precision Agriculture	The author has discussed the IoT communication protocols, open issues, and challenges in IoT-based smart agriculture. The author doesn't provide the existing framework or architecture for smart agriculture. It lacks to provide the solution for using the G-IoT communication protocols.
(Kour & Arora 2020)	2020	Smart Agriculture	The author has explained the various communication protocols, goals, IoT OS, architecture and discussed the smart agriculture applications like crop monitoring, irrigation, farm management, and disease detection with various future directions but lacks in providing an efficient and sustainable green framework in smart agriculture as a solution for a future world.
(Saad *et al.* 2020)	2020	Water Management	The author has concentrated on the water management problem and presents techniques that strive to optimize water consumption and improve the overall quality of agriculture. The capacity to do energy-efficient real information processing for controlling and making decisions during the watershed management system is highly suggested.
(Farooq *et al.* 2019)	2019	Smart Farming	The researchers address the numerous IoT issues and security requirements for smart agriculture with open policies in various cities. The author should raise the concept of green communication, green disposal, reducing energy consumption, and improving sustainability.
(Ferrag *et al.* 2019)	2019	Fog enabled IoT	The authors should configure the proxy servers at the fog computing layer and use the anonymous key agreement protocol in smart farming.
(Boursianis *et al.* 2022)	2022	UAV and IoT in smart Agriculture	The authors have analyzed the uses of UAVs in numerous situations, such as pesticides and weed control, plant growth monitoring, irrigation, fertilizers, and crop diseases to illustrate the importance of UAVs in smart agriculture. The author should introduce the concept of optimizing algorithms with energy efficiency and can also use solar energy to reduce the power consumption in UAV devices.

machinery. Data-driven decisions, soil health compaction, autonomous weed and pest classification, and more are possible with these Agbots. AI algorithms teach these Agbots to identify pests, recognize fruits, and count them (Agbots 2020). To make the right decisions, predictions, forecasts, and predictive models in agriculture, a lot of data must be processed throughout the lifetime (Mohanraj *et al.* 2016). Cloud or fog computing can handle massive amounts of data to make the right judgements and automate the agricultural process with G-IoT (Bibri 2018).

2.3 *Industry applications of AI/ML in the precision agriculture*

In this section we show the industrial applications of AI/ML in the precision agriculture. Table 2. Provides a list of the software platforms exist in precision agriculture which employ Big Data Analytics, Mobile Computing, Cloud Computing, IoT and AI/ML. We summarize these applications with their strength and also mention whether these applications adapt G-IoT.

Table 2. Software applications of G-IoT in smart agriculture.

Applications	Adapted G-IoT Yes/No	Strength	Description
CropIn (*Cropin*, n.d.)	Yes	The proposed framework provides monitoring and detection of pest, risk management and livestock management.	An application which provides the alert logs to detect the pesticides, satellite information, traceability and output productivity, geo-tagging.
AgNext (*AgNext – digitizing food quality* 2019)	Yes	It provides AI based technology to count the leaf of tea and provide international market trends.	This application provides the quality, traceability and data initiatives with image analysis, spectral analysis and sensor analysis.
Cropx Technologies (*CropX technologies* 2022)	Yes	Smart irrigation and optimize nitrogen management by optimizing soil analysis.	Smart irrigation solution which provides the application-based platform for varieties of pulses, rice and wheat.
Biz4intellia (*An End-to-End IoT Solutions provider* n.d.)	Yes	All in one platform for soil analysis and smart farming.	A framework which provides the solution of both soil and crop management. It uses ML and big data to provide the soil analysis, diseases detection, energy management, livestock tracking etc.
Semios (*Know more, worry less.*™ 2020)	Yes	Provide predictive analytics solution using big data and ML.	A platform having a tool which tampers for yield improvement. It is used to assess plant growth and detect diseases.
Sensefly (*Agriculture* 2022b)	Yes	It monitors the field only with drones.	This platform provides a monitoring system using drones. It provides emergency and early growth assessment through pre-harvest prediction.
Farm Management Software (*FarmLogs:*.)	No	The framework provides the current price in the market and price fluctuation.	This platform maintains the crop, facilitating the planning and monitors of the field condition.
Plantix (PEAT GmbH. n.d.)	No	For the detection of diseases.	A smartphone application, uses ML algorithm to detect the diseases present in any plant.
Farmapp (*Farmapp.* N.d.)	No	The platform is used to compare previous and current measures and provides predicted values of farm activities like diseases, pest etc.	A software application which records the data quickly into the server and provides the report of solutions in a chart and graphs.

3 DISCUSSION

3.1 *Statistical and meta-analysis*

A total of 197 articles were identified through the process of screening and eligibility in various repository such as SCOPUS, IEEE xplore etc. These articles were then refined using the search query "green AND IoT AND internet AND of AND things AND farming AND agriculture". From this refinement, a final database of 36 articles was identified. These 36 articles were accepted for inclusion in the meta-analysis, as depicted in Figure 4.

Figure 4.   Sequence diagram of the search database and meta-analysis.

As most of the articles published on IoT based smart agriculture, IoT has increased rapidly to provide the good quality harvesting by controlling the irrigation, automate the production of agriculture and make day-to-day work easy but it also produces the carbon emission, energy use and e-waste. To transform the IoT based application to Green-IoT which aims to reduce the carbon footprints and enhance the model. In Figure 5, it summarizes the statistical analysis of research articles in IoT based smart agriculture using AI/ML techniques. As shown in the Figure 6, It has been estimated that scope of the articles in G-IoT has been increased in the upcoming years and whole community has shifted their interest towards the green communication for sustainable environment in most state-of-the-art approaches for a diverse range of applications. By statistical analysis, we conclude that the G-IoT based application are still at young age and has lot of scope in diverse application in future.

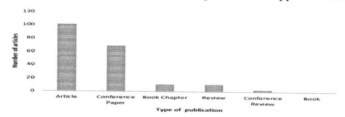

Figure 5.   Number of research articles, conferences and review articles increasing from 2013–2021.

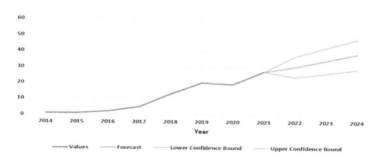

Figure 6.   Prediction of Green-IoT based applications.

3.2 Comparison of IoT and G-IoT applications

The utilization of intelligent devices at the foundation of IoT devices that demand powerful connection, a dependable interface, and real-time analysis and data analysis is termed as the next phase of IoT (Neagu *et al.* 2019). Cutting-edge digital revolution, along with effective approaches and strategies like 5G technology, smart storage, smart agriculture and smart healthcare, will transform our future environment, making it healthier, smarter, and greener while maintaining a high quality of service. The number of network-connected gadgets is expected to grow in the future. As a result, Green IoT architecture must adapt. Current studies do not address thoroughly on smart agriculture policies and methodologies for enabling greener smart agriculture. To the best knowledge of the authors, previous research has not focused to analyzing technology and framework in G-IoT based smart agriculture, such as enabling ICT, decreasing energy consumption, decreasing CO_2 emissions, minimizing waste management, and enhancing sustainability. We offered green energy methods to power IoT devices in this article. Despite huge research efforts, G-IoT in smart agriculture technology is still in its initial stages as depicted in Figure 7.

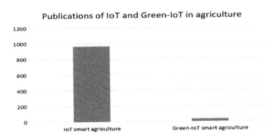

Figure 7.   Comparison of IoT and G-IoT in smart agriculture.

Table 3.   Experimental analysis of energy efficient protocols/ algorithms in G-IoT.

Ref.	Focus area	Protocols / algorithms	Parameters	Energy efficiency	Throug- hput	Network lifetime
(Abdul-Qawy & Srinivasulu, 2019)	Green IoT enabled energy efficient protocol	SEES	Residual Energy, nearest relay node.	62%	114%	42%
(Behera *et al.* 2019)	Residual energy in IoT applications	Enhanced – LEACH	Network diameter, total number of nodes, network energy, energy dissipation.	+ 64%	+60%	+66%
(Mahmud & Cho 2019)	FANET routing protocol in Green UAVS	(EE-Hello) energy –efficient scheme.	Mobility model, number of UAVs, source/ destination pairs, loss model, channel capacity, routing protocols, simulation time, packet size / rate.	AODV (25%) and OLSR (23%)	+25%	–
(Shahid *et al.* 2020)	Fog computing using caching mechanism	Load-balancing algorithm	Current energy level, computational node capacity, active nodes and computational availability.	92.6 percent less energy and 82.7 percent less energy than the basic caching method.	–	Increased.

3.3 *Key issues and challenges in G-IoT*

To achieve a reliable and accurate system for smart and precision agriculture application using G-IoT and AI/ML, there are many challenges which need to be addressed. These challenges are identified as following:

3.3.1 *Real-time data and interference in noise*

G-IoT-enabled smart agriculture uses real-time sensors to collect data, enabling informed decisions and operational efficiency. However, transmitting this data across a large network presents challenges. Implementing real-time data processing in a large cloud-based infrastructure can introduce complexity to architectural design. Interference and noise can impact data volume and reliability, potentially causing data loss and reducing the overall reliability of the G-IoT platform.

3.3.2 *Quality and accuracy of data*

The presence of data noise and inaccuracies is a common occurrence in sensor data, often resulting from inherent limitations and potential faults in the data collection process. The maintenance of data accuracy and dependability is of utmost importance in the context of decision-making. The topic of discussion pertains to the calibration and maintenance of equipment or systems.

3.3.3 *Data security*

The topic of privacy and security in the context of the Internet of Things (IoT) has been elucidated in the work of (Turgut and Boloni 2017). (Asplund and Nadjm-Tehrani 2016) assert that the primary impediment to the implementation of IoT devices is security. In the realm of agricultural, the implementation of G-IoT devices encounters challenges related to potential tampering by theft, intrusion, or animal interference, hence posing difficulties in their deployment within open field environments. Machine learning applications have the potential to introduce fabricated data, so undermining the reliability and integrity of the system. Preserving privacy is a significant consideration in the context of smart agriculture, as a substantial volume of data originates from individuals who may have concerns over the potential disclosure of their private information.

3.3.4 *Optimization approach*

The optimization of resources is a crucial aspect to consider in various fields. The G-IoT resource optimisation platform has the capability to recognise various types of sensors. According to (Turgut and Boloni 2017), the sensing capabilities of agricultural systems can be utilised to monitor crops and livestock by considering factors such as the volume of data to be transferred, the capacity of cloud storage, and the number of Internet of Things (IoT) sensors, all of which can vary depending on the size of the farmland. In order to reduce expenses for farmers and enhance crop productivity and profitability, the implementation of a sophisticated Internet of Things (IoT) framework is necessary.

4 CONCLUSION

In conclusion, the use of G-IoT applications in agriculture, along with AI/ML approaches, can lead to precision agriculture and help solve the world's hunger problem. The expansion of the internet to G-IoT has enabled various agricultural applications, making agriculture smarter and more efficient. G-IoT allows farmers to remotely monitor farm activities and manage farms efficiently using renewable sources of green energy. However, there are still key issues and challenges that need to be addressed for large-scale deployment in agriculture practices.

Future research in G-IoT should focus on understanding the characteristics and service needs of different IoT applications, designing power-efficient components, considering green IoT architecture to optimize energy usage while meeting service goals, and exploring the potential of virtualized sensor as a service (SNaaS) for consumer access and management of private IoT.

REFERENCES

Abdul-Qawy, A.S.H. and Srinivasulu, T., 2019. SEES: a scalable and energy-efficient scheme for green IoT-based heterogeneous wireless nodes. *Journal of Ambient Intelligence and Humanized Computing*, 10(4), pp. 1571–1596.

Adão T, Hruška J, Pádua L, Bessa J, Peres E, Morais R, Sousa J. 2017. Hyperspectral imaging: A review on UAV-Based sensors, data processing and applications for agriculture and forestry. *Remote Sensing*. 9 (11):1110. https://doi.org/10.3390/rs9111110

AgNext – Digitizing Food Quality | AI-based Food Assessment Technology. *Agnext [Internet]*. https://agnext.com/

Agriculture.AgEagle Aerial Systems Inc [Internet]. [accessed 2023 Oct 16]. https://www.sensefly.com/industry/agricultural-drones-industry

AgriTech Stories,Videos, Articles, Interviews, Reviews & News | AgriTechTomorrow. wwwagritechtomorrowcom [Internet]. https://www.agritechtomorrow.com/

An End-To-End IoT Solutions Provider. *Biz4Intellia [Internet]*. [accessed 2023 Oct 16]. https://www.biz4intellia.com/

Arshad R, Zahoor S, Shah MA, Wahid A, Yu H. 2017. Green IoT: An Investigation on Energy Saving Practices for 2020 and Beyond. *IEEE Access*. 5:15667–15681. https://doi.org/10.1109/access.2017.2686092

Asplund M, Nadjm-Tehrani S. 2016. Attitudes and perceptions of IoT security in critical societal services. *IEEE Access*. 4:2130–2138. https://doi.org/10.1109/access.2016.2560919

Best Farm Management Software. Bushel Farm [Internet]. [accessed 2023 Oct 16]. https://farmlogs.com

Bibri SE. 2018. The IoT for smart sustainable cities of the future: An analytical framework for sensor-based big data applications for environmental sustainability. *Sustainable Cities and Society*. 38:230–253. https://doi.org/10.1016/j.scs.2017.12.034

CropIn. SaaS-based AgriTech | Smarter Agriculture Technology Solutions. *CropIn [Internet]*. https://www.cropin.com/

CropX Agronomic Farm Management System. *CropX [Internet]*. [accessed 2023 Oct 16]. https://www.cropx.com/

Farmapp.farmappwebcom [Internet]. [accessed 2023 Oct 16]. https://farmappweb.com

Farooq MS, Riaz S, Abid A, Abid K, Naeem MA. 2019. A Survey on the Role of IoT in Agriculture for the Implementation of Smart Farming. *IEEE Access [Internet]*. [accessed 2020 Mar 28] 7:156237–156271. https://doi.org/10.1109/ACCESS.2019.2949703

Farooq MS, Riaz S, Abid A, Umer T, Zikria YB. 2020. Role of IoT Technology in Agriculture: A Systematic Literature Review. *Electronics*. 9(2):319. https://doi.org/10.3390/electronics9020319

Ferrag MA, Shu L, Yang X, Derhab A, Maglaras L. 2020. Security and Privacy for Green IoT-based Agriculture: Review, Blockchain solutions, and Challenges. *IEEE Access*:1–1. https://doi.org/10.1109/access.2020.2973178

GmbH P. Plantix | Best Agriculture *App*. *Plantix [Internet]*. https://plantix.net/en/

GREENIQ.GREENIQ [Internet]. [accessed 2023 Oct 16]. https://greeniq.eu

Khanna A, Kaur S. 2019. Evolution of Internet of Things (IoT) and its significant impact in the field of Precision Agriculture. *Computers and Electronics in Agriculture [Internet]*. [accessed 2019 Sep 26] 157:218–231. https://doi.org/10.1016/j.compag.2018.12.039

Kour VP, Arora S. 2020. Recent Developments of the Internet of Things in Agriculture: A Survey. *IEEE Access*. 8:129924–129957. https://doi.org/10.1109/access.2020.3009298

Ku M-L, Li W, Chen Y, Ray Liu KJ. 2016. Advances in Energy Harvesting Communications: Past, Present, and Future Challenges. *IEEE Communications Surveys & Tutorials [Internet]*. [accessed 2019 Jun 29] 18 (2):1384–1412. https://doi.org/10.1109/comst.2015.2497324

Maddikunta PKR, Hakak S, Alazab M, Bhattacharya S, Gadekallu TR, Khan WZ, Pham Q-V. 2021. Unmanned Aerial Vehicles in Smart Agriculture: Applications, Requirements, and Challenges. *IEEE Sensors Journal*.:1–1. https://doi.org/10.1109/jsen.2021.3049471

Neagu, G., Ianculescu, M., Alexandru, A., Florian, V., & Rădulescu, C. Z. (2019). Next generation IoT and its influence on decision-making. An illustrative case study. *Procedia Computer Science, 162*, 555–561.

Ruan J, Wang Y, Chan FTS, Hu X, Zhao M, Zhu F, Shi B, Shi Y, Lin F. 2019. A Life Cycle Framework of Green IoT-Based Agriculture and Its Finance, Operation, and Management Issues. *IEEE Communications Magazine [Internet]*. [accessed 2020 Jan 16] 57(3):90–96. https://doi.org/10.1109/mcom.2019.1800332

SemiosBIOTechnologies. Semios [Internet]. https://semios.com

The World Bank. 2018. *Agriculture and Food. World Bank [Internet]*. https://www.worldbank.org/en/topic/agriculture.

Turgut D, Boloni L. 2017. Value of Information and Cost of Privacy in the Internet of Things. *IEEE Communications Magazine*. 55(9):62–66. https://doi.org/10.1109/mcom.2017.1600625

© 2025 The Author(s), ISBN 978-1-032-73865-9

Towards a serverless computing and edge-intelligence architecture for the Personal-Internet-of-Things (PIoT)

Ankur Gupta & Surbhi Gupta

Model Institute of Engineering and Technology, Jammu, J&K, India

ABSTRACT: Serverless computing has emerged as a compelling paradigm for the deployment of applications at the edge facilitating performant and lightweight services. Services deployed at the edge typically consume lesser compute and storage resources while offering relatively lower latency. Hence, a serverless architecture is a natural fit for the requirements of the Personal-Internet-of-Things (PIoT), an emerging sub-domain of the IoT in which data acquisition, analysis and computation is performed in the context of an individual user and her environment. For such environments privacy preservation, resource-light models, fast detection of emergency situations and low response times for services is paramount. In addition users demand hyper-personalized service delivery requiring embedded intelligence for real-time interventions. This paper proposes a serverless computing architecture with embedded intelligence at the edge for the PIoT which effectively meets the stated requirements. A proof-of-concept using AWS Lambda and tiny AI is implemented and initial results presented, establishing the effectiveness of the proposed framework.

1 INTRODUCTION

The Personal-Internet-of-Things (PIoT) (Gupta *et al.* 2022) is an exciting sub-domain of the IoT, encompassing sensors, digital artefacts, consumer electronics, digital assistants, smart home devices, internet-enabled vehicles etc. which operate within the context of a particular user or family. These sensors, devices and applications are typically supported by a decent home network and a personal cloud for computation and storage. The primary requirement for the PIoT (Sahoo *et al.* 2021) is to provide a privacy-preserving, yet hyper-personalised (Maayan, 2020) tech-driven environment for individual users or families leading to intelligent automation and on-demand service consumption. This realization of the vision for the PIoT is still some distance away, due to vendor/device heterogeneity and lack of standards for interoperability besides privacy-aware cloud service delivery models. PIoT caters to major emerging trends in consumer behavior and expectations (Ahmed *et al.* 2023; Zhang *et al.* 2023a). Users have begun to place a high-premium on their privacy including how their data is collected and managed. Besides, they expect high-levels of personalization, referred to as hyper-personalization (Jain *et al.* 2021). These two expectations are non-complementary usually representing a trade-off. To deliver a high level of personalization, user-data needs to be tracked and analyzed. Hence, new approaches are needed to cater to this trade-off while satisfying both requirements.

The Internet of Things (IoT) and its many sub-variants including the PIoT are characterized by a large number of devices/sensors generating massive amounts of time-series data requiring significant computing power to process. Serverless computing and edge intelligence have emerged as a promising solution to address the scalability and performance issues while lowering the cost of central computing power neede to process the information. Serverless computing, also known as Function-as-a-Service (FaaS), is a cloud computing

DOI: 10.1201/9781003466383-12

model in which a third-party service provider manages the infrastructure and automatically allocates computing resources as needed to execute code in the form of discrete functions. By hosting these discrete functions on the edge, further latency optimization can be achieved. Although servers are still used in the process, the details of the servers and their management are abstracted away from the developers. Due to its capacity to save operational costs, increase scalability, and boost developer productivity, serverless computing has gained traction, especially when data need not be persistently stored for further processing.

Thus, serverless computing and edge intelligence are valuable tools for building scalable, cost-effective, and agile IoT platforms (Yao *et al.* 2023). Furthermore, this paper investigates the benefits of these deployment models in reducing latency, increasing agility, and improving privacy provisions in the PIoT environment. Hence, the main contributions of this paper are:

- The present work highlights the specific requirements for the PIoT necessitating the development of customized frameworks to be developed beyond traditional IoT architectures.
- The study proposes a model for the PIoT which leverages the power of edge computing, combining it with serverless computing paradigm to provide real-time processing and analysis of PIoT data, while ensuring privacy and reducing latency and storage requirements.
- The proposed framework is evaluated experimentally to showcase significant improvements in latency, reduced data storage requirements, and enhanced data privacy, laying the foundation for developing performant PIoT models in future.

The rest of the paper is organized as follows: section 2 provides an overview of related work in the domain of PIoT and related environments such as smart homes. Section 3 proposes the system model for a server-less computing and edge-intelligence based framework. Experimental results from a custom simulator are provided in Section 4, while section 5 concludes the paper.

2 BACKGROUND AND RELATED WORK

In recent years there is increased convergence of consumer electronics, IoT, edge computing, AI and cloud computing to deliver enhanced experiences to end users and help improve the quality of human lives. This has led to high-levels of automation and intelligence in smart home environments with a focus on health, well-being and automation of mundane tasks through robots and software bots. Hence, hyper-personalization as a trend is here to stay. This also implies increased data collection from the user and her immediate vicinity, leading to privacy concerns. Thus, models of tech deployment which enable hyper-personalization to be delivered in a privacy-preserving manner are going to be increasingly built to meet the requirements of increasingly sophisticated individual users. A prominent industry report by MarketsandResearch.com (Research and Markets 2018) highlights the significant potential of the PIoT market, the players, potential applications and usage scenarios, new market opportunities through an in-depth coverage. The study identifies connected consumer devices, wearable computing, smart homes, connected vehicles and V2X platforms and related services, AR/VR/XR technologies, mobile edge computing and analytics services as building blocks for the PIoT ecosystem.

The concept of Personal-Internet-of-Things is nascent with not much literature available. One of the earlier mentions of PioT is available in Siow *et al.* 2016, where the authors use inexpensive edge nodes to create a repository of user linked data in a smart home environment, demonstrating the feasibility of the serverless approach for such environments. A recent study (Gupta *et al.* 2022) on the Personal-Internet-of-Things (PIoT) highlighted that user privacy, data security, and device security must be ensured to realize the PIoT vision and enable exciting new applications. An architecture supporting secure and privacy-preserving PIoT interactions for hyper-personalization and smart automation was also proposed. An early attempt to setup a personalized IoT environment using a DIY model is presented in Ambe *et al.* 2019. The study proposes an IoT Un-Kit experience, a co-design strategy that engages individuals in creating personally relevant IoT apps and in-home solutions. Other solutions focusing on individual aspects such as personalised healthcare have also been proposed. In (Gupta *et al.* 2023) authors propose a framework for the Personalized Health of Things (PHoT). The PHoT architecture proposed in the study automates diagnostic and therapy processes, modifies patient behavior and health status in real time, and aims to provide more personalized healthcare services. AI approaches to handling personal data of users in smart home environments are discussed in Zhang *et al.* 2023b and issues such as privacy, trust and explainability for AI algorithms dealing with personal data are emphasized in the study. A collection of articles related to AI approaches for networking and communication challenges in PIoT and related environments is available in Jin *et al.* 2020. An interesting application leveraging an individuals PIoT data for user identification in digital forensics is presented in Hutchinson 2023. The study highlights the need to provide greater privacy controls for users while ensuring that the data can be correlated to determine identity when required by law enforcement agencies. Recently the risks associated with the personal-internet-of-things has been highlighted in a comprehensive study (Di Gangi *et al.* 2023) indicating that experts and ens users carried a high risk perception about the pervasiveness of the IoT devices in their immediate vicinity. The study further indicates a significantly higher potential of hacking of low cost sensor devices by malicious users, granting them access to sensitive personal data of the users. Hence, the risk perception emerged as a major consideration hampering the broad-based adoption of the PIoT framework.

Thus, related work is limited which addresses the PIoT as a domain and the term has not yet gained traction. However, individual aspects aggregating towards realizing the vision of the PIoT have been researched for instance smart homes, wearable computing, connected vehicles etc. More studies are therefore needed which address the end-to-end needs of the PIoT ecosystem holistically.

3 MYIOT: SYSTEM MODEL

To address the issues discussed above, this paper proposes MyIoT, a system model delivering high-performance edge services for an individual user or a family in a hyper-connected smart home and personal computing environment. The schematic for the proposed system model is presented in Figure 1.

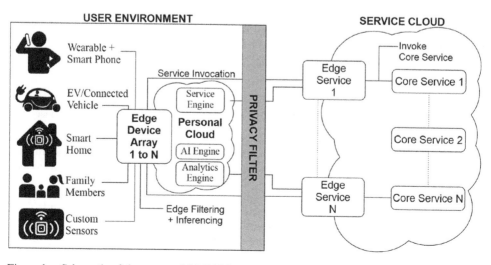

Figure 1.   Schematic of the proposed MyIoT framework.

The major sub-systems of the Proposed framework are:

3.1 *User environment*

In the Personal Internet of Things (PIoT) architecture, the user environment characterizes the entire sensing, analyzing and computing environment in the context of an individual user and her immediate family. This environment comprises a multitude of sensors ranging from wearables to smart devices, smart home sensors, appliances/gadgets, connected vehicles, connected devices, connected healthcare and ambient intelligence as shown in figure one. Thus, sensor diversity is a major characteristic of the PIoT environment. Till sensor interoperability and open standards for sensor data become pervasive across the industry, solutions for catering to specific sensor types and data formats need to be devised. The proposed PIoT architecture caters to this need by positioning an array of low-cost edge devices (some with GPU capabilities), catering to each sensor type. These edge devices receive, filter, perform inferencing based on TinyAI algorithms and send the filtered data for further processing at the personal cloud. The personal cloud is a collection of services without storing any persistent data, since the data storage is managed by the end service providers. For instance, smart phone data is managed by the phone providers cloud or a public cloud, while connected vehicle data will be managed by the vehicle providers servers. In some cases the edge devices may directly invoke edge cloud services hosted at the Service Cloud, for instance in the case of an emergency or as part of application logic or where data storage in the personal cloud is not needed. For example in the case of CCTV monitoring an crime detection, only the frames of interest may be shared with the cloud services for storage and evidence, while a majority of the frames with no point of interest can be filtered and dropped at the edge device. This obviates the need of persistent storage for many applications and it is entirely possible to attain a completely serverless model for the PIoT in the near future.

3.2 *Cloud service environment*

The cloud service environment delivers the service aspect of the PIoT and houses edge services for delivering performant services to multiple PIoT users by optimizing the service latency and response times. Only when major processing is needed and records need to be maintained are core cloud services invoked and provisioned. Clearly, the focus is on offloading a majority of the computation and data traffic to the edge while reducing the traffic flow between the user environment and the cloud core services, till possible. The personal cloud hosts the engines responsible for service invocation as required, AI Engine for applying AI models to data for hyper-personalization needs and an analytics engine to make sense of data and take actions based on analysis of received data.

4 SIMULATION RESULTS

AWS IoT platform to simulate multiple IoT devices and generate data streams based on real-world data, with the following steps:

- User Scenarios: 3 user scenarios were considered requiring data analysis and further actions to be taken based on the data by interacting with external third-party services as per configured rules.
- IoT Devices: 25 virtual IoT devices were simulated for smartwatches, phones, wearables, and other sensors such as fire/smoke detectors, smart devices, garage doors, CCTV cameras and the like, using the AWS IoT Device Simulator to generate data streams based on real-world sensor readings. All sensors were configured to collect readings every 5 minutes, except for CCTVs which send a fixed size data stream continuously.

- Edge Devices: Simulated edge nodes were setup using AWS Greengrass. These devices process and analyze data locally before sending it to the personal cloud/services cloud.
- Personal Cloud: An AWS service instance was created for implementing AI models and inferencing, analytics and service invocation, besides removing sensitive or personal information in case not performed by the edge device (for specific sensor types) to mimic the privacy filter.
- Edge Services: Multiple AWS Lambda functions were setup to model edge services hosted either by the sensor vendors or to perform specific tasks on behalf of the user by invoking external cloud services.
- Cloud Services: Finally, multiple AWS VM instances were instantiated to host central cloud services including emergency response or third-party services such as online grocery ordering etc.

All the above AWS elements were co-located in the AWS Mumbai region. Using the setup, measurements for multiple parameters and interactions were obtained:

- Sensor to Edge Latency
- Edge Processing/Inferencing
- Cloud Edge Service Latency
- Cloud Core Service Latency
- Server Storage Optimization

Figure 2(a) provides the sensor to edge latency measurements averaged hourly for 25 sensors, indicating very little variance over a 24 hour period. Average time taken for edge processing/inferencing for 3 sensors, a user wearable, a CCTV camera and a generic sensor is shown in Figure 2(b) over a 1 hour period (12 instances, one every 5 minutes). Edge processing involves three operations including dropping of duplicate/similar readings and inferencing includes any AI model or rules executed to initiate actions as per rules. Maximum inferencing is performed for the CCTV data stream.

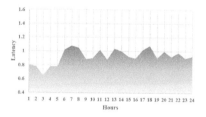

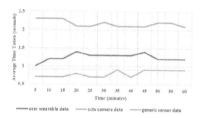

(a) Average sensor to edge latency measurements for 25 sensors

(b) Average time taken for edge processing/inferencing for 3 sensors

Figure 2. Average latency and edge processing time.

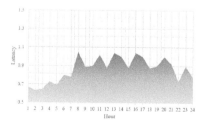

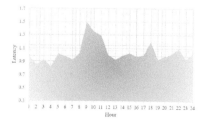

(a) Server to cloud edge latency

(b) Cloud Edge to Cloud Core Latency

Figure 3. Average latency.

61

Figure 3(a) encapsulates the average hourly latency for the Server to cloud edge service latency (modeled as AWS lambda functions). Here the latency varies between 0.63 to 1.07 milliseconds over a 24 hour period indicating a manageable overhead. The variations in latency occur due to periods of varying traffic and load on the AWS platform and its associated services. Figure 3(b) depicts the average hourly latency for the edge to core cloud service (hosted at the AWS IoT Core) to model third-party services. Here the latency ranges from 0.82 to 1.35 milliseconds. For simplicity it is assumed that the actual service processing time is constant for a particular service.

Figure 4 shows the traffic optimization achieved for the three sensors used in Figure 2. For the CCTV stream, by dropping duplicate frames at the edge, server optimization of upto 60% was attained, while it was a more modest 15% and 19% for the other two sensors (similar readings with insignificant variation).

Finally, Figure 5 compares the service provisioning time when data is analyzed and actions initiated at the local edge versus using the cloud services. Results show that the edge devices perform very well when performing computations in the context of the user and service performance improvement of over 50% is attained when inferencing is performed locally at the edge (user's home) compared to at the edge of the cloud or through third-party edge services

Thus, the simulation results establish the efficacy of the proposed system model, which delivers performance while managing the overheads.

5 CONCLUSION

MyIoT encompasses a serverless computing and edge intelligence model to deliver the vision for the PIoT, which could gain significant traction in the coming years as users seek out a tech-driven

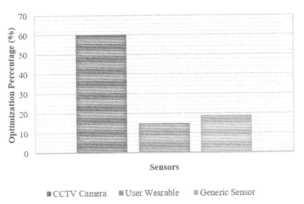

Figure 4.   Traffic optimization at the edge for 3 different sensors (percentage of frames dropped as duplicates).

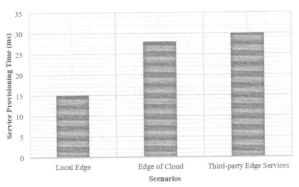

Figure 5.   Comparison of service provisioning time.

environment to live and operate optimally in. The serverless model ensures that user data is not stored on third-party cloud servers and used in an unauthorised manner, while edge services with TinyAI build intelligence by analyzing user data devoid of any PII, ensuring privacy. The user data is stored internally on the user's personal cloud and managed as per user requirements. Thus, the user gains control over how her data is collected, analyzed and consumed while delivering the required personalized services. The present work addresses the significant challenge of delivering personalization with privacy which systems catering to individual users, such as the PIoT, need to address going forward. Future work in the domain will need to focus on:

- Exploring federated learning, secure multiparty computation, and differential privacy in the context of PIoT.
- Investigating resource management strategies that dynamically allocate and optimize computing resources to meet the varying demands of edge-intelligence tasks in the PIoT.
- Developing context-aware personalization techniques that leverage methods for capturing and utilizing contextual information, such as location, user behavior, and environmental factors, to provide personalized services in real-time for an entire family of users and even guests.
- Exploring the integration of machine learning, deep learning, and reinforcement learning algorithms for adaptive and context-aware personalization in the PIoT.

REFERENCES

Abdussalam Ali Ahmed, Mohamed Belrzaeg, Yasser Nassar, Hala J El-Khozondar, Mohamed Khaleel, and Abdulgader Alsharif. (2023). A comprehensive review towards smart homes and cities considering sustainability developments, concepts, and future trends. *World J. Adv. Res. Rev*, 19(1), 1482–1489.

Aloha Hufana Ambe, Margot Brereton, Alessandro Soro, Min Zhen Chai, Laurie Buys, and Paul Roe. (2019). Older people inventing their personal internet of things with the IoT un-kit experience. In *Proceedings of the 2019 CHI Conference on Human Factors in Computing Systems*, pages 1–15.

Ankur Gupta, Purnendu Prabhat, and Bisma Gulzar. (2022). Personal-internet-of-things (piot): A vision for hyper-personalization delivered securely. In *2022 IEEE Delhi Section Conference (DELCON)*, pages 1–6. IEEE

Biswa PS Sahoo, Saraju P Mohanty, Deepak Puthal, and Prashant Pillai. (2021). Personal internet of things (piot): What is it exactly? *IEEE Consumer Electronics Magazine*, 10(6), 58–60.

Eugene Siow, Thanassis Tiropanis, and Wendy Hall. (2016). Piotre: Personal internet of things repository.

Fengjiao Zhang, Zhao Pan, and Yaobin Lu. (2023). AIOT-enabled smart surveillance for personal data digitalization: Contextual personalization-privacy paradox in smart home. *Information & Management*, 60 (2), 103736

Geetika Jain, Justin Paul, and Archana Shrivastava. (2021). Hyper-personalization, co-creation, digital clienteling and transformation. *Journal of Business Research*, 124, 12–23.

Gilad Maayan. (2020). Hyper personalization: Customizing service with AI. Eyl ul, 18, 2020.

Paul M Di Gangi, Barbara A Wech, Jennifer D Hamrick, James L Worrell, and Samuel H Goh. (2023). Risk perceptions about personal internet-of-things: Research directions from a multi-panel Delphi study. *Journal of Cybersecurity Education, Research and Practice*, 2022(2)

Research and Markets.Personal internet of things (IoT) and connected devices: Applications and services in wearables and IoT devices, connected vehicles, connected healthcare, ambient intelligence, and quantified self 2018–2023.

Shinelle Hutchinson. (2023). *Leveraging Personal Internet-of-Things Technology To Facilitate User Identification in Digital Forensics Investigations*. PhD thesis, Purdue University Graduate School.

Surbhi Gupta, Mohammad Shabaz, Ankur Gupta, Abdullah Alqahtani, Shtwai Alsubai, and Isaac Ofori. (2023). Personal healthcare of things: A novel paradigm and futuristic approach. *CAAI Transactions on Intelligence Technology*.

Xuyi Yao, Ningjiang Chen, Xuemei Yuan, and Pingjie Ou. (2023). Performance optimization of serverless edge computing function offloading based on deep reinforcement learning. *Future Generation Computer Systems*, 139, 74–86.

Yong Jin, Honghao Gao, Tao Hu, and Xinrong Li. (2020). Special issue on AI-driven smart networking and communication for personal internet of things: Part II. *International Journal of Wireless Information Networks*, 27, 207–208.

Next Generation Computing and Information Systems – Gupta. (Ed.)
© 2025 The Author(s), ISBN 978-1-032-73865-9

IOT and developed deep learning based road accident detection system and societal knowledge management

K.R. Ananth
Department of Computer Science (AI&DS), Nandha Arts and Science College, Erode, Tamil Nadu, India

Sarfraz Fayaz Khan
SAT – Algonquin College, Ottawa, Canada

Abhishek Agarwal
Vermont State University, USA

Mohammed Wasim Bhatt
Model Institute of Engineering and Technology, Jammu, J&K, India

Ahmad Murtaza Alvi
College of Finance, Department of Business Administration, Saudi Electronic University, Riyadh, Saudi Arabia

Sheshang Degadwala
Department of Computer Engineering, Sigma University, Vadodara, Gujarat, India

ABSTRACT: Business organizations and the research community try to precisely detect occurrences and assist in the case of a disaster. Most development systems are hardware-based, making them pricey and unavailable in every vehicle. A vehicle's sensors can be destroyed in various ways, including minor accidents or fixed interactions. In instances, the sensors are incapable of detecting an accident. Intelligent phone sensors are a great alternative because of their dependability and availability. The study involves creating a smartphone application that continually reads sensor data and sends it to the cloud. The crash is discovered by threshold analysis. The critical novelty of this research is creating a scheme that alerts nearby hospitals and ambulances when an accident occurs. The system will have more minor inaccuracies, precisely identify accidents, and perform better than earlier techniques using four sensory inputs. This paper introduced novel types of deep learning for accident detection.

1 INTRODUCTION

Cities are becoming increasingly congested with tourists, residents, and cars. Increased vehicle use has increased traffic. According to the last WHO estimate, 1.35 million people die yearly in traffic accidents, and 50 million are wounded (Khan *et al.* 2018). Road crashes are now the ninth leading cause of death. However, the International Road Safety Association (ASIRT) predicts that, unless substantial improvements occur, they will soon increase to the fifth leading cause of death. Furthermore, the societal costs of road accidents are significant. The International Road Safety Association expects that road accidents will consume one to two percent of each country's yearly budget. Even in sophisticated nations with road safety measures, the global yearly number of driving deaths has recently climbed. However, the issue remains that low- and middle-income nations have the most incredible pressure on tolls and road injuries (Arora *et al.* 2022). There are several meanings of IoT. For example,

DOI: 10.1201/9781003466383-13

defines the Internet of Things as "a network of products each with embedded sensors" linked to the Internet. According to Internet meanings, things are a physical, cyber system that connects physical items to the cyber domain. Sensors have limited computing and storage capacity, which might pose problems, particularly regarding security and dependability. Some of these challenges have been addressed using cloud computing (More and Raisinghani 2017; Shaikh and Zeadally 2016).

2 REVIEW OF LITERATURE

Several concepts and tactics for dealing with road safety, vehicle communications, and post-accident rescue operations may be found in the literature. This study concentrates on the most practical approaches and strategies: software and hardware-based solutions. It primarily focuses on accident detection technology that makes use of a variety of sensory inputs (Ali and Alwan 2015). The review of the current systems connected to traffic dangers and road accidents in this part reveals their strengths, shortcomings, and limits.

2.1 Part I) Smartphone-based systems

A substantial amount of research on this topic can be found to solve the challenge of low-cost, resilient solutions to detect and notify automobile accidents based on mobile technology. It debuted a crash alarm system in (Rajkiran and Anusha 2014; Sane *et al.* 2016; Tushara and Vardhini 2016) and that used accelerometers and GPS data from mobile devices to identify accidents. Using an integrated accelerometer and gyroscope, the writers developed a smartphone-based application that detects unintended accidents and alerts the adjacent emergency. The planned strategy, is focused on reducing reaction time and does not consider vehicle accidents.

2.2 Part II) Hardware-based Systems

As mentioned, road accidents are the primary cause of fatalities; thus, more study is needed to discover and swiftly begin rescue efforts. The likelihood of fatality is lowered if the timeframe between the crash and dispatch of the rescue team is shortened, and this has inspired several researchers to minimize reaction time. The authors presented a crash tracking system in (Fogue *et al.* 2014), which employs a microcontroller to regulate all processes. Messages are delivered to a determined cellphone number. The response team is then dispatched to the location by the Emergency Center. Despite its many benefits, the system has several disadvantages.

3 PROPOSED METHOD

This study proposes an Optimized Accident Detection and Reporting System, or OADRS. The new OADRS crash tracking and reporting system utilizes modern Android smart phone capabilities and thus reduces overall cost because no special hardware is required. The main responsibility of the perception layer is to interact with smartphone sensors in the suggested architecture, and gathering data from the sensors is the primary aim of the perception layer in OADRS architecture (Chaturvedi and Srivastava 2018). There are set threshold values; if an accident happens, the sensor data offer a superior deal than the threshold value. When criteria are met, an alarm is generated and sent to the car's driver. The hospital will not be called if the motorist disregards the signal to avoid false reporting. The cloud service will alert the nearest hospital within 10 seconds (Bhatt and Sharma 2023).

The chief aim of the proposed system is to provide an architecture that can handle 5 cases: 1) direct communication of the vehicle to the infrastructure, 2) automated sharing of accident data, 3) increase the precision of an accident detection, 4) reduce fake reports and messages, 5) establish a profitable system. The activities of the presented architecture. The most significant variables of this study include multi-smartphone sensors like the accelerometer and GPS for detection of crashes. in several different sections. The first part deals with the accident detection components.

3.1 Phase one: Accident detection components (Deep learning inputs)

Accident identification is used to avoid occurrences that cause damage or injury and minimize the number of people killed in traffic accidents. The detection process determines the presence of an accident. Accident detection component as discussed above are such Figure 1.

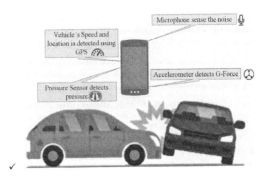

Figure 1. Components of accident detection.

3.2 Phase two: Phase of the notification

When an accident is detected, effective communication and dispatch are vital. The system locates an accident using the smartphone's GPS when it is spotted. The cloud has a hospital database and employs a mapping tool to locate the closest hospital. A message containing the location's data and owner information is sent to the hospital. The obtained data is saved in the current database.

3.3 Phase three: Database

Database of cars: An automobile database provides all vital information regarding registered vehicles. Information of owner, name, address, and the number of automobiles is saved in the cloud to resolve any incident. The samples from the automobile database are given in Table 1.

Table 1. Samples of cars database.

ID of car	Name of car	Number of cars	Name of owner	ID of owner
C 1	Suzuki	RAZ 3825	Bilal Khaled	36512-4530645-9
C 2	Landover	MP 3509	Shahid Khalid	33103-9963108-6
C 3	Toyota	LAL 76 4320	Ali Hosseini	12145-1519307-7

Database of Hospital: The system must be aware of all adjacent hospitals to notify them of an emergency. The cloud must locate and pass the message to the nearest hospital. Table 2 displays the information that has been saved.

Table 2. Database of hospital.

ID of hospital	Name of hospital	Address of hospital	Number of hospitals
H 1	Usmani Rd Panjab	Jahan	+92-42-95231543
H 2	Kohistan Rd Mumbai	Health	+92-51-2455613
H 3	Abid Majeed Rd Delhi	Soldierly	+92-51-9260376

Clustering can be utilized in these networks to mitigate these detrimental consequences. Moving autos in the same direction are regarded as part of a cluster. It mitigates the harmful impact of recurrent clustering. This essay considers the link between the size of the cluster, speed of the vehicle, traffic, and the CW's size on the hand and their influence on output efficiency and closure losses. Another issue in creating such networks is selecting the appropriate cluster.

This procedure was used to choose the cluster based on the weighted method. Furthermore, the average signal strength received according to Friies' rule is recognized throughout the preceding experiments (1).

$$E(p_{re}) = \frac{P_{tr}G_{tr}G_{re}\lambda_c^2}{16\pi^2 d^2} \tag{1}$$

In this regard, d denotes the distance of the transmitter with the receiver, as well as Ptr, Gtr, and Gre of the transmitter and reception antennas (previously specified). Equation will be used to calculate the wavelength of the wireless signal (2).

$$\lambda_t = \frac{c}{f} \tag{2}$$

(2) denotes the speed of light as c and the frequency of the wireless channel as f (5.850 GHz to 5.926 GHz), with one channel especially tailored to boost automobile safety and other channels for car-specific applications. To simplify, we'll suppose the frequency is in the 5.890 GH range. Regarding the channel considerations, the probability density function can be computed as below for the delivered signal:

$$F_{re}(p_{re}) = \frac{1}{E(p_{re})} e^{-\frac{P_{re}}{E(P_{re})}} \tag{3}$$

The emergency message is decrypted if the received signal intensity is higher than the lowest sensitivity (P min). If this is not done, emergency communication will be lost (Yao et al. 2021). The receiver calculates the signal intensity when the emergency message is received. Consequently, the possibility of receiving an emergency message within the distance L tr is calculated using the equation (4).

$$pr_{succ(d)} = pr\{p_{re} > p\_min|l_{ir} = d\} = 1 - pr\{p_{re} \leq p\_min|l_{ir} = d\}$$

$$= 1 - \int_0^{P\_min} \frac{1}{E(p_{re})} e^{-\frac{pre}{E(pre)}} d(p_{re}) = \exp\left[-\frac{p_{min}}{E(p_{re})}\right]$$

Even though the processing units placed in automobiles are capable, this cannot eliminate the delay caused by vehicle propagation since wireless networks' limited resources produce this delay. As a result, the emission delay (T_{tr}) is the time elapsed between back-off and retransmission in the MAC layer.

According to the equation, the unscreened applicants' number for channel access is determined as Nc if an automobile plans to transmit an emergency message.

$$N_c = n\tau + 1 \tag{4}$$

Back-off is meant to avoid interference when autos issue emergency notifications. The forwarding probability of the collision.

In this equation, w_{min} denotes the least conceivable CW size, which is often believed to be 32. The average number of transmitted entries will be specified in the equation concerning the maximum transmitting value (l_{re})

$$E(N) = \sum_{N=1}^{l_{re}} N\left(1 - Pr_{\frac{w}{0}}\right)^{N-1} Pr_{\frac{w}{0}} \tag{5}$$

CW doubles its maximum value $w_{max} = 2^m w_{min}$, if there is no interference channel. Mistake a system parameter, which is presumed to be 5 in this case. As w_{con} approaches its maximum value of w_{max} $m + 1$, the CW will be used as the maximum in the following computations. As a result, the equation defines the average latency for a successful post η represents the back-off slots' length. The data transmission length (T_{data}) is defined as the product of the packet's size message (L_{size}) and the data transfer rate (M) and can be calculated as equation (9).

$$T_{data} = \frac{L_{size}}{M} \qquad (6)$$

Combining equations, in the form of equation (7).

$$T_{trs} = T_{back} + \frac{L_{size}}{M} \qquad (7)$$

In this part, a parameter called M8 will be added to define the relay node in the cluster to link the nodes inside the cluster to each other. The capabilities of Neighbor and Node Relay will be fully utilized. The maximum delay of T_{max} is 100 ms based on the reference number. Also, to compute the delay factor in one step, the parameter D^* is introduced that has a symmetrical connection with the T_{tr} emission delay variable. Thus, if T_{tr} is 0, D^* will be 1, and vice versa. The minor D^*, the larger the delayed T_{tr} release. As a result, in general, the parameter D^* will be specified by equation (8).

$$D^* = \begin{cases} 0; \ T_{tr} > T_{max} \\ 1 - \dfrac{T_{tr}}{T_{max}}; else \end{cases} \qquad (8)$$

The parameter D^* will have a value between 0 and 1. Furthermore, suppose the T_{tr} diffusion delay in one phase is more significant than T_{max}. In that case, D^* will be 0, the worst-case situation. The parameter value $M^* \alpha$ and β indicate the propagation's reliability and latency properties. The parameter M^*'s of changes is similarly between 0 and 1.

4 SIMULATION AND RESULTS

The problem of capturing vehicle distances and speeds on straight and Steep Mountain bumps must be solved (Jairath *et al.* 2022; Lei *et al.* 2022). It is determining traffic bottlenecks, traffic, and any accidents over high terrain is difficult since it looks like a series of amplifiers are necessary. Specifying the Smart City's first parameters with the Internet of Things setup and infrastructure is required. Table 3 depicts the starting parameters of a Smart City with the Internet of Things setup and infrastructure, as specified by a reference paper in this field (Rida *et al.* 2015).

Table 3. Smart city with internet of things configuration and infrastructure parameters.

Network Dimension	500×500 m^2 or 1000×1000 m^2
Each packet's size	1000 byte
Numbers of Vehicles	300
Cars with the slowest speeds (in mountainous roads)	30.0 km/h
Cars with the slowest speeds (in tunnels)	25.0 km/h
The range of each vehicle's radio signal.	20m
The energy of each vehicle's sender	0.02 Jul
The energy received by each vehicle	0.04 Jul
Per-second fuel usage	2 sec
Network port count	4000
Paths' lengths	5.0m
Modulation Diagram	OFDM
Data transmission power	21 dB
Transfer speed	6 MB/s

OFDM channels use QPSK and 64QAM in the roads and tunnels, respectively. At first, PDR will be shown PDR is calculated by an equation (9).

$$PacketDrop_{Calculation} = \frac{SentPacket\_Number}{RecievedPacket_{Number}} \tag{9}$$

i is the index of summation according to the above two relations. When network resources are shared, the outcome reveals a high rate of PDR. PDR is evaluated at zero intervals just one vehicle is in the covered zone and resources (simulation start). This does not imply that the PDR will be assessed just for an automobile, even if a car fills the area (Abbas *et al.* 2021; Rida *et al.* 2014). The equation calculates the average end-to-end latency.

$$n - to - n_{delay} = \frac{decrease\ in\ delay}{4}, \quad unit = ms \tag{10}$$

The simulation's throughput is determined depending on runtime, input condition, and vehicle unavailability to the network coverage region. Average throughput computed using equation (4.4).

$$Throughput_{Avg} = \left(\sum_{i=1}^{4} \frac{RecievedPacket_{Number} \times PacketSize}{Total\ Simulation\ Time} \right) \times 100 \tag{11}$$

According to an analysis, the ID 1 vehicle was withdrawn from the envelope in 1.1 seconds, whereas the vehicle with ID 2 began sending data in 20 seconds. The results of two-time delays demonstrate a significant decrease in bit rate. The outcomes are as follows: The protocol's delivery and data loss rates were compared to AODV and AOMDV, before a recommended technique, SPA- (S, P), was implemented for enhanced resilience in each segment (Javed *et al.* 2018; Mazhar *et al.* 2023).

The suggested technique has a higher packet delivery rate than AODV and AOMDV protocols and is more suited to packet loss rates. As a result, the permeability is comparatively more elevated, and data transmission and reception are faster (Mazhar *et al.* 2023). A sustained routing comparison is presented in the Figure below (2). Through the obtained results, the suggested routing stability protocol beats AODV and AOMDV.

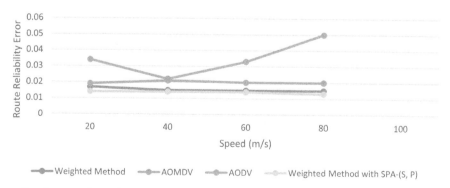

Figure 2. A comparison of routing stability.

This section has been added in the OFDM channels, the modulation type is separated into two parts: The figure shows a bit error rate (Figure 3).

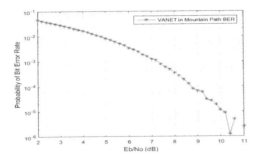

Figure 3.   Proposed method bit error rate.

The bit error rate appears to be reducing. The graph is lowering and reducing, indicating the chance of bit error rate on hilly pathways is low by E b/No and dB. Low doesn't imply terribly; rather, it indicates less inaccuracy in the data, which might include disruptions such as noise (Soni and Jain 2018; Soni *et al.* 2021). As seen in Figure 4, the latency is reduced once the bit error rate is reduced.

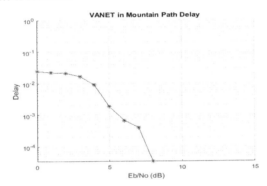

Figure 4.   Delay or latency rate.

The delay is expressed in E_b/N_o units, indicating that it is minimized. Reduced bit error rates and delays will result in shorter latencies. The network's principal energy was a total of 200 Jules. Energy is utilized at 100 Jules from the beginning to the conclusion. At start of Simulation #2 that reached 11 in ending of simulation, demonstrating development. The bit error rate began at four and ended at two. The delay was there from the start of Simulation 1 and eventually dropped to zero. Nonetheless, the ups and downs of high-altitude access and tunnels have implications.

5   CONCLUSION

In recent years, the number of vehicles in metropolitan areas has increased substantially. Accidents have grown as a result of increased traffic. Despite the introduction of different accident detection devices to the market, a significant number of fatalities occur. We demonstrated how using various sensors may help more precisely recognize a traffic incident. The proposed system immediately identifies an accident, locates the closest hospital. Then, it forwards the emergency help message to the relevant section of the hospital. This technology makes a decision based on the obtained data of smartphone sensors that detect data around the vehicle's status. We proved that our suggested method minimizes false alerts. One disadvantage of this study is that we conducted the system's primary assessment in a simulated circumstance. We are working to make the system more responsive while boosting its privacy and security level. Even so, the method demands a comprehensive investigation.

REFERENCES

Abbas, S., Talib, M.A., Ahmed, A., Khan, F., Ahmad, S. and Kim, D.H., 2021. Blockchain-based authentication in internet of vehicles: A survey. *Sensors*, *21*(23): 7927.

Ali, H.M. and Alwan, Z.S., 2017. *Car accident detection and notification system using smartphone.* Saarbrucken: Lap Lambert Academic Publishing.

Arora, V.K., Sharma, V. and Sachdeva, M., 2022. On QoS evaluation for ZigBee incorporated Wireless Sensor Network (IEEE 802.15. 4) using mobile sensor nodes. *Journal of King Saud University-Computer and Information Sciences*, *34*(2): 27–35.

Bhatt, M.W. and Sharma, S., 2023. An Object Recognition-Based Neuroscience Engineering: A Study for Future Implementations. *Electrica*, *23*(2).

Chaturvedi, N. and Srivastava, P., 2018. Automatic vehicle accident detection and messaging system using GSM and GPS modem. *Int. Res. J. Eng. Technol.(IRJET)*, *5*(3): 252–254.

Fogue, M., Garrido, P., Martinez, F.J., Cano, J.C., Calafate, C.T. and Manzoni, P., 2013. A system for automatic notification and severity estimation of automotive accidents. *IEEE Transactions on mobile computing*, *13*(5): 948–963.

Jairath, K., Singh, N., Shabaz, M., Jagota, V. and Singh, B.K., 2022. Performance analysis of metamaterial-inspired structure loaded antennas for narrow range wireless communication. *scientific programming*, *2022*.

Javed, F., Khan, S., Khan, A., Javed, A., Tariq, R., Matiullah and Khan, F., 2018. On precise path planning algorithm in wireless sensor network. *International journal of distributed sensor networks*, *14*(7): 1550147718783385.

Khan, A., Bibi, F., Dilshad, M., Ahmed, S., Ullah, Z. and Ali, H., 2018. Accident detection and smart rescue system using Android smartphone with real-time location tracking. *International Journal of Advanced Computer Science and Applications*, *9*(6): 341–355.

Lei, Y., Vyas, S., Gupta, S. and Shabaz, M., 2022. AI based study on product development and process design. *International Journal of System Assurance Engineering and Management*: 1–7.

Mazhar, T., Irfan, H.M., Haq, I., Ullah, I., Ashraf, M., Shloul, T.A., Ghadi, Y.Y., Imran and Elkamchouchi, D.H., 2023. Analysis of Challenges and Solutions of IoT in Smart Grids Using AI and Machine Learning Techniques: A Review. *Electronics*, *12*(1): 242.

Mazhar, T., Irfan, H.M., Khan, S., Haq, I., Ullah, I., Iqbal, M. and Hamam, H., 2023. Analysis of Cyber Security Attacks and Its Solutions for the Smart Grid Using Machine Learning and Blockchain Methods. *Future Internet*, *15*(2): 83.

More, A. and Raisinghani, V., 2017. A survey on energy efficient coverage protocols in wireless sensor networks. *Journal of King Saud University-Computer and Information Sciences*, *29*(4): 428–448.

Rajkiran, A. and Anusha, M., 2014. Intelligent automatic vehicle accident detection system using wireless communication. *Int. J. Res. Stud. Sci. Eng. Technol*, *1*(8): 98–101.

Rida, I., Al Maadeed, S. and Bouridane, A., 2015, August. Unsupervised feature selection method for improved human gait recognition. *Proc. of IEEE 23rd European Signal Processing Conference (EUSIPCO)*: 1128–1132.

Rida, I., Almaadeed, S. and Bouridane, A., 2014, December. Improved gait recognition based on gait energy images. *Proc. of 26th International Conference on Microelectronics (ICM)*: 40–43.

Sane, N.H., Patil, D.S., Thakare, S.D. and Rokade, A.V., 2016. Real time vehicle accident detection and tracking using GPS and GSM. *Int. J. Recent Innov. Trends Comput. Commun*, *4*(4): 479–482.

Shaikh, F.K. and Zeadally, S., 2016. Energy harvesting in wireless sensor networks: A comprehensive review. *Renewable and Sustainable Energy Reviews*, *55*: 1041–1054.

Soni, M. and Jain, A., 2018, February. Secure communication and implementation technique for sybil attack in vehicular Ad-Hoc networks. *Proc. of Second International Conference on Computing Methodologies and Communication (ICCMC)*: 539–543.

Soni, M., Rajput, B.S., Patel, T. and Parmar, N., 2021. Lightweight vehicle-to-infrastructure message verification method for VANET. *Proc. of International Conference on Data Science and Intelligent Applications: Proceedings of ICDSIA*: 451–456. Springer: Singapore.

Tushara, D.B. and Vardhini, P.H., 2016, March. Wireless vehicle alert and collision prevention system design using Atmel microcontroller. *Proc. IEEE International Conference on Electrical, Electronics, and Optimization Techniques (ICEEOT)*: 2784–2787.

Yao, Q., Shabaz, M., Lohani, T.K., Wasim Bhatt, M., Panesar, G.S. and Singh, R.K., 2021. 3D modelling and visualization for vision-based vibration signal processing and measurement. *Journal of Intelligent Systems*, *30*(1): 541–553.

Next Generation Computing and Information Systems – Gupta. (Ed.)
© 2025 The Author(s), ISBN 978-1-032-73865-9

Healthcare CAD model for hierarchical processing of surgical navigation system

Niladri Maiti
School of Dentistry, Central Asian University, Tashkent, Uzbekistan

Arun Kumar Maurya
Bit College of Pharmacy, Dehradun, India

Shri Ganesh V. Manerkar
Department of Information Technology, Goa College of Engineering, Farmagudi, Ponda – Goa. South Goa, India

Mohammad Shabaz
Model Institute of Engineering and Technology, Jammu, J&K, India

Sheshang Degadwala
Department of Computer Engineering, Sigma University, Vadodara, Gujarat, India

Raj A. Varma
Symbiosis Law School (SLS), Symbiosis International (Deemed University) (SIU), Vimannagar, Pune, Maharashtra, India

ABSTRACT: To avoid the "feeling out of control" caused by the redundant information of the human-computer interface to the doctor, based on the cognitive model of the doctor during the operation, the human-machine interaction criterion in the augmented reality surgical navigation system is proposed. Aiming at the multi-sensory information fusion problem of AR navigation interface, a human-computer interaction paradigm based on eGOMS model is proposed. To objectively evaluate the human-computer interaction usability of the AR navigation system, the analytic hierarchy process was used to construct the human-computer interaction usability evaluation index system, and the fuzzy theory was used to comprehensively assess the human-computer interaction usability of the AR navigation system. The evaluation result of the system is good, and it can be applied in practice.

1 INTRODUCTION

Anterior cruciate ligament (ACL) rupture cannot be treated effectively, which will seriously affect patients' quality of life (Saxena *et al.* 2013; Yang *et al.* 2020). Therefore, arthroscopic ACL reconstruction surgery is the most common surgical procedure. However, since retaining the ACL remnant will increase the difficulty of positioning the bone tract, traditional arthroscopic ACL reconstruction surgery requires complete removal of the ACL remnant (Liu 2020). Therefore, the ACL reconstruction with residual remnants is the development trend of ACL reconstruction surgery (Soni *et al.* 2018; Yang and Chen 2009). Augmented Reality (AR) technology is an effective method to enhance the perception of a minimally invasive surgical environment (Li *et al.* 2021). It can assist doctors in accurate target positioning while preserving the remnants and improving postoperative rehabilitation effect and patients' quality of life; it can also provide doctors with rich anatomical structures, expand the doctor's surgical field of vision, and reduce the difficulty of surgery (Wang and Xu 2020).

DOI: 10.1201/9781003466383-14

Augmented reality navigation systems have the following three characteristics: (1) virtual-real fusion; (2) precise "alignment"; (3) real-time interaction. Although the surgical navigation system has been used for more than 20 years, the current interactive technology still has shortcomings in terms of stability, perception and feedback, limited by hardware technology (Lu *et al.* 2010). Therefore, augmented reality navigation systems, especially those based on an optical perspective, are very different from traditional human-machine interfaces. Its information channels are more diverse, so the interaction mode is very different from the traditional human-machine interface (Zhao *et al.* 2013). In addition, during surgery, doctors are under high stress. Therefore, smooth and comfortable human-computer interaction is essential. On the other hand, the redundant information in the human-machine interface may distract the doctor's energy and make the doctor feel "out of control" (Lee *et al.* 2011).

2 OPTICAL FLUOROSCOPY-BASED SURGICAL NAVIGATION SYSTEMS FOR ACL RECONSTRUCTION

2.1 *Surgical navigation system*

2.1.1 *System structure*

As shown in Figure 1, this paper proposes an augmented reality navigation system based on HoloLens. The system includes a CT machine, C-arm, PC, Micron Tracker, and HoloLens. The relationship between the nine modules of the ACL navigation system based on augmented reality technology is as follows:

- The position tracking module provides the entire system with the position information of each tracking target (calibrator, surgical target, calibration block and HoloLens). The position information is collected by Micron Tracker and transmitted to other modules through the UDP protocol;
- The C-arm space registration module can realize the registration of X-ray images and the surgical space, and the registration results can provide parameter support for the 2D-3D registration module; 2D-3D registration module can realize the registration of the pre-operative CT image and the intraoperative X-ray image;
- The virtual-real registration module can recognize the spatial registration between the virtual space and the actual scene; as a result, information such as virtual models, path planning, etc. accurate scene fusion display;
- 3D reconstruction module can realize 3D reconstruction of preoperative CT images and provide a 3D data model for 2D-3D registration, path planning and visual presentation.

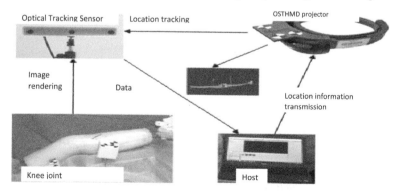

Figure 1.   Augmented reality system structure of ACL reconstruction system.

2.1.2 *Navigation process*

The usage flow of the ACL navigation system based on augmented reality technology is as follows:

1) Preoperative treatment
   - Offline calibration of the HoloLens workspace.
   - CT scan of the target area.
   - Segmentation and reconstruction of preoperative CT data were performed to obtain femur and tibia, STL models.
   - On the segmented STL model, perform surgical path planning.

2) Intraoperative Navigation
   - Take X-rays, perform spatial calibration on the C-arm, and transmit the calibration results and X-ray images to the 2D-3D registration module.
   - In the 2D-3D registration module, import the preoperative CT data and the segmented STL model, adjust the initial position of the STL model and X-ray, perform 2D-3D image registration, and transmit the registration results to the visualization Reality Module.
   - Online calibration of HoloLens.
   - Enhanced display, navigation tracking.

2.2   *Spatial registration technology*

2.2.1   *C-arm calibrator based on tapered helix*
As shown in Figure 2, to improve the spatial registration accuracy of point pair registration, a C-arm calibrator based on a conical helix is used to calibrate the C-arm. In the vicinity of the surgical target, this calibrator has higher calibration accuracy than the traditional dual-plane calibrator (Yu *et al.* 2012).

2.2.2   *Splat-based 2D-3D image registration*
To achieve the spatial unification of preoperative CT images and digital X images a DRR projection-based 2D/3D automatic registration method strategy is adopted. DRR reconstruction, similarity criterion, and parameter optimization algorithms are essential components of the 2D/3D registration process (Collazos and Merchan 2015).

2.2.3   *Virtual and real calibration based on space compensation*
To achieve fast online calibration, the calibration process is decomposed into offline calibration and online calibration by analyzing the workspace of the HoloLens.

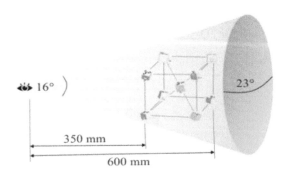

Figure 2.   Workspace of HoloLens.

2.3   *Human-computer Interaction Technology*

2.3.1   *Human-computer interface*
For scene development in Unity3D, the human-computer interaction interface of the system is shown in Figure 3. The setting of the navigation tracking method can be confirmed by voice or gesture, which does not involve channel integration.

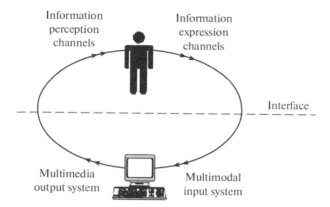

Figure 3. Human-computer interaction interface.

2.3.2 *Human-computer interaction model*

In this paper, the eGOMS model is used to analyze human-computer interaction tasks and realize the integration of multi-channel information. According to the eGOMS model, the interaction model of the virtual-real alignment of the calibration block and the visibility setting of the virtual-real fusion target (Looi and See 2010).

3 GUIDELINES FOR HUMAN-COMPUTER INTERACTION BASED ON COGNITIVE MODELS

3.1 *Human-computer interaction guidelines for surgical navigation systems*

This article uses Microsoft HoloLens as the AR display device. Microsoft HoloLens can not only complete the superposition of virtual and real scenes based on optical perspective but also perform voice and gesture interaction (Lokhande *et al.* 2021). Therefore, the human-computer interaction rules of the AR surgical navigation system mainly include:

1) Consistency of virtual and real registration: virtual objects and natural scenes should be effectively integrated to minimize the distortion of virtual objects, tracking errors and parameter errors of user perspective.
2) User's sense of security: The biggest problem of the surgical navigation system based on VST-HMD is that the doctor cannot look directly at the actual scene with glasses, so there is a certain sense of fear.
3) Effectiveness of information: Information should be classified and managed to avoid redundant information distracting users' energy.

4 USABILITY EVALUATION OF HUMAN-COMPUTER INTERACTION

4.1 *Usability evaluation index based on AHP*

According to the ISO9241-11 standard proposed by the International Organization for Standardization, usability refers to the effectiveness, efficiency and satisfaction of a specific user using a product to achieve a particular goal under a specific context of use. Some researchers have conducted usability evaluations for virtual reality systems. However, there are no usability studies on AR navigation system interaction. Therefore, combined with the human-computer interaction criteria proposed in Section 3.2, refer to the virtual reality interface usability evaluation method (Li *et al.* 2021; Silva and Medeiros 2021; Tao *et al.* 2018).

The consistency test indicators CR and CR_i are: 0.0371, 0, 0, 0.0061, 0, 0.0028, respectively. All levels of consistency test indicators are less than 0.1, so all passed the test, where i = 1, 2, 3, 4, 5 If λ and λ_i are the largest eigenroots corresponding to matrices A and B_i, respectively, and x and x_i are the eigenvectors corresponding to matrices A and B_i, respectively. Normalize ε and ε_i respectively to get vectors θ and θ_i. If x_i represents the element of the vector θ, and y_{ij} represents the element of the vector θ_i, the weight of each layer is combined and calculated, and the total weight of each secondary index can be obtained as x_{ij} = $y_{ij} \times \theta_i$. The weight of each indicator is shown in Table 1 and Figure 4.

Table 1. Availability indicator weights.

Index	Weight x_i	Index	Weight y_{ij}	Comprehensive weight x_{ij}
Task performance	0.2345	Error rate	1.00002	0.2339
Appropriateness	0.3608	Key information rate	0.6298	0.2265
		suggestive	0.3708	0.1336
Ease of use	0.1598	Symbolic Manipulation	0.5109	0.0818
		Complexity	0.3359	0.0533
		Operation depth	0.1548	0.0243
		total operation		
Learnability	0.1029	New knowledge rate	0.5837	0.0597
		Study-time	0.4165	0.0428
Interactivity	0.1452	Naturalness	0.2637	0.0384
		Immediacy	0.2149	0.0313
		predictability of results	0.5223	0.0759

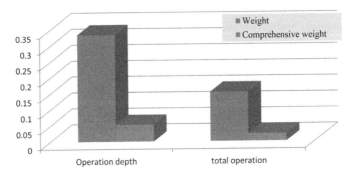

Figure 4. Operational depth ratio.

4.2 Comprehensive evaluation based on fuzzy theory

The fuzzy evaluation method is based on fuzzy mathematics, applying the synthesis principle of undefined relationships, quantifying some factors with unclear boundaries and challenging to quantify, and comprehensively evaluating the affiliation status of the considered object from multiple aspects (Wang *et al.* 2011; Yao *et al.* 2021; Zhou *et al.* 2021).

1) Determine the set of evaluation factors and the set of evaluation grades.
    The performance of human-computer interaction is divided into five grades, V = {excellent, good, fair, poor, and inferior}.
2) Single-factor evaluation
    When using fuzzy theory for a comprehensive evaluation, it is necessary to quantify fuzzy sets through membership function. According to the evaluation index of usability evaluation characteristics, this paper adopts the trapezoidal/semi-trapezoidal membership function. Given by experts based on experience, the boundaries of each indicator are shown in Table 2.

Table 2. Usability evaluation index boundary value.

Index	Fuzzy Evaluation Division				
	Very good	Better	Generally	Poor	Difference
Error rate/%	95~100	90~95	81~90	71~80	0~71
Key information rate/%	85~100	75~85	41~70	21~40	0~21
Suggestive	4.5~5	4~4.5	3~4	1.5~3	0~1.5
Symbolic Manipulation Complexity	0~75	71~100	101~120	121~150	>150
Operation depth/%	1	1~3	3~5	5~7	>7
Total operation	1~5	3~5	5~8	8~12	>11
New knowledge rate/%	0~10	11~25	25~50	51~70	71~100
Study time/h	0~0.5	0.55~2	2~6	6~10	>10
Naturalness	4.5~5	4~4.50	3~4	1.5~3	0~1.5
Immediacy	4.5~5	4~4.50	3~4	1.5~3	0~1.5
Predictability of results	4.5~5	4~4.50	3~4	1.5~3	0~1.5

The rate of new knowledge, the rate of crucial information, the symbol complexity of the system, the depth of operation, and the total amount of operation work can be calculated according to the definition, and the calculation results are 18%, 85%, 95%, 3, and 4, respectively. According to the membership function and the evaluation results, the measurement values of the usability evaluation index belonging to the evaluation level are shown in Table 3.

Table 3. Measure values of the usability evaluation index belonging to the evaluation level.

Index	Index result	Evaluation level measure value r_{ij}				
		Very good	Better	Generally	Poor	Difference
Error rate/%	90	0.25	1	0.8	0	0
Key information rate/%	86	1.02	1	0	0	0
Suggestive	4.75	1.02	0.45	0	0	0
Symbolic operation complexity/%	96	0.18	1	0.85	0	0
Operation depth	3	0	1	1	0	0
Total operation	4.5	0.52	1	0.52	0	0
New knowledge rate/%	17	0.48	1	0.53	0	0
Study time/h	1.25	0.52	1	0.48	0	0
Naturalness	3.82	0	0.85	1	0.21	0
Immediacy	3.45	0	0.45	1	0.61	0
Predictability of results	4.45	0.84	1	0.21	0	0

4.3 *Comprehensive evaluation results*

Combined with the data in Table 3, the corresponding fuzzy comprehensive evaluation matrix is obtained. According to the weight vector of each index in the results of Section 4.1, the weighted average operator is used as the fuzzy synthesis operator to carry out a fuzzy synthesis.

$$S_i = W_i \times R_i \qquad (1)$$

$$W_i = y^i \qquad (2)$$

$$R_i = \begin{bmatrix} r_i^0 \\ r_i^1 \\ \vdots \\ r_i^s \end{bmatrix} \qquad (3)$$

First, perform fuzzy synthesis on the secondary indicators: where i = 1,2,3,4,5. W_i represents the weight of the secondary indicator, R_i represents evaluation level measure, S represents the number of second-level indicators in the i-th first-level index (Ojha and Rajan 2019; Shajideen and Preetha 2018). Then perform fuzzy synthesis on the first-level hands, and the comprehensive evaluation vector of fuzzy synthesis is:

$$R = [S_1, S_2, S_3, S_4, S_5] \tag{4}$$

$$S = W \times R \tag{5}$$

The result of the fuzzy comprehensive evaluation vector is:

$$s = [0.1945, 0.3875, 0.2257, 0.00390]$$

According to the principle of maximum membership degree, the usability evaluation level of AR navigation system human-computer interaction is better. This is because it has better task performance, appropriateness, ease of learning, ease of use and interactivity.

5 CONCLUSIONS

Through the doctor's cognitive model analysis in the perioperative period and the investigation of the human-computer interaction criteria of the virtual reality system, the design criteria of the human-computer interaction of the surgical navigation system based on OST-AR technology are proposed. To objectively evaluate the human-computer interaction usability of the AR navigation system, this paper shows the usability evaluation index system of human-computer interaction by using AHP and proposes five first-level evaluations of task performance, appropriateness, learnability, ease of use and interactivity. Indicators and 11 secondary indicators such as error rate, key information rate, promptness, symbol operation complexity, operation depth, total operation, new knowledge rate, learning time, naturalness, directness, and predictability of results. Finally, the fuzzy theory is used to comprehensively evaluate the human-computer interaction usability of the AR navigation system.

REFERENCES

Collazos, C.A. and Merchan, L., 2015. Human-computer interaction in Colombia: bridging the gap between education and industry. *IT Professional, 17*(1): 5–9.

De Medeiros, F.P.A., 2021, June. Digital legacy post mortem-data mortality as part of digital life-an analysis from the perspective of human computer interaction researches in Brazil. In *2021 16th Iberian Conference on Information Systems and Technologies (CISTI)* (pp. 1–6). IEEE.

En, L.Q. and Lan, S.S., 2010, April. Social gaming—analysing Human Computer Interaction using a video-diary method. In *2010 2nd International Conference on Computer Engineering and Technology* (Vol. 3, pp. V3-509). IEEE.

Lee, M.K., Tang, K.P., Forlizzi, J. and Kiesler, S., 2011, March. Understanding users' perception of privacy in human-robot interaction. In *Proceedings of the 6th international conference on Human-robot interaction* (pp. 181–182).

Li, H., Shabaz, M. and Castillejo-Melgarejo, R., 2023. RETRACTED ARTICLE: Implementation of python data in online translation crawler website design. *International Journal of System Assurance Engineering and Management, 14*(1): 484–484.

Li, Z., Shu, H., Song, W., Li, X., Yang, L. and Zhao, S., 2021, November. Design and implementation of traditional kite art platform based on human-computer interaction and WebAR technology. In *2021 2nd International Conference on Intelligent Computing and Human-Computer Interaction (ICHCI)* (pp. 209–212). IEEE.

Liu, Y., 2020, December. Human-computer interface design based on design psychology. In *2020 International Conference on Intelligent Computing and Human-Computer Interaction (ICHCI)* (pp. 5–9). IEEE.

Lokhande, M.P., Patil, D.D., Patil, L.V. and Shabaz, M., 2021. Machine-to-machine communication for device identification and classification in secure telerobotics surgery. *Security and communication networks*, *2021*: 1–16.

Ojha, D. and Rajan, R.G., 2019, June. Histogram based human computer interaction for gesture recognition. In *2019 3rd International conference on Electronics, Communication and Aerospace Technology (ICECA)* (pp. 263–266). IEEE.

Saxena, S., Saxena, P. and Dubey, S.K., 2013, August. Various levels of human stress & their impact on human computer interaction. In *2013 International Conference on Human Computer Interactions (ICHCI)* (pp. 1–6). IEEE.

Shajideen, S.M.S. and Preetha, V.H., 2018, December. Hand gestures-virtual mouse for human computer interaction. In *2018 International Conference on Smart Systems and Inventive Technology (ICSSIT)* (pp. 543–546). IEEE.

Soni, M., Jain, A. and Patel, T., 2018, November. Human movement identification using Wi-Fi signals. In *2018 3rd International Conference on Inventive Computation Technologies (ICICT)* (pp. 422–427). IEEE.

Tao, Y., Yao, Z. and Liang, B., 2018, December. Human-Computer Interaction Using Fingertip Based on Kinect. In *2018 IEEE 4th Information Technology and Mechatronics Engineering Conference (ITOEC)* (pp. 888–893). IEEE.

Wang, D. and Xu, Z., 2020, August. Bibliometric analysis of the core thesis system of Interaction Design Research on Human-Computer Interaction. In *2020 International Conference on Big Data and Social Sciences (ICBDSS)* (pp. 101–105). IEEE.

Wang, X., Li, Z. and Bai, J., 2011, August. Non-contact human-computer interaction system based on gesture recognition. In *Proceedings of 2011 International Conference on Electronic & Mechanical Engineering and Information Technology* (Vol. 1, pp. 125–128). IEEE.

Xiang Lu, J., Wang, P., huai Yu, S. and de Lu, C., 2010, August. An improved spectral subtraction algorithm based on auditory masking in voice human-computer interaction. In *2010 IEEE International Conference on Mechatronics and Automation* (pp. 1938–1941). IEEE.

Yang, S., Liu, W. and Xu, C., 2020, November. Research on fashion design based on human-computer interaction technology in the era of big data. In *2020 5th International Conference on Information Science, Computer Technology and Transportation (ISCTT)* (pp. 365–368). IEEE.

Yang, X. and Chen, G., 2009, March. Human-computer interaction design in product design. In *2009 First International Workshop on Education Technology and Computer Science* (Vol. 2, pp. 437–439). IEEE.

Yao, Q., Shabaz, M., Lohani, T.K., Wasim Bhatt, M., Panesar, G.S. and Singh, R.K., 2021. 3D modelling and visualization for vision-based vibration signal processing and measurement. *Journal of Intelligent Systems*, 30(1): 541–553.

Yu, X., Yan-Ning, W., Bo-Tao, G., Hui-Hui, W. and Lu-Fen, T., 2012, December. Study on Human-Computer Interaction System Based on Binocular Vision Technology. In *2012 Second International Conference on Instrumentation, Measurement, Computer, Communication and Control* (pp. 1541–1546). IEEE.

Zhao, X., Zhou, C. and Huang, W., 2013, October. Smart home power management system design based on human-computer interaction model. In *Proceedings of 2013 3rd international conference on computer science and network technology* (pp. 1247–1250). IEEE.

Zhou, Y., Hu, X. and Shabaz, M., 2021. Application and innovation of digital media technology in visual design. *International Journal of System Assurance Engineering and Management*: 1–11.

Next Generation Computing and Information Systems – Gupta. (Ed.)
© 2025 The Author(s), ISBN 978-1-032-73865-9

Applicability of eye tracking technology in virtual keyboard for human-computer interactions

Heena Wadhwa, Keshvi Dhir & Lokesh Deswal
Chitkara University Institute of Engineering & Technology, Chitkara University, Punjab, India

ABSTRACT: The vast field of Human-Computer Interaction, or HCI, which is primarily concerned with the interactions between people and computers, used to be solely concerned with scientific testing in the past. But nowadays it also designs user-focused interfaces and aids in the creation of smart surroundings. We have discussed the detailed applications of HCI. However, a lot of difficulties arise in implementing environment friendly solutions as a result of these advancements in HCI. It has been related to eudaimonia, which is the concept of feeling meaningfulness, realizing one's potential. Moreover, practically any area may now apply HCI, including industrial design, psychology, sociology, and computer science. Further, we analyzed the evolution of HCI over the past decade. Developments in HCI research have also led to the invention of the eye tracking control-based system. The use of an eye-tracking system to enhance learning processes and facilitate accurate and efficient visual recognition of geospatial data has also been covered in this study. In addition, a typing experiment is carried out where participants are given access to a virtual keyboard in order to better understand how this technology might improve users' accuracy and speed when typing. The Technology Acceptance Model (TAM), which is used in this experiment, allows us to determine whether or not consumers are adopting the suggested technology based on responses to the TAM questionnaire. The majority of users deemed this system to be helpful, simple to use, and understandable, based on the experiment's outcomes and the TAM questionnaire. The investigation revealed that while female participants' average accuracy was lower than male participants', male participants' average speed was higher. Later in the paper, we have elaborated challenges of HCI, a few of which are ethics, privacy, security, accessibility, learning and creativity, etc.

1 INTRODUCTION

HCI has been transformed from focusing on a single specialty area (scientific testing) to generating new user interface designs which include a community of tech professionals, researchers, designers, and creators, working together for better user experience. HCI has covered almost all forms of information technology. Today, in this field we give importance to designing interactive computer interfaces and computing machines so that the user can have a better experience, easy-to-understand software, smart environment, etc. The original technical focus of HCI was and is on the concept of usability. Technologies like voice-based and Internet of Things (IoT) cognitive interaction have increased a lot, having the ability or skill to do everything. The interface is one of the main components that can upgrade the overall user experience. There are various interface-related features such as touch, click, voice, display size, color contrast, and brightness. Major job roles in HCI is usability engineer, human factor specialist, designer, researcher, etc. HCI makes software and gadgets more useful for us by designing and implementing interactive interfaces which focus on different interface aspects as described in Figure 1. All these aspects make sure that the user can adjust things according to their needs creating a better experience for them.

DOI: 10.1201/9781003466383-15

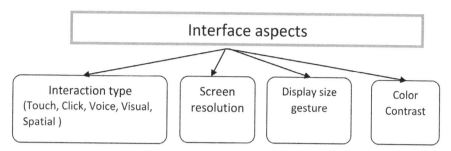

Figure 1. Interface aspects to interact with users.

HCI has transformed a lot compare to its early days because of research, one great example of it is World Wide Web (WWW) (Keskin *et al.* 2023). WWW is a result of HCI research, through which we can store web pages or websites in a web server and connect to it through local computers. Through research, we have developed applications, such as drawing, text editing, hypertext, animation systems, gesture recognition system, eye tracking system, biometric tracking systems, cloud computing, speech recognition, etc.

Another approach is design thinking, which is a practice that originated from the design discipline and aims towards solving problems through a human-focused approach (Culén *et al.* 2014). This practice concerns real-world problems, which can be solved through empathy, prioritizing users, their needs, innovative solutions, and synthetic reasoning. It also makes sure that the innovation is useful, realistic, noticeable, and addresses human values. HCI designers and practitioners follow an approach that concerns user requirements. A deeper exploration of similarities or differences between design thinking and HCI design can reveal new ways for practitioners in shaping innovative technologies for the future. Few Institutes around the world organize different workshops in which we discuss new ideas, recently facing problems, how to solve them, new methods, the idea behind HCI, its history, and much more for researchers, designers, practitioners, beginners, others interested in innovation to help them understand the issues involved in designing and implementation of HCI.

Because of this transformation in HCI and the rapid increase in technology, people with different abilities can use it in their everyday life. Due to this people lacking knowledge, and formal training can easily interact with computers and need not worry about the complications that arise during the early stage of learning any computing system. Hence, we can say that user-friendly interfaces and computing systems make sure that everything is easy and clear for everyone.

2 APPLICATIONS OF HCI

Nowadays, the internet and advanced technology have changed every area of our life from waking up to sleep at night we experience HCI technology throughout the day, one simple example is a light switch (McCormick *et al.* 1976). A person does not need to own a computer system or a laptop or a smartphone to have an impact on HCI in their lives. Examples of HCI in our daily lives are railway ticket-selling machines, health trackers, smart TVs, wearable systems, etc. Today companies or industry which is reliant on technology or computing machines needs HCI in their daily routine. Employees can work more efficiently and fast if the computing machines would be well-designed, easy to use, and easy to understand. HCI plays an important role in safety-control systems and many more.

In this world a lot of people have disabilities so it becomes very difficult for them to understand computing machines. HCI plays a very important role in this case, as HCI offers a very secure, easy-to-understand, efficient, and usable environment. So by focusing on user-centered techniques and functions people with disabilities will also be able to use computing machines comfortably. In the early days of technology computing machines used to be very expensive and not everyone could afford them. But by looking at the benefits of technology, designers, researchers, manufacturers, and providers made sure that everyone could get access to smart devices by making technology cheap without compromising its quality.

Nowadays very few people read manuals as they are lengthy and hard to understand for a beginner. And in this case, HCI makes people learning and accessing computer systems easy (Majaranta *et al.* 2014). HCI makes user-friendly and easy-to-use interfaces and computing systems so that a beginner can understand everything within a few minutes.

3 EVOLUTION OF HCI

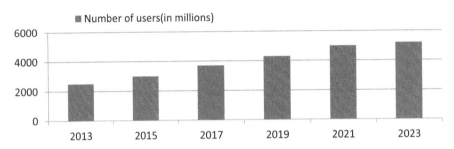

Figure 2.   Number of users increased over past years.

The evolution of HCI in designing new interactive, efficient, easy to use and easy-to-understand interfaces has evolved a lot in the past few decades by focusing on graphical user interfaces (GUIs) as described in Figure 2. Because of the evolution of HCI the number of users using internet and technology has increased rapidly. With the help of GUI, users can now have a whole new experience while dealing with technology, using touch screens have now become a part of our lives by using desktops, smartphones, and tabs (Murad *et al.* 2019). Now the designers are also focusing on Voice User Interfaces (VUIs) to interact with users but our design system is still focused mainly on GUIs hence, for now, it may not be fully possible to design, implement and focus on VUI (Murad *et al.* 2019). A lot of studies, research, and experiments have been done to improve HCI. Various articles and research papers have been published by practitioners, researchers, designers, and users about various problems, their solutions, etc these articles and research papers can help us a lot in evolving HCI (Chen *et al.* 2006). We can go through them and can experiment with a few effective methods, then implement them. Over the years it has been seen that the "living laboratory" has created new ways of experimenting with new methods and developing HCI by analytical analysis, stats, coding, and hypothesis generation on data (Alavi *et al.* 2020).

One other approach is that we can use design patterns as design tools to solve individual problems. We can also use pattern-oriented designs for different related design problems, which help designers in creating visionary designs (Seffah et al. 2012). The design pattern's goal is that how patterns can be reused, and provide proven and valid design solutions (Seffah et al. 2012). As the technology and HCI both evolved new learning solutions and technologies were introduced, which help humans in every way such as gesture-based interaction, a new learning solution for autistic children, and much more.

3.1   *Gesture-based interaction*

It is a type of interaction in which users interact with the computing system through a set of gestures. In this type of interaction, the user uses physical gestures for instance, swiping, tapping, pinching, and scrolling, instead of using input devices such as a keyboard, mouse, etc. HCI aims to make interactions smooth and to do that their most significant technique is gesture recognition, this system makes sure that it is easy to use, provides mobility, and requires less hardware (Kumar *et al.* 2018). Gesture-based applications can be very useful for elderly users as with their aging they lose their sensory, motor, and cognitive abilities (Chen et al. 2013). So they can easily interact with the help of movements of their hands, face, and other body parts because of the advantage of simplicity. With the help of HCI, we can provide new learning solutions for children with Autism Spectrum Disorder (ASD). Research has concluded that the use of computer technology combined with HCI in the education of children with ASD has positive and beneficial effects (Millen *et al.* 2010). Augmented Reality (AR) technology can be used to create a familiar environment in which autistic children can learn their learning content in real-time through audio and video (Sharma et al. 2022). However, the HCI community has not been paying much attention to providing usable and efficient solution to autistic children but a few solutions is still there for instance robot-supported learning solution. These solutions come with many challenges and researchers are trying to overcome the problems.

4 EYE TRACKING CONTROL SYSTEM

Eye tracking systems are used to measure the activity of our eye, it measures the position of our eyeballs, gaze direction, and the movement of the eye can be measured with different technologies (Zhang *et al.* 2017). Eye trackers can help HCI researchers to understand visual and display-based information and this information can be used in improving usability interfaces, various products, websites, apps, etc (Joseph et al. 2020). Eye tracking technology is available for 100s of years but for the past few years, the development in this field has increased a lot. Zhang *et al.* (2017) have mentioned 4 techniques for it: infrared-oculography (IROG), sclera search coil method (SSC), electro-oculography (EOG), and video-oculography (VOG).

This system is now commonly used in medical and psychological research to understand human visual behavior more effectively. This can also be very helpful for people with disabilities and in our workplaces as a security system (Majaranta et al. 2014). Nowadays users prefer biometric systems more as their security system and with the growing interest in this tool in medicine, market research, web design, marketing, security, digital media, product design, defense, etc it has become more affordable than before. A few years back eye trackers used to be very expensive and not everyone could afford them but with the rapid increase in demand for eye trackers in various fields, low-cost eye-tracking devices became available in the market for example GazePoint (GP3) HD 150 Hz, HTC Vive Pro Eye, smart eye aurora – 60/120 Hz, EyeTech VT3 mini – 40/60 Hz, Smart Eye AI-X, etc.

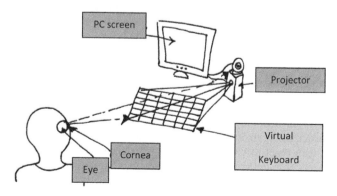

Figure 3.   Eye-tracking technology and virtual keyboard.

The user may control his personal computer (PC) by tapping the pictures of the keys on the virtual keyboard, which can be projected and interacted with on the desktop. Here, we create a virtual keyboard control system that tracks the user's eye movements in addition to their finger movements and converts them into keystrokes (Figure 3). Users may simultaneously use their fingers and eyes to operate the virtual keyboard (Lin *et al.* 2008).

This system can also help users in their learning methods with the help of interactive and revolutionized ways of teaching. McCormick et al. (1976) in their work indicate that more than 80% of human beings learn things faster through visual representation. Thus the visual power of learning is much better than other cognitive processes. The user experience of HCI and eye-tracking technology can be used in a much more effective way to enhance digital learning. In this system, the eye movement data is so important but in the past observing the movements through the eyes was not objective enough, less accurate, etc. The main component of cognitive psychology is the information processing model (Kumar *et al.* 2021; Singh *et al.* 2021; Wu et al. 2012). It is a framework used by cognitive psychologists to describe the processes of the human brain.

5 EXPERIMENTAL RESULTS

A typing experiment is conducted in which we are evaluating how fast participants react or move through visual changes. A virtual keyboard is provided to the participants. In this experiment we are using TAM, introduced by Davis et al. (1989), it is an information system that tells us how the user accepts and rejects any proposed system or technology. In TAM we are using a TAM questionnaire to get the results, in it we need to evaluate 5 items, PU, EOU, ATU, BI, and PI according to the system adopted by Davis et al. (1989) and Venkatesh et al. (2000). We are using the TAM questionnaire at the end of this experiment to know if the proposed technology is comfortable, beneficial, and easy to use for the participants.

In this experiment, more than 1000 participants were part of this experiment and survey. Out of which 64.3% were males and 35.7% were females, with some experience involved in using computing devices. Participants were students of age group from 17 to 25 years old. At first, the participants were introduced to the virtual keyboard and then spent 2–3 minutes with it by themselves to get familiar. After that participants were told only to look at the virtual keyboard to avoid any errors and increase the system's accuracy. After the experiment participants filled out the TAM questionnaire for the proposed virtual keyboard system. This experiment involves 5 simple steps.

– Go to the website link: https://www.typing.com/student/tests.
– Enter the button "Start 1-minute test".
– Now type the letter which is in "blue" color on the screen and make sure to be as fast as you can and avoid any typing errors.
– Then after 1 minute you can check speed and accuracy.
– After that, participants completed their surveys.

The result of this experiment is defined in Table 1. The speed is calculated in Words Per Minute (WPM) and the accuracy is in percentage (%). According to the results, the average speed of male participants (36.5 wpm) is greater than female participants (28.4 wpm), indicating that males can react to visual changes faster than females. The average accuracy of males is 91% whereas that of females is 96%, indicating that females can type with few errors than males. The difference in average accuracy and speed between males and females is shown in Figure 4.

Table 1.   Typing experiment results.

Gender	Average Speed (wpm)	Average Accuracy (%)
Male	36.5	91
Female	28.4	96

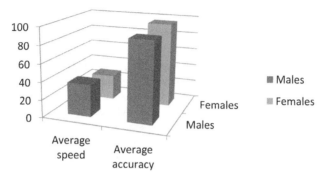

Figure 4.   Average speed and accuracy results.

According to the results of TAM questionnaire, 44.6% of participants use this system in their daily routine or at work while 42.9% use it sometimes. 91.1% of participants think that this system is easy to use and easy to understand. 66.1% would like to use this system in future also. 73.2% participants think that this virtual keyboard typing practice system can help them in improving their performance at work. The results of the TAM questionnaire is described below in Table 2.

Table 2.   Results of the survey.

Decision	Does it improve your performance at work?	Is this system easy-to-use and easy-to-figure?	Do you use it in your daily routine?	Would you like to use it in future?
Yes	73.2	91.1	44.6	66.1
No	7.1	7.1	12.5	26.7
Maybe	19.6	1.7	42.9	7.1

6   FUTURE CHALLENGES OF HCI

There are various challenges and problems that HCI had to face through the years which are mentioned in the following section.

6.1   *Human-technology symbiosis*

With the rise in the development of AI, everything is getting automated so it becomes very crucial that humans control everything, not various algorithms and computers. Therefore, we should create technology which is supporting users in their daily lives and respecting humans, their rights, and establishes trust. To that we need three factors which are transparency, understandability, and accountability, which will build a fair relationship of trust between humans and machines. Interactions in changing and smart environments have been transformed from conventional to focusing on the user's emotions, touch, and gestures meaning shifting from HCI to HEI (human-environment interaction) (Stephanidis *et al.*

2019). A major concern in this approach is the disappearance of computers as they become visible in distinctive devices, creating a smart environment as the computers become part of the furniture (tables, walls, etc) (Stephanidis *et al.* 2019). As technology keeps growing the need for adapting these changes in our everyday life is also important and concerning that these technological changes do not disturb our environment (Stern 1993).

The important role of HCI is to make sure that there is smart interaction between both worlds, creating a hybrid world that consists of physical and digital worlds both. Having smart robots working with and for us, computers embedded in our environment.

6.2 *Ethics, privacy, and security*

The goal is to provide these advanced HCI technologies to a large number of people to have a potential impact and to do that it becomes very important to take care of human privacy, ethics, and security. So, the question arises, 'How do we create and implement guidelines, privacy, security, and standards for HCI and AI technologies that are universally acceptable and also recognize human rights?' (Kisselburgh *et al.* 2020). The answer to this question is research, experiments, and cyber security. Nowadays as everything is digital cybercrimes have taken a rise and people have become more cautious of their privacy and security. Today one can track every information of the other through cyber-attacks and that's why maintaining cyber security becomes very important. Technology provides answers to this problem with various systems like surveillance cameras, IOT-connected home appliances, smart door locks, facial recognition systems, etc.

6.3 *Well-being, health, and eudaimonia*

Technological advances offer various opportunities for humans to have a more advanced and less expensive healthy lifestyle and become more fit. Eudaimonia is a concept of feeling meaningfulness, realizing one's potential, feeling positive, and creating a balance between skills and challenges. Because of these various advanced technologies humans have created 'Personal Medical Devices' (PMDs) that one can wear, interact with, carry, or attach to it, to monitor their health and generate medical data. PMDs not only refer to wearable devices but there are a lot of self-monitoring applications in smartphones that can improve the overall health of the user because they are easy to use, easy to understand, and low cost (Culén *et al.* 2014).

7 CONCLUSION

Virtual keyboard typing practice system can be very beneficial for those who need to type in their jobs or studies. As through this system users can know where they lack, so they can keep practicing to improve their typing speed and accuracy dramatically. Typing fast with great accuracy can save a lot of an individual's time, allowing them to be more productive with few errors in their work places and studies. To know if this application is really helpful or not for students and working professionals, a survey is conducted with the help of TAM model. To understand user's acceptance of this application TAM questionnaire was used in this survey. TAM questionnaire includes 5 essential points which tell us how the user feels about the proposed system, if it is useful for them or not, if it is beneficial for them in their daily lives or not, what do they think about the proposed system. In the above typing experiment and survey, we concluded that the participants liked the proposed system and would like to continue using it to enhance their speed with great accuracy. According to the results most participants think this system is useful, beneficial, easy-to-use and easy-to-understand, so they would like to use it in future for themselves to improve their skills.

REFERENCES

Abascal, J., Barbosa, S.D., Nicolle, C. and Zaphiris, P., 2016. Rethinking universal accessibility: a broader approach considering the digital gap. *Universal Access in the Information Society*, 15(2), pp.179–182.

Alavi, H.S., Lalanne, D. and Rogers, Y., 2020. The five strands of living lab: a literature study of the evolution of living lab concepts in HCI. *ACM Transactions on Computer-Human Interaction (TOCHI)*, 27(2), pp.1–26.

Chen, C., Panjwani, G., Proctor, J., Allendoerfer, K., Aluker, S., Sturtz, D., Vukovic, M. and Kuljis, J., 2006. Visualizing the Evolution of HCI. In *People and Computers XIX—The Bigger Picture: Proceedings of HCI 2005* (pp. 233–250). Springer London.

Chen, W., 2013. Gesture-based applications for elderly people. In *Human-Computer Interaction. Interaction Modalities and Techniques: 15th International Conference, HCI International 2013, Las Vegas, NV, USA, July 21–26, 2013, Proceedings, Part IV 15* (pp. 186–195). Springer Berlin Heidelberg.

Culén, A.L. and Følstad, A., 2014, October. Innovation in HCI: What can we learn from design thinking?. In *Proceedings of the 8th nordic conference on human-computer interaction: Fun, fast, foundational* (pp. 849–852).

Davis, F.D., 1989. Perceived usefulness, perceived ease of use, and user acceptance of information technology. *MIS quarterly*, pp.319–340.

Joseph, A.W. and Murugesh, R., 2020. Potential eye tracking metrics and indicators to measure cognitive load in human-computer interaction research. *J. Sci. Res*, 64(1), pp.168–175.

Keskin, M. and Kettunen, P., 2023. Potential of eye-tracking for interactive geovisual exploration aided by machine learning. *International Journal of Cartography*, pp.1–23.

Kisselburgh, L., Beaudouin-Lafon, M., Cranor, L., Lazar, J. and Hanson, V.L., 2020, April. HCI ethics, privacy, accessibility, and the environment: A town hall forum on global policy issues. In *Extended Abstracts of the 2020 CHI Conference on Human Factors in Computing Systems* (pp. 1–6).

Kumar, A., Kaur, A., Sharma, B. and Mantri, A., 2018. Gesture Based Human Computer Interaction Model for Mixed Reality System.

Kumar, A., Sharma, S., Goyal, N., Singh, A., Cheng, X. and Singh, P., 2021. Secure and energy-efficient smart building architecture with emerging technology IoT. *Computer Communications*, 176, pp.207–217.

Lin, C.S., Lin, C.H., Lay, Y.L., Yeh, M.S. and Chang, H.C., 2008. Eye-controlled virtual keyboard using a new coordinate transformation of long and narrow region. *Optica Applicata*, 38(2).

Majaranta, P. and Bulling, A., 2014. Eye tracking and eye-based human–computer interaction. In *Advances in physiological computing* (pp. 39–65). London: Springer London.

McCormick, E.J., 1976. Human factors in engineering and design. *(No Title)*.

Millen, L., Edlin-White, R. and Cobb, S., 2010, March. The development of educational collaborative virtual environments for children with autism. In *Proceedings of the 5th Cambridge Workshop on Universal Access and Assistive Technology, Cambridge* (Vol. 1, No. 7).

Murad, C., Munteanu, C., Cowan, B.R. and Clark, L., 2019. Revolution or evolution? Speech interaction and HCI design guidelines. *IEEE Pervasive Computing*, 18(2), pp.33–45.

Seffah, A. and Taleb, M., 2012. Tracing the evolution of HCI patterns as an interaction design tool. *Innovations in Systems and Software Engineering*, 8, pp.93–109.

Sharma, B., Mantri, A., Singh, N.P., Sharma, D., Gupta, D. and Tuli, N., 2022, October. EduSense-AR: A Sensory Learning Solution for Autistic Children. In *2022 10th International Conference on Reliability, Infocom Technologies and Optimization (Trends and Future Directions)(ICRITO)* (pp. 1–4). IEEE.

Singh, G., Mantri, A., Sharma, O. and Kaur, R., 2021. Virtual reality learning environment for enhancing electronics engineering laboratory experience. *Computer Applications in Engineering Education*, 29(1), pp.229–243.

Stephanidis, C., Salvendy, G., Antona, M., Chen, J.Y., Dong, J., Duffy, V.G., Fang, X., Fidopiastis, C., Fragomeni, G., Fu, L.P. and Guo, Y., 2019. Seven HCI grand challenges. *International Journal of Human–Computer Interaction*, 35(14), pp.1229–1269.

Stern, P.C., 1993. A second environmental science: human-environment interactions. *Science*, 260(5116), pp.1897–1899.

Venkatesh, V. and Davis, F.D., 2000. A theoretical extension of the technology acceptance model: Four longitudinal field studies. *Management science*, 46(2), pp.186–204.

Wu, C.I., 2012. HCI and eye tracking technology for learning effect. *Procedia-Social and Behavioral Sciences*, 64, pp.626–632.

Zhang, X., Liu, X., Yuan, S.M. and Lin, S.F., 2017. Eye tracking based control system for natural human-computer interaction. *Computational intelligence and neuroscience*, 2017.

Next Generation Computing and Information Systems – Gupta. (Ed.)
© 2025 The Author(s), ISBN 978-1-032-73865-9

A digital twin framework for smart contract-based DeFi applications in the metaverse: Towards interoperability, service scaleup & resilience

Ankur Gupta*, Surbhi Gupta* & Saurabh Sharma*
Department of CSE, Model Institute of Engineering and Technology, Jammu, J&K, India

ABSTRACT: Decentralized Finance (DeFi) is expected to be the backbone for the metaverse facilitating B2B, B2C, and C2C commerce at scale. Blockchain and smart contracts are key drivers of the DeFi vision, enabling secure and non-repudiable financial transactions based on the fulfillment of pre-defined conditions. However, smart contract business logic can be faulty resulting in erroneous transactions and loss of capital. Testing smart contracts and ensuring 100% correctness also remains a challenge. Finally, to realize the interoperability requirement of the metaverse as mandated by the Metaverse Standards Forum (MSF), service providers shall be required to quickly scale and provide similar services across a plethora of metaverse platforms and ecosystems. Service providers also need to ensure the high availability of their DeFi services and remain resilient in the face of attacks. This research paper presents A Digital Twin Framework for smart contract-based applications for the metaverse, allowing DeFi service providers to create and maintain multiple active and passive twins of their services across metaverse platforms. Passive twins of the service can be used by prospective partners to test their smart contract execution and validate the outcomes. Passive twins can be activated dynamically to improve availability, ensure hyper-scaleup, or respond to attacks. Simulation results establish the effectiveness of the proposed framework laying the foundation for meeting the expectations for the future metaverse.

1 INTRODUCTION

The metaverse is fast emerging as a tangible reality with pieces of technology required to realize its vision becoming available (Gupta *et al.* 2023). This promised digital universe is an evolving ecosystem driven by the fusion of high-performance computing, low-latency networks, and immersive AR/VR/XR technologies (Chen *et al.* 2022; Singh *et al.* 2021) which are fast becoming mainstream and helping diverse industries imagine and experiment with exciting use-case scenarios. At the heart of this transformative landscape lies the promise of decentralized finance (DeFi) (Auer *et al.* 2023), a financial paradigm shift that leverages blockchain technology and smart contracts to reshape how we perceive, interact with, and experience economic transactions within the metaverse. The economic impact of the metaverse, driven by DeFi is expected to be a few trillion dollars over the next 2 decades.

The metaverse, with its immersive virtual environments, interconnected digital economies, and thriving virtual communities, presents an unprecedented canvas for the evolution of financial systems(Dincelli & Yayla 2022). In this realm, users can traverse vast virtual landscapes, interact with diverse avatars, own and trade digital assets, and participate in a spectrum of economic activities that mirror real-world counterparts. The metaverse's

---

*Corresponding Authors: ankur.gupta@mietjammu.in, surbhi.cse@mietjammu.in and saurabh.cse@mietjammu.in

     DOI: 10.1201/9781003466383-16

inherent potential to redefine commerce, entertainment, and social interactions is under-pinned by blockchain technology, the decentralized ledger that enables trustless transactions and transparency, and smart contracts (Gupta *et al.* 2023), self-executing agreements that automate and enforce the rules of engagement.

DeFi, which has already disrupted traditional finance (Zhang *et al.* 2023), is poised to take center stage within the metaverse, offering a compelling vision of a financial ecosystem characterized by autonomy, accessibility, and inclusivity. Unlike the conventional financial system, which is governed by intermediaries and centralized authorities, DeFi leverages the power of blockchain and smart contracts to provide individuals with direct control over their assets and financial activities. It also opens new avenues for economic participation, enabling users to earn income through virtual jobs, create and trade virtual goods and ser-vices, and even establish their own decentralized autonomous organizations (DAOs)(Taulli 2022). The Metaverse Standards Forum (Metaverse Standards Forum), defines interoper-ability between all metaverse platforms as a key requirement for the future development of disparate yet collaborating platforms giving users the freedom to navigate the metaverse seamlessly. Thus, DeFi too would need to seamlessly work across platforms allowing the user the freedom to engage in financial transactions on the go, with guarantees of security, and privacy built in. To facilitate this basic requirement, DeFi or smart contracts need to be pervasive across the metaverse fabric (La Barbera 2023), be easily testable, scalable, resilient, and provide assurances of correct execution irrespective of where and how they are invoked. This represents a technical challenge for smart contract developers and DeFi service provi-ders. This research paper provides a distributed framework for smart contracts based on the concept of digital twins (Barricelli *et al.* 2020), allowing DeFi service providers to easily scale their services through replication and edge deployment, providing seamless access to finan-cial transactions for users, while ensuring security and transparency. The significant con-tributions of the present study are listed below:

- The present study proposes the novel concept of Smart Contract Digital Twins to repre-sent replicated smart contracts as edge services for different metaverse platforms.
- The present study provides a new framework for scaling DeFi services across the meta-verse ensuring security (offered as authenticated service endpoints at the edge), inter-operability (working across different metaverse platforms), resilient (to attacks as services can be scaled up through replication) and performant (low-latency as it is edge-based and offers load balancing).
- The present work addresses an important challenge in ensuring the correctness of smart contracts by allowing DeFi users the ability to test the output of smart contract execution by using edge services in passive mode. This ensures that the final output is not committed to the central blockchain. Currently, there is no mechanism for end-users to verify the correct execution of smart contract business logic.
- Simulation results also show that the framework is feasible, easily scalable, and perfor-mant, thus addressing a major requirement for enabling metaverse cross-platform DeFi use cases.

The rest of the manuscript is organized as follows: Section 2 discusses the related work in the domain. Section 3 presents the detailed system model along with the functional elements and workflow, along with the salient features and advantages, while Section 4 presents the experimental setup and simulation results obtained with their discussion. Finally, Section 5 concludes the paper.

2 RELATED WORK

Decentralized Finance (DeFi) has gained traction since the emergence of blockchain, smart contracts, and Decentralized Oracle Networks (DONs) along with the rise of Web3 and the

Metaverse (Liew 2022). In fact, DeFi is touted as the foundational element for realizing the commercial potential of the metaverse (Turi 2023). The Metaverse Standards Forum (Metaverse Standards Forum) is a consortium of leading corporate houses helping set an equitable and inclusive road map for metaverse development. It identifies interoperability as a major requirement for future metaverse platforms. Thus, all central elements of the metaverse should ideally operate seamlessly across multiple and diverse metaverse platforms built and operated by different organizations. This requirement has implications for DeFi service providers as well as they would need to provision their services efficiently across geographically diverse metaverse service providers while ensuring security, transparency, resilience, performance, and scalability (Banaeian Far *et al.* 2023). One of the challenges with smart contracts is accurate and predictable execution (Mense & Flatscher 2018). End users are not able to test smart contracts execution prior to availing services, leaving them vulnerable to incorrect execution of business logic (Khan *et al.* 2021). Other challenges with smart contracts are captured in (Kaur *et al.* 2023). An early suggestion to address this through the use of proxy smart contracts is suggested in (Gupta *et al.* 2023), but this solution is specific to one blockchain type. Current research in enabling seamless DeFi service composition, discovery and consumption across the metaverse can best be described as nascent.

To overcome the stated requirements and associated challenges this research paper proposes a DeFi framework for the metaverse which is based on digital twins for smart contracts, hosted as edge services, providing easy access to metaverse entities to engage in financial transactions using a custom or standard business logic encapsulated in the smart contracts. Further, users can test smart contract logic prior to availing the actual DeFi service. Some considerations for effectively using digital twins in the metaverse setting as discussed in (Banaeian Far *et al.* 2023). The proposed framework addresses these security concerns besides providing additional features to ease cross-metaverse operations.

3 SYSTEM MODEL

Figure 1 depicts the detailed system model for the proposed Digital Twin based DeFi Service Model (DTDSM) based on providing Service-Access End Points (SEPs) across Metaverse Platforms. These SEPs are mapped to DeFi Edge Services (DES) hosted at the edge of the DeFi Service Cloud (DSC) and designed to provide low-latency DeFi services to users across diverse metaverse platforms.

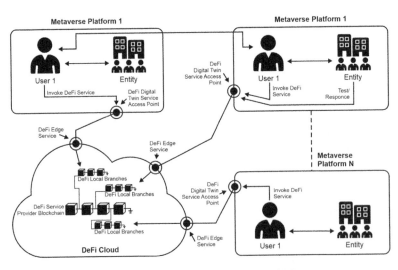

Figure 1.   Proposed digital twin based DeFi Service Model (DTDSM).

The DeFi Service Cloud (DSC) hosts the blockchain providing immutable and non-repudiable service consumption and payment transactions related records for users across diverse metaverse platforms. The Blockchain model is hierarchical with forks in the main blockchain to track transactions and smart contract executions for individual metaverse platforms. While data from the main blockchain is synced periodically with the forked chains, the data from the forked chain to the main chain is committed only when the associated smart contract is in "active mode". In "passive mode" the smart contract performs a dummy transaction on the forked chain to return the indicative result to the parties involved in the transaction. This feature is useful from a testing perspective. The Service-Access End Points (SEP) hosted by a metaverse platform are authenticated and issued authentication tokens which any metaverse entity can verify on demand. This ensures that entities enter into trusted interactions with hosted DeFI Service-Access End Points. All hosted smart contracts can be tested/verified by the platform, entities, and users before entering into the final transaction. The concept of proxy-smart contract has been proposed in an earlier work (Gupta *et al.* 2023) and has shown to work well to overcome issues with incorrect smart contract execution based on the execution of flawed business logic. Based on the service load, the DeFi Service Provider can scale up the DeFi Service-Access End Points and the corresponding Edge Services on the DeFi Service Cloud, thereby ensuring load management and providing deterministic quality-of-service guarantees. Further, in the face of denial-of-service attacks on the DeFi services or metaverse platform, having multiple instances will help defray the overall service impact for the end users. The salient features of the Proposed Digital Twin based DeFi Service Model are highlighted below:

- **Interconnected Metaverse Platforms**: The proposed framework helps realise the interoperability vision articulated by the Metaverse Standards Forum, where users can interact, transact, and engage in various collaborative activities.
- **Service Access Points**: The system facilitates the establishment of service access points across multiple metaverse platforms. These access points act as gateways or interfaces to access DeFi services seamlessly from within and outside the metaverse.
- **DeFi Cloud Infrastructure**: The DeFi Cloud serves as the backbone of the system. It hosts a range of DeFi services, edge services, and Digital Twins of Smart Contracts for secure, transparent and scalable solutions.
- **Digital Twins of Smart Contracts**: Digital Twins of Smart Contracts are replicas of smart contracts hosted remotely on the DeFi Cloud with transactions recorded on forked branches and automated forward syncing. These replicas are synchronized with the originals and enable faster execution, scaling, and fault tolerance.
- **DeFi Service Providers**: DeFi service providers offer various financial services, such as lending, borrowing, trading, yield farming, and more, to users across the metaverse platforms. Their service access points are authenticated and testable by prospective collaborators before actual use.
- **Multi-Blockchain Model**: The DeFi service provider maintains multiple blockchains forked from the main blockchain with each forked branch for transactions from a single metaverse for easier reporting. When actual transactions are conducted, only then the final transaction is committed to the main blockchain from the forked branches.

The proposed model is efficient in terms of latency, scalability, and reliability. Users can seamlessly access DeFi services within their metaverse environments, or while navigating other metaverse platforms. Additionally, DeFi service providers can expand their services to new platforms. The framework achieves lower latency in smart contract execution by hosting Digital Twins of Smart Contracts as edge services on the DeFi Cloud. This is essential for time-sensitive DeFi activities, including high-frequency trading. Digital Twins of Smart Contracts can be horizontally scaled to accommodate a growing user base and increased transaction volume, ensuring the system can handle increased demand. Replicated smart contract instances on the DeFi Cloud provide fault tolerance. In case of a failure or down-time in one instance, others seamlessly take over, ensuring continuous service availability. The framework allows for compatibility across multiple metaverse platforms, making it easier for users to transition between different virtual worlds while using familiar services and maintaining context.

4 SIMULATION RESULTS

The metaverse platforms have been simulated by selecting AWS instances across different geographical regions (USA, Europe, Asia, Australia, and New Zealand) with their asso-ciated zones, regional edge caches, and edge locations. Table 1 captures the configuration.

Table 1. AWS Configuration describing the location of components of the proposed framework.

Cloud Region (Metaverse Platform Location)	US West (USA)	US East (USA)	London (Europe)	Mumbai (Asia)	Sydney (ANZ)
Regional Edge Caches	California	Ohio	London	Mumbai	Sydney
Edge Locations (For DeFi Edge Services)	Los Angeles	New York	London	Mumbai	Sydney
DeFi Cloud Location	Mumbai	Mumbai	Mumbai	Mumbai	Mumbai

For simulating edge services, we make use of AWS CloudFront Edge Services using Lambda Edge Functions which provide low-latency services across multiple geographically dispersed locations. For the purposes of simulation, we have considered the DeFi Cloud to be hosted at a single geographical location. However, future optimizations may include offering decentralized DeFi services for further performance optimization.

Figure 2 captures the average latency time for Edge Service Invocation from within the metaverse platform by an entity in the same metaverse. It can be seen from the figure that the average latency for edge service invocation ranges from 0.83 milliseconds to 1.96 milli-seconds over a 24-hour period of operation during which 10,000 transactions were simulated by programs running on AWS instances representing user-entity interactions in the metaverse.

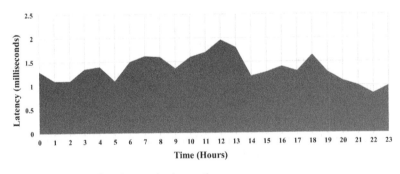

Figure 2. Average latency for edge service invocation.

Figure 3 a captures the average latency for the Edge-DeFi Cloud communication in the same AWS region, while Figure 3 b captures the average latency for the Edge-DeFi Cloud latency in a different region. The DeFI Cloud is hosted in the Mumbai AWS region. It can be seen that the average latency for same-region communication lies between 1.7 to

2.8 milliseconds while the average latency for Edge-DeFi cloud latency ranges from 2.4 to 5.7 milliseconds over a 24-hour period.

The end-to-end transaction time between two entities in a metaverse platform now comprises:

$$T_{txn} = DeFi_{Invo-Time} + Edge_{Serv-Latency} + Edge_{Core-Latency} + SC_{Exe-Time} + Commit \qquad (1)$$

In the above equations, T_{txn} represents the total transaction time in the system. $DeFi_{Invo-Time}$ represents the time taken for invocation of DeFi service. $Edge_{Serv-Latency}$ and $Edge_{Core-Latency}$ denotes the Edge service and Edge core latencies respectively. The term $SC_{Exe-Time}$ denotes the Smart Contract execution time. Further, $Commit$ indicates the time taken for blockchain commit, possibly in the forked blockchain or main blockchain.

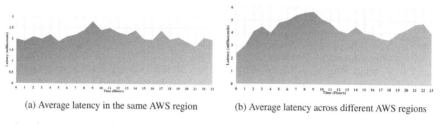

(a) Average latency in the same AWS region　　(b) Average latency across different AWS regions

Figure 3.　Average latency for the Edge-DeFi Cloud communication in the same and different regions.

Figure 4 depicts the average transaction time-varying over a 24-hour window of operation across time zones and geographical locations. A maximum variance of 78% is observed across transaction times due to varying factors such as traffic load variations, cross-region communication and whether the digital twin smart contracts were invoked in active or passive mode (random distribution).

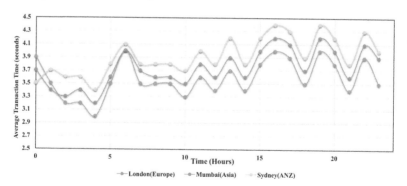

Figure 4.　Average transaction time over a 24-hour window of operation across time zones and geographical locations.

Using a two-level blockchain with duplicate smart contract execution adds significant overhead to the overall processing, but given the uncertainty surrounding correct smart contract execution, the overheads are justified.

5　CONCLUSION

This research paper presents a novel framework for designing DeFi services for the metaverse based on the concept of Digital Twins, essentially replicas of smart contracts and other financial services. These replicas allow users to verify the correct execution of smart contract-based DeFi services before actually using them, overcoming challenges in traditional smart contracts. Further, the multi-instance edge model is quickly scalable, resilient to attacks and provides low-latency operation due to its location at the metaverse edge. A hierarchical blockchain model supports the Digital Twin DeFi Services model ensuring that only active transactions are committed while the passive transactions used for testing are not made a part of the actual transaction records. Future work shall involve testing the framework at scale for diverse use cases and optimizing the performance further through intelligent global placement/re-deployment of edge services.

REFERENCES

Auer, R., Haslhofer, B., Kitzler, S., Saggese, P., & Victor, F. (2023). *The echnology of Decentralized Finance (DeFi). Bank for International Settlements*, Monetary and Economic Department.

Banaeian Far, S., & Imani Rad, A. (2022). Applying digital twins in metaverse: User interface, security and privacy challenges. *Journal of Metaverse*, 2(1), 8–16. 2022.

Banaeian Far, S., Imani Rad, A., & Rajabzadeh Asaar, M. (2023). Blockchain and its derived technologies shape the future generation of digital businesses: a focus on decentralized finance and the metaverse. *Data Science and Management*, 6(3), 183–197. 2023.

Barricelli, B. R., Casiraghi, E., Gliozzo, J., Petrini, A., & Valtolina, S. (2020). Human digital twin for fitness management. *IEEE Access*, 8, 26637–26664. 2020.

Chen, Z., Wu, J., Gan, W., & Qi, Z. (2022). Metaverse security and privacy: An overview. In *2022 IEEE International Conference on Big Data (Big Data)* (pp. 2950–2959). December 2022.

Dincelli, E., & Yayla, A. (2022). Immersive virtual reality in the age of the metaverse: A hybrid-narrative review based on the technology affordance perspective. *The Journal of Strategic Information Systems*, 31 (2), 101717. 2022.

Gupta, A., Gupta, R., Jadav, D., Tanwar, S., Kumar, N., & Shabaz, M. (2023). Proxy smart contracts for zero trust architecture implementation in decentralized oracle networks based applications. *Computer Communications*, 206, 10–21.

Gupta, H. U., Khan, S., Nazir, M., Shafiq, M., & Shabaz, M. (2023). Metaverse security: Issues, challenges and a viable ZTA model. *Electronics*, 12(2), 391. January 12 2023.

Kaur, G., Lashkari, A. H., Sharafaldin, I., & Lashkari, Z. H. (2023). Smart contracts and DeFi security and threats. In *Understanding Cybersecurity Management in Decentralized Finance: Challenges, Strategies, and Trends* (pp. 91–111). Springer, 2023.

Khan, S. N., Loukil, F., Ghedira-Guegan, C., Benkhelifa, E., & Bani-Hani, A. (2021). Blockchain smart contracts: Applications, challenges, and future trends. *Peer-to-peer Networking and Applications*, 14, 2901–2925. 2021.

La Barbera, S. (2023). Navigating the virtual frontier: The convergence of decentralized finance and the metaverse.

Liew, V. K. (2022). Web3 Made Easy: A comprehensive guide to web3: Everything you need to know about Web3, Blockchain, DeFi, Metaverse, NFT and GameFi.

Mense, A., & Flatscher, M. (2018). Security vulnerabilities in ethereum smart contracts. In *Proceedings of the 20th International Conference on Information Integration and Web-based Applications & Services* (pp. 37–380).

Metaverse Standards Forum. The metaverse standards forum.

Singh, M., Singh, S. K., Kumar, S., Madan, U., & Maan, T. (2021). Sustainable framework for metaverse security and privacy: Opportunities and challenges. In *International Conference on Cyber Security, Privacy and Networking* (pp. 329–340). Cham, September 2021.

Taulli, T. (2022). *Decentralized Autonomous Organizations (DAOs) Governance for Web3* (pp. 81–96). Apress, Berkeley, CA, October 29 2022.

Turi, A. N. (2023). *Financial Technologies and DeFi: A Revisit to the Digital Finance Revolution*. Springer Nature.

Zhang, Y., Chan, S., Chu, J., & Shih, S.-h. (2023). The adaptive market hypothesis of decentralized finance (DeFi). *Applied Economics*, 55(42), 4975–4989. 2023.

Next Generation Computing and Information Systems – Gupta. (Ed.)
© 2025 The Author(s), ISBN 978-1-032-73865-9

A novel approach to glass identification using ensemble learning for forensics

Priyansh Sanghavi, Mahir Mehta, Tarang Ghetia & Yogesh Kumar
School of Technology, Pandit Deendayal Energy University Gandhinagar, Gujarat, India

ABSTRACT: Criminal cases, including burglaries, robberies, hit-and-runs, murders, assaults, ram raids, criminal damage, and car thefts, may all employ glass as evidence. When anything made of glass is broken, the shards are likelier to go in the direction of whoever or whatever caused the break. In order to conduct effective criminological investigations, it is important to get information from databases containing criminal evidence. The paper gives a detailed study of glass identification based on the KNN algorithm. The identification method is based on the examination of many feature columns such as Refractive Index (R. I.), Sodium (Na), Magnesium (Mg), Potassium (K), and other substances that serve as essential forensic evidence. An ensemble strategy comprising several machine learning models was used to improve the accuracy and reliability of glass detection. In addition to KNN, SVM, Logistic Regression, Naive Bayes, and Random Forest classifiers have been integrated into the ensemble framework. The ensemble methodology leverages soft voting to combine the predictions of these diverse models, thereby achieving a more robust and effective glass identification system. The proposed ensemble technique demonstrates superior performance compared to individual models through extensive experimentation and validation, resulting in enhanced accuracy, precision, and recall. Such advances in glass identification hold significant promise for digital forensics, allowing forensic experts to make more informed determinations in criminal investigations.

1 INTRODUCTION

Criminological research served as a foundation for studying the glass classification problem. If the glass left at the crime scene is properly recognized, it can be used as evidence. It is typical practice in casework to compare glass pieces found at the site of a crime with glass pieces linked to a suspect(Mehta *et al.* 2022). These shards of glass might sometimes be imperceptibly small. These minute glass pieces must be recognized and examined in case they are relevant to a forensic investigation (Colomban 2013a).

The oxide concentrations for various elements have been provided via the quantitative examination of glass. However, only the oxides of calcium, silicon, aluminum, magnesium, sodium, and potassium typically exist at concentrations high enough to be measured (Hickman 1981). Therefore, the elements in our dataset, such as sodium (Na), magnesium (Mg), aluminum (Al), calcium (Ca), barium (Ba) etc. are what we have taken into consideration for the feature column in this paper. Furthermore, a key component in differentiating between various types of glass and their uses is the Refractive Index (RI). This dataset includes glasses divided into seven distinct categories depending on their function. Tableware, pendant lighting, window glass (both float and non-float), top lights, and vehicles (both float and non-float) (Mathur and Surana 2020).

DOI: 10.1201/9781003466383-17

The primary purpose of this study is to properly classify a single glass shard by its major component measurement. The K-Nearest Neighbor classifier(Mashael S. Aldayel 2012), one of the most popular data mining methods for Pattern detection and classification were the basis of our solution for the glass problem. Recent studies have compared voting to other single classifiers and found that it performs better than the rest (Rincy and Gupta 2020).

In the Ensemble technique, each specified classification algorithm is trained on the dataset. Each model learns to make predictions based on the features (e.g., refractive index, chemical composition) and their relationships to glass types. After training, each model is used to predict from the dataset (Dong *et al.* 2020; Rincy and Gupta 2020). These predictions represent the model's assessments of the glass type for each observation. After that, soft voting is done in which the ensemble technique aggregates the projections from all individual models, and the predictions are combined in a weighted manner. The ensemble technique calculates a weighted average or combines the probability scores from the individual models. This weighted combination of predictions determines the final classification. The class with the highest weighted sum or the highest probability score is chosen as the ultimate classification for a given observation.

2 RELATED WORKS

In recent times, a significant amount of research has been conducted to investigate the possibility of different algorithmic methods to streamline the identification process of eyeglasses. This study contains machine-learning techniques applied to the same glass detection dataset, such as bagging, kernel density estimation, AdaBoost, and fuzzy clustering (Ruying and Rongcang 2009). One notable researcher (Pandey and Jain 2017), has used the K-nearest neighbors (KNN) technique as an approach to address this formidable job.

The main goal of these initiatives is to improve the accuracy and precision level in the glass detection models. Accurate detection of glass is paramount across various applications, including automated quality control in industrial processes and security and surveillance systems. To accomplish this objective, scholars acknowledge the need to integrate additional classification algorithms into constructing the glass detection models(Cholakova *et al.* 2016).

Integrating a K-closest neighbors (KNN) classifier with a feature-weighted nearest neighbor approach, based on the principles of a chi-square statistical test, is a successful strategy for improving its precision (Dino and Abdulrazzaq 2019). This unique concept was proposed by the authors (Rincy and Gupta 2020), and it has the potential to considerably increase the performance of glass detection models based on the K-nearest neighbors method. This strategy, which assigns weights to features based on their relevance and importance, allows the classifier to prioritize the most valuable attributes of the input data, resulting in improved accuracy in glass detection.

Including feature weighting into the K-nearest neighbors (KNN) method is theoretically sound. It has practical applications. Glass detection datasets sometimes include a profusion of information, some of which may need to be more relevant in distinguishing distinct glass types or states. The use of feature weighting, driven by statistical tests such as the chi-square test, may effectively discover and prioritize the most discriminatory characteristics, hence augmenting the prediction capabilities of the model (Oladipo *et al.* 2020).

Moreover, it is crucial to underscore the utmost importance of ensuring the precision and validity of the findings derived from these scholarly pursuits. The precise categorization of glass has significant consequences for practical applications in the real world. In a manufacturing setting, accurately identifying and classifying various kinds of glass helped optimize quality control procedures, resulting in fewer faults and lowered production costs. In security and surveillance, the precise detection of glass may significantly enhance the efficacy of recognizing possible security risks or criminal activities (El-Khatib, Abu-Nasser and Abu-Naser 2019).

The primary objective of these research endeavors is to enhance the overall precision and dependability of glass detection models. Therefore, these efforts are in accordance with the overarching goal often pursued in research endeavors: to progress the current knowledge and understanding within a certain discipline. The primary objective in this instance pertains to the augmentation of the precision of glass categorization, a work that has extensive implications (Goswami and Wegman 2016).

3   DATASET DESCRIPTION

The UCI Machine Learning Repository offers a plethora of additional datasets for use with machine learning. Glass may be identified by its components in the glass identification dataset. Ten input characteristics and one output attribute (Type of Glass) are included in the 214 data points that make up this dataset. The proper oxide percentage by weight is used to define the properties (German 1987).

3.1   Attribute information

Properties such as "S.NO.", "RI (Refractive Index)", "Na", "Mg", "Al", "Si", "K", "Ca", "Ba" and "Fe", (calcium, barium, and iron) may be found in the set. Features like these are used as inputs in a classification task.

Table 1.   Glass dataset

S.No.	RI	Na	Mg	Al	Si	K	Ca	Ba	Fe	Type of Glass
1	1.51761	12.81	3.54	1.23	73.24	0.58	8.39	0	0	One
2	1.51761	13.89	3.6	1.36	72.73	0.48	7.83	0	0	One
3	1.51618	13.53	3.55	1.54	72.99	0.39	7.78	0	0	One
4	1.51766	13.21	3.69	1.29	72.61	0.57	8.22	0	0	One
5	1.51742	13.27	3.62	1.24	73.08	0.55	8.07	0	0	One
6	1.51596	12.79	3.61	1.62	72.97	0.64	8.07	0	0.26	One
7	1.51743	13.3	3.6	1.14	73.09	0.58	8.17	0	0	One
8	1.51756	13.15	3.61	1.05	73.24	0.57	8.24	0	0	One
9	1.51918	14.04	3.58	1.37	72.08	0.56	8.3	0	0	One
10	1.51755	13	3.6	1.36	72.99	0.57	8.4	0	0.11	One

Depending on their output qualities, different types of glass fall into one of seven categories. These categories are related to the function or production method of the glass, and they are not numbered. There are seven categories: "float processed windows", "non-float processed windows", "vehicle windows", "vehicle windows that haven't been float processed", "headlamps", "containers" and "tableware".

The dataset's primary goal is to classify each kind of glass into one of the seven categories based on their characteristics; hence, it contains details on a wide range of glass characteristics.

Input and output characteristics were discussed at the top. The input variables' values will establish the kind of glass used; the model will be trained with these values, and tests will be run to establish the Glass classification of the record. A sample of this data is shown in Table 1.

4   METHODOLOGY

Data preparation and classifier application are the two main sections of this study. The purpose of the pre-processing phase is to get the dataset ready for analysis. In the second stage, we use single and multiple classifiers to build a highly accurate prediction model to solve the glass identification problem. Multiple classifiers will be used to achieve this goal (Dong *et al.* 2020; Rincy and Gupta 2020). The flow diagram used to solve this problem is shown in Figure 1 below.

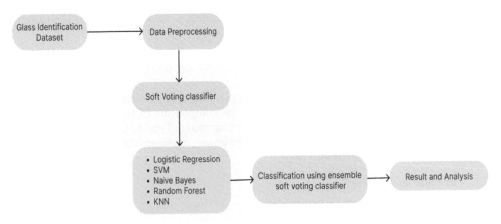

Figure 1.   Flow diagram to be followed to solve this problem.

4.1   Data pre-processing

Despite the adherence to the appropriate structure of a dataset, it may nevertheless need pre-processing before using a data mining method. This will allow for a higher-quality analysis due to the study. There are a lot of different approaches to preparing data. Since there are no missing numbers, the data on glass is in a relatively clean state.

4.2   Feature selection

Feature analysis is done to identify the most relevant compounds influencing glass type. Various feature selection approaches may be used to ascertain the crucial variables required for classification, including correlation analysis, recursive feature removal, and feature significance derived from ensemble models.

4.3   Ensemble model selection

A method based on "ensemble learning" has been selected. In this instance, various machine learning models will be used to form the ensemble. These models aim to increase confidence in glass identification.

4.4   Support vector machine algorithm

The SVM method has a kernel and linear sections. SVM can solve classification and regression problems. SVM is better at finding deep patterns in complex datasets than other methods. For classification, we calculate the characteristic-based hyperplane that separates the two classes. Hyperplanes categorize decision boundaries. Comparing data points on opposing sides of the plane reveals distinct types(Colomban 2013b).

Support vector machines aim to maximize data point-hyperplane margin(Rincy and Gupta 2020).

Linear equation: y = a + b. We obtain $ax1x2 + b = 0$ if we replace x with x1 and y with x2. It can be shown that $wx + b = 0$ if x is defined as (x1,x2) and w as (a,1).

This equation is based on vectors in just two dimensions. However, this holds for any arbitrary number of dimensions. This is the hyperplane's equation.

$$h(x_i) = \begin{cases} +1 \, if \, w \cdot \varkappa + b \geq 0 \\ -1 \, if \, w \cdot \varkappa + b < 0 \end{cases} \tag{1}$$

Points that are located above or on the hyperplane are allocated to class +1, whereas points that are located below the hyperplane are assigned to class −1.

4.5 K-Nearest neighbors algorithm

This approach can be used to solve regression and classification issues similarly. We're employing classification here. KNN selects the desired number of examples (K) closest to the query by measuring the distances between each example in the data and the query. It then selects the label that appears the most frequently or averages the labels in the case of regression (Mashael S. Aldayel 2012).

The Euclidean metric will be used as the distance metric of choice.

$$d(x, x^1) = \sqrt{(x - x_1')^2 + \dots (x_n - x_n^1)^2} \tag{2}$$

Finally, the input x is allocated to the most probability class.

$$P(y = j | x = x) = \frac{1}{k} \sum_{i \in A} I\left(y^{(i)} = j\right) \tag{3}$$

4.6 Random forest algorithm

Random Forest, or Ensemble machine learning, connects several classifiers to tackle a difficult problem. The majority of voting predicts data outcomes. Prediction class voting occurs among trees. The majority-voted class wins. Random Forests sometimes use the Gini index to decide the order of nodes on a decision tree branch when categorizing data.

$$Gini = 1 - \sum_{i=i}^{c} (p_i)^2 \tag{4}$$

The Gini of each branch on a node is calculated using the class and probability to establish the most probable branch. Entropy may also be used to establish the relationships between nodes in a decision tree (Dong et al. 2020).

$$Entropy = \sum_{i=1}^{c} -p_i * \log_2(p_i) \tag{5}$$

4.7 Logistic regression algorithm

This supervised machine learning approach is similar to linear regression in that it is used for classification instead of regression. It categorizes the independent variables into binary form and displays their linear relationships (Dong et al. 2020).

The basic logistic model uses the formula:

$$\ln(P) = \ln\left(\frac{P}{1-P}\right) \tag{6}$$

So, Logistic Regression formula is given by:

$$\ln\left(\frac{P}{1-P}\right) = \beta_0 + \beta_1 x \tag{7}$$

$$=> P = \frac{e^{\beta_0 + \beta_1 x}}{1 + e^{\beta_0 + \beta_1 x}} \tag{8}$$

4.8 Naïve Bayes algorithm

The Naive Bayes technique is a probabilistic classification approach frequently used in machine learning for classification difficulties, such as text categorization, spam detection, sentiment analysis, and others (Rincy and Gupta 2020). Baye's Theorem may be expressed in a straightforward mathematical manner as follows:

$$P(A|B) = P(B|A) * P(A)/P(B) \tag{9}$$

The marginal probability of an occurrence, $P(A)$, is termed the prior, and the probability we're interested in, $P(A|B)$, is called the posterior.

4.9 Soft voting ensemble

The decisions of numerous classifiers are combined using the multiple classifiers voting technique. It involves creating a classifier for each of the smaller, equal subsets of the training dataset that are created. Soft voting has been used in which the individual models' predictions are weighted based on their confidence scores, and the final prediction is made by considering these weighted contributions. The accuracy of these classifiers has been improved when voting is applied to classification algorithms(Rincy and Gupta 2020).

5 RESULT AND DISCUSSION

We used the Ensemble Learning approach, where we used multiple models for classifying glasses according to their categories. As a result of this study, we found an accuracy of 95.29% using the Ensemble method for the glass dataset.

Multiple studies show that using multiple classifiers on a dataset yields far better results. Therefore, the classification decision is based on the combined outputs of numerous models, making it more accurate than using a single classifier to identify the glass type.

Using data on the glass's elemental composition and refractive index, we trained several machine learning algorithms to make an identification. We used five different machine learning methods.

Table 2 below details the precision of the various machine-learning models used in the ensemble method. Random Forest was the best model for making predictions in this data set, achieving an accuracy of 96.92%. Ensemble voting yielded an accuracy of 95.29 percent.

Table 2. Classification accuracy details in glass dataset.

Model	Accuracy	Precision	Recall	F1-Score
K-Nearest-Neighbor	89.23%	86%	86%	85%
Support Vector Machine	83.85%	98%	95%	96%
Gaussian Naïve Bayes	72.31%	78%	72%	72%
Logistic Regression	93.85%	96%	94%	95%
Random Forest	**96.92%**	**99%**	**98%**	**99%**
Ensemble Voting	95.29%	96%	95%	96%

We have used different evaluation metrics to evaluate the performance of different models. These metrics provide a more holistic assessment of a categorization model's performance instead of relying only on accuracy. Prediction accuracy refers to the proportion of positive situations that happen. The F1-score is computed by taking the harmonic mean of the accuracy and recall. Recall, in contrast, quantifies the proportion of true positive cases properly detected by the model. Uneven class distribution may be effectively addressed by ensuring a balance between recall and accuracy, which offers significant advantages.

6 CONCLUSION

Criminal evidence databases must be accessed to perform extensive and reliable criminological research. Data mining extracts relevant information from datasets by finding obvious and believable patterns. Classification problems are often solved with a k-nearest neighbor. Time and again, studies have shown that voting with several classifiers performs better than voting alone. KNN first helped criminologists determine glass class in this article. We used the Ensemble approach. Combining the most successful classifiers on the glass dataset plus soft voting yields Random Forest's greatest accuracy (96.92%) and ensemble voting's substantial result.

Additional research will improve the model's usefulness and practicality. Adding glass attributes like density and heat resistance to the model may improve system accuracy. New attributes must fix the dataset's imbalance, which is its main issue. A rigorous categorization validation of the glass shows an increase in accuracy. We divided the training and testing data at 70:30 in this study. However, applying cross-validation can improve outcomes. The normal classifier may perform better by splitting the dataset into n bins. Finally, we may use non-ML methods like ANN to enhance the model in the future.

REFERENCES

Cholakova, A., Rehren, T. and Freestone, I.C. (2016) 'Compositional identification of 6th c. AD glass from the lower Danube', *Journal of Archaeological Science: Reports*, 7, pp. 625–632. Available at: https://doi.org/10.1016/j.jasrep.2015.08.009.

Colomban, P. (2013a) 'The destructive/non-destructive identification of enameled pottery, glass artifacts and associated pigments—A brief overview', *Arts*, 2(3), pp. 77–110. Available at: https://doi.org/10.3390/arts2030077.

Colomban, P. (2013b) 'The destructive/non-destructive identification of enameled pottery, glass artifacts and associated pigments—A brief overview', *Arts*, 2(3), pp. 77–110. Available at: https://doi.org/10.3390/arts2030077.

Dino, H.I. and Abdulrazzaq, M.B. (2019) *Facial Expression Classification Based on SVM, KNN and MLP Classifiers*.

Dong, X. *et al.* (2020) 'A survey on ensemble learning', *Frontiers of Computer Science*. Higher Education Press, pp. 241–258. Available at: https://doi.org/10.1007/s11704-019-8208-z.

El-Khatib, M.J., Abu-Nasser, B.S. and Abu-Naser, S.S. (2019) *Glass Classification Using Artificial Neural Network, International Journal of Academic Pedagogical Research*. Available at: www.ijeais.org/ijapr.

German, B. (1987) *Glass Identification – UCI Machine Learning Repository*. Available at: https://doi.org/https://doi.org/10.24432/C5WW2P.

Goswami, S. and Wegman, E.J. (2016) 'Comparison of Different Classification Methods on Glass Identification for Forensic Research', *Journal of Statistical Science and Application*, 4(2). Available at: https://doi.org/10.17265/2328-224x/2015.0304.001.

Hickman, D.A. (1981) *A classification scheme for glass, Forensic Science International*. Elsevier Sequoia S.A.

Mashael S. Aldayel (2012) *K-Nearest Neighbor Classification for Glass Identification Problem*. [IEEE].

Mathur, H. and Surana, A. (2020) 'Glass classification based on machine learning algorithms', *International Journal of Innovative Technology and Exploring Engineering (IJITEE)*, (9), pp. 2278–3075. Available at: https://doi.org/10.35940/ijitee.H6819.0991120.

Mehta, N. *et al.* (2022) *A Comprehensive Study on Cyber Legislation in G20 Countries, Communications in Computer and Information Science*. Available at: https://doi.org/10.1007/978-3-031-23095-0_1.

Oladipo, F.O. *et al.* (2020) *The State of the Art in Machine Learning-Based Digital Forensics*. Available at: https://ssrn.com/abstract=3668687.

Pandey, A. and Jain, A. (2017) 'Comparative analysis of KNN algorithm using various normalization techniques', *International Journal of Computer Network and Information Security*, 9(11), pp. 36–42. Available at: https://doi.org/10.5815/ijcnis.2017.11.04.

Rincy, T.N. and Gupta, R. (2020) *Ensemble learning techniques and its efficiency in machine LEARNING: a survey, 2nd International Conference on Data, Engineering and Applications (IDEA)*.

Ruying, S. and Rongcang, H. (2009) *Data Mining Based on Fuzzy Rough Set Theory and its Application in the Glass Identification*.

Next Generation Computing and Information Systems – Gupta. (Ed.)
© 2025 The Author(s), ISBN 978-1-032-73865-9

Deep learning-based cognitive digital twin system for wrist pulse diagnostic and classification

Jitendra Kumar Chaudhary

Associate Professor, School of Computing, Graphic Era Hill University Bhimtal Campus, Uttrakhand, India

T. Parimalam

Associate Professor and Head, Department of Computer Science, Nandha Arts and Science College, Erode, Tamilnadu, India

Faisal Yousef Alghayadh & Ismail Keshta

Computer Science and Information Systems Department, College of Applied Sciences, AlMaarefa University, Riyadh, Saudi Arabia

Mukesh Soni

Dr. D. Y. Patil Vidyapeeth, Pune, Dr. D. Y. Patil School of Science & Technology, Tathawade, Pune, India

Sheshang Degadwala

Department of Computer Engineering, Sigma University, Vadodara, Gujarat, India

ABSTRACT: In analyzing and recognizing wrist pulse signals, it isn't easy to mine the nonlinear information of wrist pulse signals using analysis methods such as time and frequency. Traditional machine learning methods require the manual definition of features and cannot perform self-learning of features. A cognitive digital twin technique for pulse analysis and recognition based on threshold-less recursive graph and CNN is proposed. The wrist pulse signal is converted into a threshold-free recursive graph based on the nonlinear dynamics' theory. The VGG-16 CNN automatically extracts the nonlinear features of the recursive graph, and a pulse condition classification model is established. Experimental finding several that the classification of the proposed method accuracy can reach 98.14%, as compared with the existing pulse classification methods. This study offers a novel concept and strategy for classifying pulse signals, and it has application to the objectification of pulse diagnosis.

1 INTRODUCTION

Pulse diagnosis is one of the components of traditional Chinese medicine and has important clinical value. Additionally, there aren't many reliable objective diagnostic markers (Jiagang *et al.* 2022). Therefore, it is necessary to apply modern computer technology to study the quantitative identification method of pulse signal, to realize the standardization and objectification of pulse diagnosis. Fourier transform, and the frequency domain features are extracted; the time-frequency domain analysis method can simultaneously describe the time and frequency domain characteristics of the signal, and the commonly used methods are wavelet transform, Hilbert-Huang transform, etc. (Yifan and Li-Yun 2022). Research shows that the pulse signal has nonlinear characteristics, and the first three methods will inevitably lose nonlinear information in the process of analyzing the pulse signal. At present, the nonlinear analysis methods of pulse signal mainly include approximate entropy (Viktor *et al.* 2022), Lyapunov index (Gautam *et al.* 2021), Existing approaches such as BP neural network, linear discriminant analysis, Bayesian classifier, Support Vector Machine (SVM), convolutional neural network (CNN) are primarily employed in the investigation of pulse signal pattern identification. Conventional machine learning methods need to define and extract pulse characteristics manually, and it is extremely challenging to identify the ideal feature set to discriminate between various types of pulse circumstances since there exists a complicated nonlinear relationship among the pulse types and pulse features (Kshirsagar and Londhe 2017). The CNN transfers information layer by layer by simulating the human neuron structure to achieve automatic feature extraction. These features have been proven in many studies to be more representative than traditional hand-extracted features, resulting in better classification and recognition results. This paper proposes a pulse signal analysis and recognition method based on threshold-less recursive graph and convolutional neural network (Yutao *et al.* 2022). The method converts the pulse signal into a corresponding threshold-free recursive graph through nonlinear analysis, so that its nonlinear features are mapped into a two-dimensional plane. Through the multi-layer convolution calculation of the convolutional neural network, the self-learning and classification of the pulse signal characteristics is realized, so that it can distinguish different pulse types (Liu 2022).

2 RESEARCH METHODS

2.1 System overview

The overall process of the proposed pulse signal analysis and identification method is illustrated in Figure 1. The pulse signal is converted into a threshold-free recursive graph, and then the VGG-16 CNN is utilized to extract and classify the pulse signal recursive graph, and establish a pulse signal classification model.

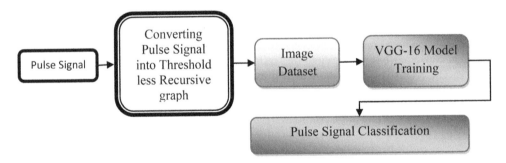

Figure 1.   Pulse signal classification flowchart.

2.2 Threshold-free recursive graph of pulse signal

Recursion graphs are a useful tool for analyzing a system's nonlinear dynamic properties because they can show how the signal is internally organized and intuitively convey the signal's complexity and instability. This method was presented in (Xuebin 2022), and the first step is to reconstruct the signal in phase space. As per the embedding theory, choose the appropriate embedding dimension m and delay time τ to reconstruct the phase space, and then the reconstructed vector is:

$$X_i = \left(x_i, x_{i+\tau}, \cdots x_{i+(m-1)\tau} \right), \quad i = 1, 2, \cdots, N \tag{1}$$

The recursion graph can intuitively map the motion state in the high-dimensional phase space of the signal to a two-dimensional plane, thereby reflecting its nonlinear dynamic characteristics. When the distance $D_{i,j}$ between X_i and X_j is less than ε, it means that the states between X_i and X_j are very similar, that is, the motion state appears recursive, and the value of $R_{i,j}$ is 1 and represented by black dots in the figure; on the contrary, when the distance $D_{i,j}$ between X_i and X_j is greater than ε, it means that the motion state between them is very different. Currently, the value of $R_{i,j}$ is 0 and it is represented by a white dot in the figure (Herzberg *et al.* 2021). The recursive graph and the threshold-free recursive graph of the pulse signal. Although the recursive graph can express the pulse's recursive character-istics in an intuitive way, it lacks detailed features; while the threshold-free recursive graph can effectively retain detailed information and has richer nonlinear dynamic features (Ambaw *et al.* 2020). As a result, there are more recursive points and a bigger circular region formed. There is a curve extending from the upper left vertex to the lower right vertex in the figure, which reflects the overall waveform characteristics of the pulse signal from rising to falling.

2.3 *CNN and VGG-16 network*

CNN is a deep neural network with convolutional structure, which has a wide range of applications in speech recognition and image segmentation. Convolutional neural networks usually consist of convolutional layers, down sampling layers, and fully connected layers.

The convolution layer contains multiple convolution kernels. Through local connection & weight sharing, the convolution kernel convolves the input data to obtain the features w.r.t. the data. The first convolutional layer extracts some preliminary data features to form multiple feature maps, which are then used as input to the next convolutional layer to obtain deep abstract features. The mathematical expression of the convolutional layer is:

$$x_j^{(l)} = f\left(x_i^{(l-1)} * W_{ij}^{(l)} + b_j^{(l)}\right) \tag{2}$$

In the formula: "$*$" represents the convolution operation; $x_j^{(l)}$ represents the output of the lth layer after the jth convolution kernel operation, that is, the input of the $l + 1$ layer; $W_{ij}^{(l)}$ represents the convolution kernel; $b_j^{(l)}$ denotes the bias value; $f(\cdot)$ denotes the activation function.

Down sampling layer is also called pooling layer. The dimension of the feature vector obtained after convolution increases, and if it is directly used for training, it will increase the computational load and complexity of the network. Since the 3 fully connected layers in VGG-16 have many parameters, they were originally designed for 1000 classification cate-gories, and this paper is only for two classifications. Therefore, this paper replaces the ori-ginal 3 fully connected layers with 2 fully connected layers to reduce the number of channels in the fully connected layer (Wen and Mario 2019). In the improved VGG-16, 15 weight layers are included, which are 13 convolutional layers and 2 fully connected layers. Dropout is added after the two fully connected layers to suppress overfitting caused by too large convolution parameters, making the network more generalizable (Pengju *et al.* 2021).

3 EXPERIMENTS AND RESULT ANALYSIS

3.1 *Experimental data*

The pulse diagnosis data of patients with coronary heart disease mainly come from Shanghai Zhongshan Hospital Renji Hospital, Municipal Traditional Chinese Medicine Hospital, Shu Guang Hospital, a total of 1612 cases; the pulse diagnosis data of healthy people are researched by Chinese Medicine Information, Shanghai University (Chang *et al.* 2022).

Comprehensive research experiment of four diagnoses of Chinese medicine personnel and students were collected and obtained, a total of 268 cases. The pulse signal was collected by the ZBOX I pulse signal digital acquisition analyzer created in collaboration by Traditional Chinese Medicine, Shanghai University and Asia-Pacific Computer Information Ltd, Shanghai. The optimal pulse pressure was selected and the pulse signal was collected on this basis. The collection time was 60 s.

3.2 Data processing

3.2.1 Pulse signal preprocessing

Pulse signal is a weak physiological signal, which is easily interfered by other signals, including patient's limb shaking, breathing, mechanical vibration of equipment and power frequency interference. The sampling frequency of the sample is 720 Hz (Mahajan *et al.* 2021). According to the sampling theorem, the frequency range of the collected pulse is 0~360 Hz. Single-cycle division is performed on the filtered pulse signal to obtain a single-cycle waveform of the pulse (Wang *et al.* 2021).

3.2.2 Balanced dataset

There is a clear imbalance in the dataset, where the healthy pulse is far less than the coronary heart disease pulse. If the model is trained with unbalanced samples, the generalization ability of the classification model will be poor and overfitting will easily occur. Therefore, this paper adopts the Synthetic Minority Oversampling Technique (SMOTE) (Deshmukh *et al.* 2021) to balance the dataset. Assume $k = 5$ and the sample multiple $N = 6$, 1608 healthy samples can be generated, combined with 1612 coronary heart disease patients, the balanced data set contains a total of 3220 samples.

3.2.3 Threshold less recursive graph conversion of pulse signal

When constructing the threshold less recursive graph of pulse signal, two key parameters need to be determined: the delay time τ and the embedding dimension m. Based on statistical methods, the C-C algorithm can simultaneously determine the parameter sizes of m and τ through correlation integration (Khan *et al.* 2022). Through calculation and analysis, this paper chooses $m = 3$ and $\tau = 4$ to expand the phase space of the pulse signal. Due to the different lengths of a single cycle of different pulse signal, zero-fill is performed at the end of the pulse signal with a shorter cycle to generate a recursive graph of the same size, as shown in Figure 2(a) and (b).

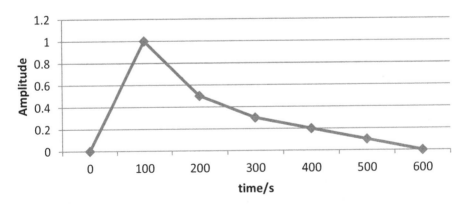

Figure 2　(a). Healthy pulse signal.

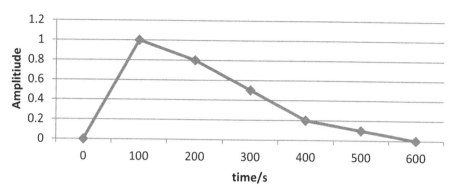

Figure 2(b).   Coronary heart disease pulse signal.

3.3 *Experimental setup and analysis*

During experimentations, participants with healthy pulses were regarded asthe "positive class" whereas those with coronary heart disease were regarded asthe "negative class". Three indicators are utilized to evaluate the classification model: Sensitivity (Sensitivity, SE), Specificity (SP), and Accuracy (Accuracy, ACC) (Khan *et al.* 2022). In the study, the improved variation of VGG-16 CNN is implemented through the Keras framework, and the SGD optimization algorithm is utilized to set the parameters as follows: the number of iterations is 2000 times, the initialization learning rate is 0.001, the momentum of network training is 0.9, and the weight decay is 0.0001, Dropout parameter is 0.5, training set: validation set: test set = 6:2:2. The GPU graphics card used in the experiment is NVIDIA Tesla V100 (Khan *et al.* 2022). Taking the two pulse recursion maps in the input of VGG-16, some feature maps automatically extracted by the CNN are obtained (as shown in Figure 3) (Asif *et al.* 2021). After the recursive graph is processed by Conv1-1, the features extracted by different convolution kernels are different, and the primary texture features are mainly extracted. The more the convolution kernel is processed, the more comprehensive the texture information of the recursive graph is obtained, and the nonlinear dynamic characteristics in the pulse signal can be more fully expressed.

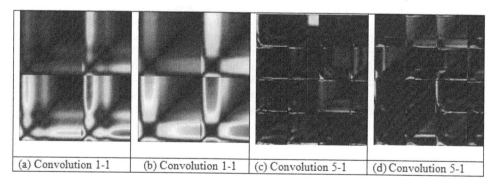

| (a) Convolution 1-1 | (b) Convolution 1-1 | (c) Convolution 5-1 | (d) Convolution 5-1 |

Figure 3.   Feature extraction results of different convolutions.

To verify the effectiveness of the threshold-free recursive graph, it is compared with the recursive graph. According to the "rule of thumb" proposed in (Soni *et al.* 2021), the threshold ε is chosen to be 10% of the phase space's maximum diameter. The two types of recurrent graphs are used as the input of VGG-16 to train the model, and the results are mentioned in Table 1. It can be observed that using the threshold less recursive graph as input can achieve higher classification accuracy, up to 97.94%, and the sensitivity and specificity are improved compared to recursive graphs Results of Different Inputs has been illustrated in Figure 4. Compared with the recursive graph, the threshold-free recursive graph retains more detailed features, and combined with VGG-16, better classification accuracy can be obtained.

Table 1.   Classification results of different inputs

Classification	Sensitivity	Specificity	Accuracy
Recursive graph + VGG-16	93.45	97.75	96.90
Threshold-free recursive graph + VGG-16	95.64	98.42	97.94

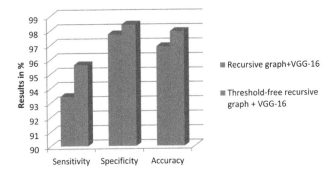

Figure 4.   Classification results of different inputs.

To further verify the proposed method effectiveness, the feature extraction and classification of threshold-free recursive graphs are explored, and compared it with the existing pulse signal classification research (Yao *et al.* 2021). In terms of accuracy, LBP and HOG are 1.71% and 1.24% higher than time-domain features respectively; for sensitivity, LBP is 3.71% higher than time-domain features; in specificity, HOG is better than time-domain features and wavelet the features are increased by 2.20% and 1.96% respectively. This shows that the threshold less recursive map of pulse signal contains rich physiological and pathological information, and the classification effect can be comparable to or even better than traditional features through feature extraction (Liu *et al.* 2022).

It can also be observed from Table 2 that the threshold less recursive graph combined with VGG-16 has the best classification effect, which is 4.04%, 2.95%, 3.01%, 2.33% higher than the time domain feature, wavelet feature, HOG, LBP, and DCNN respectively, 2.02% in terms of sensitivity, the proposed method is 5.43%, 2.57%, 4.01%, 0.61%, and 2.16% higher than the other five methods, respectively; for specificity, the proposed method is 98.18, which

Table 2.   Pulse classification results.

Classification	Sensitivity	Specificity	Accuracy
Time Domain Features + SVM	91.87	95.93	93.83
Wavelet Feature + BP Neural Network	93.42	95.87	94.68
HOG + SVM	91.44	96.93	94.26
LBP + SVM	96.19	94.32	95.78
DCNN	95.52	96.54	95.91
Threshold-Free Recursive Graph + VGG-16	97.09	98.18	97.94

is higher than that of the other five methods. LBP was 4.06% higher. Therefore, compared with other methods, the pulse classification and recognition method proposed in this paper has better classification performance.

To sum up, through the conversion of pulse signal to threshold less recursive graph, the nonlinear characteristics of pulse signal can be mapped to a two-dimensional image without setting a threshold, which can retain more nonlinear information in pulse signal. A VGG-16 CNN was constructed to train the threshold-free recursive graph data, and a pulse classification model was established. The model forms a deep network structure through the combination of 3×3 convolution kernels, which can perform self-learning of more detailed features on the threshold less recursive graph, and obtains a better classification effect.

4 CONCLUSION

In the current pulse diagnosis objectification research, most of the analysis methods such as time domain and frequency domain are used, ignoring the nonlinear characteristics of pulse signal. Considering the nonlinearity of pulse signal, using nonlinear dynamics method to study pulse signal, compared with traditional methods, it can better tap its inherent non-linear information. First, the pulse signal is converted into a threshold-free recursion map to retain more detailed features in the pulse signal. Then a CNN is set up to perform feature self-learning & training on the threshold less recursive graph to obtain a pulse signal classification model. At last, the classification and recognition of the pulse signal is carried out. The experimental results reveal that the pulse signal recognition the proposed method is effective, and can obtain higher classification accuracy than the current pulse signal classification methods. In the future work, we will further study the signal analysis method that can better represent the characteristics of pulse signal, and combine deep learning to improve the recognition rate of pulse signal.

REFERENCES

Ambaw, A.B., Bari, M. and Doroslovački, M., 2020, November. Optimizing Convolutional Neural Networks to Identify Distorted M-ary CPFSK Signals with RRC Pulse Shaped Instantaneous Frequency. In *2020 54th Asilomar Conference on Signals, Systems, and Computers* (pp. 153–156). IEEE.

Asif, M., Khan, W.U., Afzal, H.R., Nebhen, J., Ullah, I., Rehman, A.U. and Kaabar, M.K., 2021. Reduced-complexity LDPC decoding for next-generation IoT networks. *Wireless Communications and Mobile Computing, 2021*: 1–10.

Chang, S., Zhang, R., Ji, K., Huang, S. and Feng, Z., 2022. A hierarchical classification head based convolutional gated deep neural network for automatic modulation classification. *IEEE Transactions on Wireless Communications, 21*(10): 8713–8728.

Cheng, Y. and Fu, L.Y., 2022. Nonlinear seismic inversion by physics-informed Caianiello convolutional neural networks for overpressure prediction of source rocks in the offshore Xihu depression, East China. *Journal of Petroleum Science and Engineering, 215*: 110654.

Deshmukh, S., Thirupathi Rao, K. and Shabaz, M., 2021. Collaborative learning based straggler prevention in large-scale distributed computing framework. *Security and communication networks, 2021*: 1–9.

Gautam, N., Choudhary, A. and Lall, B., 2021, July. Neural networks for predicting optical pulse propagation through highly nonlinear fibers. In *2021 National Conference on Communications (NCC)* (pp. 1–6). IEEE.

Herzberg, W., Rowe, D.B., Hauptmann, A. and Hamilton, S.J., 2021. Graph convolutional networks for model-based learning in nonlinear inverse problems. *IEEE transactions on computational imaging, 7*: 1341–1353.

Khan, F., Tarimer, I., Alwageed, H.S., Karadağ, B.C., Fayaz, M., Abdusalomov, A.B. and Cho, Y.I., 2022. Effect of feature selection on the accuracy of music popularity classification using machine learning algorithms. *Electronics, 11*(21): 3518.

Khan, I., Wu, Q., Ullah, I., Rahman, S.U., Ullah, H. and Zhang, K., 2022. Designed circularly polarized two-port microstrip MIMO antenna for WLAN applications. *Applied Sciences*, *12*(3): 1068.

Khan, M.S., Khan, A.W., Khan, F., Khan, M.A. and Whangbo, T.K., 2022. Critical challenges to adopt DevOps culture in software organizations: A systematic review. *IEEE Access*, *10*: 14339–14349.

Kshirsagar, G.B. and Londhe, N.D., 2017, December. Deep convolutional neural network based character detection in devanagari script input based P300 speller. In *2017 International Conference on Electrical, Electronics, Communication, Computer, and Optimization Techniques (ICEECCOT)* (pp. 507–511). IEEE.

Leonhardt, V., Claus, F. and Garth, C., 2022. PEN: Process estimator neural Network for root cause analysis using graph convolution. *Journal of Manufacturing Systems*, *62*: 886–902.

Liu, Q., Zhang, W., Bhatt, M.W. and Kumar, A., 2022. Seismic nonlinear vibration control algorithm for high-rise buildings. *Nonlinear Engineering*, *10*(1): 574–582.

Liu, Z., 2022, April. Chlorophyll-a analysis based on hilbert huang transformation and convolutional neural network. In *2022 7th International Conference on Cloud Computing and Big Data Analytics (ICCCBDA)* (pp. 1–7). IEEE.

Mahajan, K., Garg, U. and Shabaz, M., 2021. CPIDM: a clustering-based profound iterating deep learning model for HSI segmentation. *Wireless communications and mobile computing*, *2021*: 1–12.

Qu, J., Cai, W. and Zhao, Y., 2022. Learning time-dependent PDEs with a linear and nonlinear separate convolutional neural network. *Journal of Computational Physics*, *453*: 110928.

Shou, Y., Meng, T., Ai, W., Yang, S. and Li, K., 2022. Conversational emotion recognition studies based on graph convolutional neural networks and a dependent syntactic analysis. *Neurocomputing*, *501*: 629–639.

Soni, M., Dhiman, G., Rajput, B.S., Patel, R. and Tejra, N.K., 2021. Energy-effective and secure data transfer scheme for mobile nodes in smart city applications. *Wireless Personal Communications*: 1–21.

Wang, B., Yao, X., Jiang, Y., Sun, C. and Shabaz, M., 2021. Design of a real-time monitoring system for smoke and dust in thermal power plants based on improved genetic algorithm. *Journal of Healthcare Engineering*, *2021*.

Xing, P., Dong, J., Yu, P., Zheng, H., Liu, X., Hu, S. and Zhu, Z., 2021. Quantitative analysis of lithium in brine by laser-induced breakdown spectroscopy based on convolutional neural network. *Analytica Chimica Acta*, *1178*: 338799.

Xu, X., Ma, F., Zhou, J. and Du, C., 2022. Applying convolutional neural networks (CNN) for end-to-end soil analysis based on laser-induced breakdown spectroscopy (LIBS) with less spectral preprocessing. *Computers and Electronics in Agriculture*, *199*: 107171.

Yao, Q., Shabaz, M., Lohani, T.K., Wasim Bhatt, M., Panesar, G.S. and Singh, R.K., 2021. 3D modelling and visualization for vision-based vibration signal processing and measurement. *Journal of Intelligent Systems*, *30*(1): 541–553.

Yu, W. and Pacheco, M., 2019. Impact of random weights on nonlinear system identification using convolutional neural networks. *Information Sciences*, *477*: 1–14.

Multi-class instance segmentation for the detection of cervical cancer cells using modified mask RCNN

Diksha Sambyal
Department of CS&IT, University of Jammu, India

Abid Sarwar
Department of CS&IT, Bhaderwah Campus, University of Jammu, India

ABSTRACT: Cervical cancer remains a leading cause of female mortality, underscoring the vital role of timely detection in mortality reduction. Accurate cell segmentation is pivotal for automating cervical cancer screening and enabling early diagnosis. This study offers a comparative analysis employing Mask RCNN with ResNeXt101-FPN and ResNet101-FPN for advanced multi-class instance segmentation in cervical cell analysis. Uniquely, it is the only research endeavor implementing Mask R-CNN, segmenting the whole cervical cell images into seven distinct classes aligned with the TBS reporting system. Leveraging the SIPaKMeD dataset, a stratified data partitioning scheme of 60% for training, 20% for 5-fold cross-validation, and 20% for testing was employed. Both configurations achieve similar mean Average Precision (mAP) values of 89% and 88%, alongside consistent accuracy (87%), sensitivity (89%), and specificity (88%). Results affirm ResNeXt101-FPN and ResNet101-FPN efficacy, potentially automating cervical cell analysis, reducing medical workload, and enhancing diagnostic accuracy. Future endeavors will center on innovating backbone structures. Moreover, equal emphasis will be placed on augmenting the expansive and diverse datasets to encompass a wider spectrum of cervical abnormalities.

1 INTRODUCTION

Cervical cancer is the third most common malignancy among women globally. The majority of cervical cancer fatalities, accounting for more than 85% of cases, occur in countries with low or middle incomes, where it stands as the leading cause of cancer-related deaths in women(Cohen *et al.* 2019). This glaring inequality in disease burden is primarily attributed to limited access to screening, early diagnosis, and effective treatment for pre-cancers and cancer. Over 604,000 cases of cervical cancer were diagnosed in 2020; of the 342,000 fatalities from cervical cancer, almost 90% occurred in nations with middle or low incomes (Sarwar *et al.* 2020). These statistics underscore the urgent need for enhanced screening methodologies and vaccination initiatives in resource-constrained regions to address this critical global healthcare disparity. Theoretically, by ensuring high-quality and widespread screening, the occurrence of cervical cancer could potentially be decreased by up to 90%. Routine gynecological screenings, often employing Pap tests, have proven effective in detecting the majority of cervical cancer cases. The Bethesda System (TBS), also known as The Bethesda System for Reporting Cervical Cytology, serves as a widely adopted framework for reporting Pap smear findings in cervical or vaginal cytologic diagnoses. TBS categorizes cells into seven distinct classes, including Superficial Squamous Cells, Intermediate Squamous Cells, Parabasal Cells, Basal Cells, LSIL (Low Grade Intraepithelial Lesion), HSIL (High-Grade Intraepithelial Lesion), and CIS (Carcinoma in situ) cells as per established criteria(Nayar and Wilbur 2017).

DOI: 10.1201/9781003466383-19

With the advent of deep learning and Convolutional Neural Networks (CNN)(Dongyao Jia *et al.* 2020), a new wave of segmentation models based on deep learning has surfaced, exhibiting notable performance improvements and highlighting their promise in medical image segmentation. Contemporary detection algorithms, including two-stage methods like Faster R-CNN (Ren *et al.* 2015), R-CNN, and Fast R-CNN (Girshick 2015), as well as single-stage methods such as YOLO and SSD, are frequently employed in current applications. Furthermore, models like SegNet, Mask R-CNN (He *et al.* 2017a), DeepLab, and various improved techniques have significantly elevated image segmentation performance. Specifically in cervical cancer screening, these models have revolutionized the interpretation of Pap smear images and cytologic data by enabling precise cell segmentation and classification. Their capabilities span improved localization, real-time detection, pixel-level classification, and instance segmentation—crucial for accurately identifying and characterizing various cell types according to the TBS reporting system. Their adaptability and precision in handling diverse cell structures underscore their pivotal role in augmenting the accuracy and efficiency of cervical cancer diagnostics.

2 RELATED WORK

Instance segmentation, a pivotal task in computer vision, amalgamates object detection and semantic segmentation, endowing machines with the capacity to not only recognize and categorize objects in images but also to demarcate and individuate each object instance. This capability finds a noteworthy application within the domain of medical image analysis, where the quest for precision is paramount. Over the years, a plethora of instance segmentation techniques have evolved, each contributing to the progress of this discipline.

One particularly distinguished approach is Mask R-CNN (He *et al.* 2017b), an enhanced version of the renowned Faster R-CNN model, which enhances detection of objects with a mask prediction branch, facilitating the precise demarcation of object boundaries. In the realm of medical image analysis, Mask R-CNN is harnessed for the meticulous segmentation of anatomical structures and pathological irregularities across various medical imaging modalities, thereby bolstering the accuracy of diagnostic processes and improving healthcare outcomes.For instance (Hsieh *et al.* 2020) combined Mask R-CNN, Inception V3 , and VGG16 to identify the malignant and benign of breast microcalcification clusters. The method achieved precision,specificity and sensitivity 87%, 89% and 90%, respectively.

Several researchers have ventured into the application of the MASK-RCNN method within the realm of cervical cancer cell detection. For instance, (Sagar *et al.* 2023) employed a smaller Visual Geometry Group-like Network combined with a MaskR-CNN utilizing ResNet10 as a backbone for the effective segmentation of cervical cells. Their methodology was tested using the Herlev Pap Smear dataset, yielding impressive single-cell segmentation results: an accuracy of 92.0%, recall of 91%, and ZSI of 81%. (Agustiansyah *et al.* 2022) employes a mask-RCNN architecture was for concurrent segmentation, classification, and detection of columnar areas and acetowhite lesions. Utilizing 262 positive and 222 negative cervicogram images, the model delivered notable results: 63.60% IoU for columnar area segmentation, 73.98% for acetowhite lesions, and a mAP of approximately 86.90% for columnar areas and perfect 100% for acetowhite lesions in detecting cervical cancer precursor lesions. Achieving 92% specificity and 100% sensitivity highlighted the model's accuracy in identifying cervical abnormalities.

Similarly, (Chen and Zhang 2021) presented a Mask RCNN-based approach designed to tackle the challenging task of cell segmentation, distinguishing between abnormal and normal cells within cytology images. Their work leveraged high-resolution cytology of the cervical region images and accompanying ground truth cell segmentation data from the ISBI 2014 and ISBI 2015 datasets. Notably, the Mask RCNN achieved outstanding performance, with a Dice Similarity Coefficient (DSC) of 0.92 and a False Positive Rate per pixel (FPRp) of 0.0008 at a DSC threshold of 0.8. These results indicate the efficacy of the segmentation utilizing Mask RCNN technique in the domain of cytological analysis.

None of the previously mentioned studies have leveraged the Mask R-CNN technique for multi-class instance segmentation beyond two classes. Furthermore, these studies predominantly concentrated on single cell images within the context of medical analysis. To the best of our knowledge, our research marks a pioneering effort in implementing the segmentation framework utilizing Mask R-CNN of entire images encompassing cervical cells, categorizing them into seven distinct classes as per the TBS reporting system. This novel approach, utilizing a dataset comprising whole images from pap smear slides, represents a significant advancement within the realm of medical image analysis and offers a hopeful avenue for the more comprehensive characterization and diagnosis of cervical cell abnormalities.

3 METHOD AND METHODOLOGY

3.1 *Dataset*

In this study, we utilized the publicly accessible SIPaKMeD(Plissiti *et al.* 2018) dataset, which comprises of 966 images of clustered cells obtained from Pap smear slides. Skilled cytopathologists have categorized these cells into five distinct groups based on their cellular appearance and morphology: superficial-intermediate, parabasal, koilocytes and dyskeratotic, and metaplastic cells (Plissiti *et al.* 2018). To ensure conformity with the TBS classification system, we further categorized the SIPaKMeD dataset into seven specific cell types as illustrated in Figure 1: superficial squamous cell, intermediate squamous cell, parabasal squamous cell, basal squamous cell, low-grade squamous intraepithelial lesions (LSIL), high-grade squamous intraepithelial lesions (HSIL), and CIS cell. This classification allows for a comprehensive and standardized analysis of the cellular compositions within the dataset, facilitating research and data interpretation within the field of cytopathology.

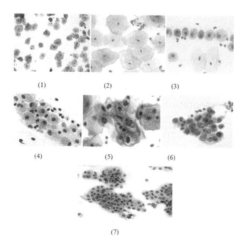

Figure 1. Seven classification categories: 1) Superficial squamous cells, 2) Intermediate squamous cells, 3) Parabasal cells, 4) Basal cells, 5) LSIL, 6) HSIL.

This systematic segmentation not only underscores the dataset's suitability for research and clinical applications but also ensures its compliance with the standards prescribed by TBS. The compatibility of the SIPaKMeD dataset with TBS highlights its significance as a valuable resource for cervical cancer diagnosis and furthers the progression of diagnostic advancements and classification models within this established framework.

For the purpose of our experiments, 60% of the dataset per class has been allocated for training, 20% is set aside for cross validation, and the remaining 20% for testing, adhering to a standard data partitioning scheme commonly employed in scientific investigations.

3.2 *Data preprocessing*

For the data preprocessing stage, we employed an AI tool, specifically the 'MakeSense.AI' annotation tool, which was tailored to label cervical cancer cells according to the TBS system. This tool efficiently streamlined activities like object detection, image classification, and segmentation, offering users the ability to delineate bounding boxes, assign labels, or trace objects. The resulting annotation information was stored in JSON files. Prior to integration into our model, we executed several image preprocessing steps to prepare the images obtained from Pap smear slides for annotation. These preprocessing steps included resizing, normalization, data augmentation, and background removal.

Our dataset encompasses a total of 966 images, each containing multiple cells. These images have been meticulously annotated, resulting in a total of 4,136 annotated images. Figure 2 provides a visual representation of the annotation process, depicting sample image with annotation labels and highlighting Squamocolumnar Junction (SCJ) Structures. These annotations cover seven distinct classes, which include Superficial Squamous Cells, Intermediate Squamous Cells, Parabasal Cells, Basal Cells, LSIL, HSIL, and CIS.

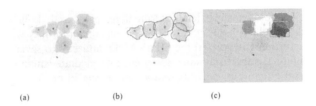

(a)   (b)   (c)

Figure 2.   Sample of Annotated Images: (a) Original Data; (b) Annotation Labels; and (c) SCJ Structures.

4   THE MODIFIED MASK-RCNN MODEL

The instance segmentation of cervical cells, following the TBS classification system, is conducted utilizing employs the well-established developed by Facebook's Artificial Intelligence Research (FAIR) Mask R-CNN (He *et al.* 2017b). This methodology represents an amalgamation of semantic segmentation and object detection techniques, signifying an advancement from the preceding Faster RCNN, Fast RCNN, and RCNN approaches. The workflow of this framework comprises two fundamental phases: (a) Generating region proposals, and (b) subsequently classifying each of these generated proposals. Figure 3 provides a visual representation of the Modified Mask R-CNN Architecture.

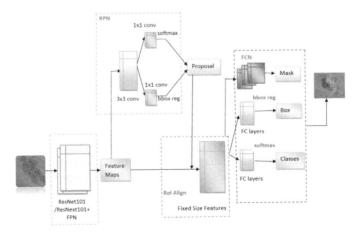

Figure 3.   Modified mask R-CNN architecture.

4.1 *Model implementation and hyperparameter configuration*

In our implementation of Mask R-CNN, we utilized a FPN as the backbone. For the bottom-up portion of FPN, we employed pre-trained ResNet-101 and ResNeXt101 models, which were initially trained on the ImageNet dataset. In the RPN, we incorporated anchors of varying sizes. Specifically, we specified the dimensions of the anchors as 128, 64, 32, 16, and 8 pixels, whereas the height of each anchor is set as a proportion relative to its width, ranging from 0.33 to 3. Consequently, the mean anchor encompasses 25 distinct sizes, and we employed 512 anchors per image for RPN training.

The network head comprises three key branches: a classification branch, a mask branch, and a detection branch. To train the network head, we used 512 RoIs per image. The classification branch produced 7 outputs, each corresponding to different cell types or abnormalities defined by the TBS classification system. The mask branch, on the other hand, typically generates binary masks. For multi-class instance segmentation, the mask branch is extended to include seven parallel sub-branches, with each sub-branch dedicated to predicting masks for one of the seven classes. For the detection branch, we adopted the default Matterport implementation to support our Mask R-CNN framework.

4.2 *Loss function*

In our methodology, the mask branch's loss function has been tailored to address the requirements of multi-class instance segmentation. With the objective of predicting masks for each of the seven distinct classes in our task, we appoint a loss function utilizing binary cross-entropy, computed individually for each class. For a given class, this loss function measures the dissimilarity between the predicted mask and the corresponding ground-truth mask. It quantifies the agreement between the predicted probability values and the true class presence, thus facilitating the accurate segmentation of objects across different categories.

5 EVALUATION METHODS

The optimal model, chosen through a rigorous 5-fold cross-validation process based on the highest mean Average Precision (mAP) achieved by the Mask R-CNN, was subsequently used for evaluation on a test dataset containing 190 pap smear slide images. Given the increased complexity inherent in our multi-class instance segmentation task for cervical cell analysis, we meticulously tailored our evaluation metrics to ensure a comprehensive assessment. The evaluation included:

(1) Class-specific mean Average Precision (mAP): Calculated independently for each of the seven classes as prescribed by the TBS system. This fine-grained analysis enables the assessment of the model's efficacy in segmenting and classifying each class individually. The overall mAP score is determined by averaging the mAP values across all seven classes.

(2) Class-specific Accuracy, Sensitivity, and Specificity per Image: For a detailed evaluation of the model's performance, metrics such as accuracy, sensitivity, and specificity were computed separately for each class. This approach provides insights into the model's performance for individual classes in terms of their accuracy, sensitivity, and specificity.

6 EXPERIMENTAL RESULTS

Figure 4 visually represents outcomes from our adapted Mask R-CNN model and offer insights into the model's performance, demonstrating its ability to identify and categorize cell types or abnormalities effectively.

The evaluation depicted in Table 1 showcased the assessment of Mask RCNN utilizing ResNeXt101-FPN and ResNet101-FPN as backbone models, resulting in promising outcomes. Both configurations exhibited robust performance, yielding similar overall mean Average Precision (mAP) values of 89% and 88%, respectively. Notably, accuracy and sensitivity attained commendable levels at 87% and 89%, while specificity stood consistently at 88% for both configurations. These findings underscore the efficacy of ResNeXt101-FPN and ResNet101-FPN backbones in multi-class instance segmentation for analyzing cervical cell images.

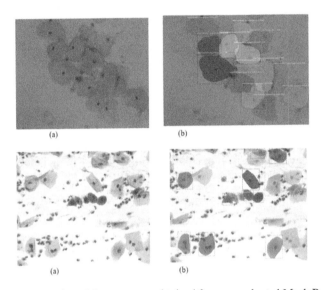

(a)   (b)

(a)   (b)

Figure 4. Visual representation of the outcomes obtained from our adapted Mask R-CNN model. (a) Original input image processed by the Mask R-CNN. (b) Color-coded depiction of the model's results showcasing bounding boxes and categorized instances.

Table 1. Assessment of Mask R-CNN using ResNeXt101-FPN and ResNet101-FPN as backbones for cervical cancer cell detection.

Model Used	Evaluation Metrics	mAP	Accuracy	Sensitivity	Specificity
ResNeXt101-FPN	Superficial	0.87	0.85	0.88	0.86
	Intermediate	0.89	0.86	0.87	0.88
	Parabasal	0.86	0.84	0.87	0.85
	Basal	0.88	0.87	0.88	0.87
	LSIL	0.90	0.88	0.89	0.90
	HSIL	0.91	0.90	0.92	0.89
	CIS	0.92	0.91	0.93	0.90
	Overall (Mean)	0.89	0.87	0.89	0.88
ResNet101-FPN	Superficial	0.86	0.84	0.87	0.85
	Intermediate	0.87	0.85	0.86	0.88
	Parabasal	0.88	0.87	0.89	0.87
	Basal	0.89	0.88	0.90	0.88
	LSIL	0.90	0.89	0.91	0.89
	HSIL	0.91	0.90	0.92	0.89
	CIS	0.92	0.91	0.93	0.90
	Overall (Mean)	0.88	0.87	0.89	0.88

Table 2. Comparative analysis of segmentation models for cervical cancer screening.

Author	Methodology	Classes Segmented	Dataset Used	Performance Achieved
(Sagar et al. 2023)	Mask RCNN with backbone ResNet10 and VGG-likeNet	Two(Abnormal, Normal)	Herlev Dataset	Acc = 92.0%,Rec = 91%, and ZSI = 81%
(Agustiansyah et al. 2022)	Mask-RCNN with ResNet 50 as a backbone.	Two (Abnormal, Normal)	Private dataset	Acc = 96.29%,F1Score = 96.67% ,Sens = 100%,Spec = 92.0%,Prec = 93.54%, mAP = 86.90%
(Hsieh et al. 2020)	Mask R-CNN with Inception V3 and VGG16	Two (benign and malignant)	Private dataset	Sens = 90%,Spec = 89%,Prec = 87%
(Chen & Zhang, 2021)	Mask-RCNN with a ResNet 50 as a backbone	Two (Abnormal, Normal)	ISBI 2014 and ISBI 2015 datasets	DSC = 92% and a FPRp = 0.08% at a DSC threshold of 0.8.
Proposed work	Mask RCNN with ResNeXt101-FPN or ResNet101-FPN as the backbones.	Seven classes as per TBS	SIPaKMeD dataset	Sens = 89%,Spec = 88%,Acc = 87%,mAP = 89%,

*Acc = Accuracy, Prec = Precision, Rec = Recall, Sens = Sensitivity, Spec = Specificity, mAP = mean Average Precision

Furthermore, as observed in Table 2, our proposed methodology, employing Mask RCNN with ResNeXt101-FPN or ResNet101-FPN as backbone models for cervical cancer cell detection, stands out amidst the methodologies outlined in the comparative analysis. While other studies excel in binary classifications (Abnormal/Normal) with accuracy ranging from 92% to 96.29%, our emphasis on delineating seven distinct classes aligned with the TBS reporting system yields a competitive accuracy of 87%. This trade-off in accuracy is offset by a higher specificity of 88% and a notable sensitivity of 89%, underscoring our models' robustness in discerning diverse cervical cell types. Noteworthy is the mean Average Precision (mAP) of 89%, signifying the precision and reliability of our approach compared to studies reporting lower mAP values. This equilibrium between sensitivity and specificity, coupled with a competitive mAP, emphasizes the clinical viability of our proposed models for precise and comprehensive cervical cancer cell detection, reflecting our commitment to a holistic and nuanced multi-class segmentation approach vital for accurate diagnosis and management in cervical cancer screening.

While our proposed methodology showcases competitive performance, it encounters limitations akin to other methodologies. The need for more expansive and diverse datasets representing a wider spectrum of cervical abnormalities remains a primary constraint. Additionally, achieving higher accuracy without compromising interpretability and resource efficiency poses challenges. Overcoming these limitations through collaborative efforts, enhanced dataset curation, and optimizing model interpretability is pivotal to augment the practical applicability of these models in clinical settings for precise cervical cancer detection and management.

7 CONCLUSION AND FUTURE WORK

This paper introduces a technique based on modified Mask RCNN tailored for segmenting whole images containing cervical cells and classifying them into seven distinct categories following the TBS reporting system. The two customized variants, employing Mask RCNN with ResNeXt101-FPN and ResNet101-FPN as the backbone architectures, were implemented, trained, and evaluated using the SIPaKMeD dataset. The findings reveal that both config-urations exhibited robust and comparable performance, achieving overall mean Average Precision (mAP) values of 89% and 88%, respectively. These consistent results underscore the reliability of both models in providing efficient segmentation and classification capabilities. Consequently, these models have the potential to significantly alleviate the workload of medical professionals and enhance diagnostic accuracy in the analysis of cervical cell images.

Future work encompasses multiple trajectories to enhance cervical cell segmentation and broaden its applications. Innovations in backbone structures and adjustments for multiclass instance segmentation stand as primary objectives to augment model adaptability and accuracy in delineating diverse cervical cell types. The method's versatility extends beyond cervical cell segmentation, holding potential for application across various biomedical imaging disciplines. Efforts to amass an extensive cervical cancer cell database will refine model training, improving overall performance. Integrating the algorithm into healthcare devices for mass cervical screening and developing AI algorithms for colposcopy could significantly enhance diagnostic capabilities and streamline the cervical cancer screening pathway. These concerted efforts aim to advance precision, accessibility, and the utility of cervical cancer screening methodologies, fostering improved healthcare outcomes.

REFERENCES

Agustiansyah, P., Nurmaini, S., Nuranna, L., Irfannuddin, I., Sanif, R., Legiran, L., Rachmatullah, M. N., Florina, G. O., Sapitri, A. I., & Darmawahyuni, A. (2022). Automated precancerous lesion screening using an instance segmentation technique for improving accuracy. *Sensors*, 22(15).

Chen, J., & Zhang, B. (2021). Segmentation of overlapping cervical cells with mask region convolutional neural network. *Computational and Mathematical Methods in Medicine*, 2021.

Cohen, P. A., Jhingran, A., Oaknin, A., & Denny, L. (2019). Cervical cancer. *Lancet (London, England)*, 393 (10167), 169–182.

Dongyao Jia, A., Zhengyi Li, B., & Chuanwang Zhang, C. (2020). Detection of cervical cancer cells based on strong feature CNN-SVM network. *Neurocomputing*, 411, 112–127.

Girshick, R. (2015). Fast r-cnn. *Proceedings of the IEEE International Conference on Computer Vision*, 1440–1448.

He, K., Gkioxari, G., Dollár, P., & Girshick, R. (2017a). Mask r-cnn. *Proceedings of the IEEE International Conference on Computer Vision*, 2961–2969.

He, K., Gkioxari, G., Dollár, P., & Girshick, R. (2017b). Mask r-cnn. *Proceedings of the IEEE International Conference on Computer Vision*, 2961–2969.

He, K., Zhang, X., Ren, S., & Sun, J. (2015). *Deep Residual Learning for Image Recognition*.

Hsieh, Y.-C., Chin, C.-L., Wei, C.-S., Chen, I.-M., Yeh, P.-Y., & Tseng, R.-J. (2020). Combining VGG16, Mask R-CNN and Inception V3 to identify the benign and malignant of breast microcalcification clusters. 2020 *International Conference on Fuzzy Theory and Its Applications (IFUZZY)*, 1–4.

Lin, T.-Y., Dollár, P., Girshick, R., He, K., Hariharan, B., & Belongie, S. (2016). Feature Pyramid Networks for Object Detection.

Nayar, R., & Wilbur, D. C. (2017). The Bethesda System for Reporting Cervical Cytology: A Historical Perspective. *Acta Cytologica*, 61(4–5), 359–372.

Plissiti, M. E., Dimitrakopoulos, P., Sfikas, G., Nikou, C., Krikoni, O., & Charchanti, A. (2018). Sipakmed: A New Dataset for Feature and Image Based Classification of Normal and Pathological Cervical Cells in Pap Smear Images. *Proceedings – International Conference on Image Processing*, ICIP, 3144–3148.

Ren, S., He, K., Girshick, R., & Sun, J. (2015). Faster r-cnn: Towards real-time object detection with region proposal networks. *Advances in Neural Information Processing Systems*, 28.

Sagar, C. V, Bhardwaj, H., & Bhan, A. (2023). Deep Learning Based Segmentation Approach for Cervical Cancer Detection Using Pap Smears. In B. K. Murthy, B. V. R. Reddy, N. Hasteer, & J.-P. Van Belle (Eds.), *Decision Intelligence* (pp. 309–317). Springer Nature Singapore.

Sarwar, A., Sheikh, A. A., Manhas, J., & Sharma, V. (2020). Segmentation of cervical cells for automated screening of cervical cancer: a review. *Artificial Intelligence Review*, 53(4), 2341–2379.

Xie, S., Girshick, R., Dollár, P., Tu, Z., & He, K. (2016). Aggregated Residual Transformations for Deep Neural Networks.

Next Generation Computing and Information Systems – Gupta. (Ed.)
© 2025 The Author(s), ISBN 978-1-032-73865-9

Dilated convolution model for lightweight neural network

Sachin Gupta
Department of CSE, Maharaja Agrasen Institute of Technology, India

Priya Goyal
Department Computer Science and Engineering, Manipal University Jaipur, India

Bhuman Vyas
Credit Acceptance, Canton, USA

Mohammad Shabaz
Model Institute of Engineering and Technology, Jammu, J&K, India

Suchitra Bala
ICT & Cognitive Systems, Sri Krishna Arts And Science College, India

Aws Zuhair Sameen
College of Medical Techniques, Al-Farahidi University, Baghdad, Iraq

ABSTRACT: Lightweight convolutional neural networks are being studied further in order to better deploy deep convolutional neural networks to edge devices, minimize the number of model parameters in deep neural networks, and reduce network complexity. On accomplish the lightweight goal, the full convolution is performed to the remaining network structure. To increase the light effect, it is merged with point-by-point convolution to build an improved dilated convolution lightweight approach. The enhanced dilated convolution is coupled with conventional convolution to lessen the attenuation of accuracy, and a lightweight fusion dilated convolution is presented. The improved dilated convolution light approach offers the most substantial lightweight benefit, according to the data. Furthermore, the lightweight fusion dilated convolution approach decreases the amount of model parameters and offers the best speed-accuracy trade-off.

1 INTRODUCTION

There have been two development trends in convolutional neural networks in recent years. On the one hand, it develops in the direction of large-scale and deep-level development; on the other hand, it grows in the order of light and high speed. With the rapid growth of the mobile Internet, people's demand for mobile edge devices is getting higher and higher, and there are more and more demands for lightweight convolutional neural networks. Running deep learning algorithms more smoothly on edge devices has become a challenge. Edge devices have little storage space and lack computing resources, which provide broad research and development platform for the lightweight method of convolutional neural networks. Atreus convolution was first systematically studied by (Zhu *et al.* 2020). The primary Atreus convolution light method applies aurous convolution directly to the network structure design, which has an excellent lightweight effect and slightly higher accuracy loss; the improved Atreus convolution lightweight method combines point-by-point with Atreus convolution.

2 RELATED WORK

Refinement of the model design to achieve a lightweight neural network is often achieved by using small convolution kernels instead of large convolution kernels, decomposing convolutions, and processing convolution channels (Yu *et al.* 2019). Modern deep neural networks usually use small convolution kernels instead of large convolution kernels for network design (Zhang *et al.* 2018). Squeeze Net is the earliest public lightweight structure design network. On the one hand, 1×1 small convolution kernels are used instead of 3×3. As a result, the kernel can reduce eight times the amount of convolution operation parameters and a large part of the convolution operation (Zhang, Zhang and Wang 2019). Shuffle Net uses the idea of grouping convolution and shuffles the feature information from different groups so that the feature information of the convolution after grouping is scrambled and mixed, which can be better circulated (Risso *et al.* 2021). The above model refinement method realizes a new high-efficiency backbone network by redesigning the macro-structure and micro-structure of the model and completes the lightweight design of the neural network from the structural level (Guo *et al.* 2021).

3 LIGHTWEIGHT METHOD OF NEURAL NETWORK BASED ON ATROUS CONVOLUTION

3.1 *Atrous convolution structure*

Whole convolution is a convolution with a unique structure, which introduces the concept of holes. By adding holes to the traditional image convolution, the purpose of expanding the convolution receptive field is achieved. For example, let H_i be the feature map of thei-th layer and G_i be the convolutional layer of the i-th layer. Then, the point in the feature map H_i, b is the point in the convolutional layer G_i.

$$(H_i * G_i)(q) = \sum_{b+c=q} H_i(b)G_i(c) \tag{1}$$

The whole convolution calculation formula is shown in Equation (2), where m is the expansion rate, and one is added to the number of holes added by ordinary convolution. When the dilation rate m is 1, the atrous convolution becomes a regular convolution.

$$(H_i * G_i)(q) = \sum_{b+mc=q} H_i(b)G_i(c) \tag{2}$$

It can be seen from the calculation formula of whole convolution and ordinary convolution that the whole convolution operation has the characteristics of equal interval sampling, and the expansion rate determines its sampling degree. The larger the expansion rate, the more extensive the whole convolution sampling range and the piece within the unit range has fewer points. The calculation formula of the receptive field is formula (3)—the stride of the i-layer convolution.

$$sf_l = sf_{l-1} + (\eta_l - 1) * \prod_{i=1}^{l-1} t_i \tag{3}$$

The whole convolution expands the receptive field without reducing the size of the feature map. Therefore, the calculation formula of the kernel size em_l of the whole convolution of the lth layer is as formula (4), where m_l represents the kernel size of the original ordinary convolution of the lth layer.

$$em_l = m_l + (m_l - 1) * (n - 1) \tag{4}$$

Substituting Equation (4) into Equation (3), the receptive field of the whole convolution kernel with the size of em_l in the lth layer for the expansion of the standard convolution kernel of size m_l is rf_expend$_l$, as shown in Equation (5):

$$rf\_expend_l = (m_l - 1) * (n - 1) * \prod_{i=1}^{l-1} t_i \qquad (5)$$

3.2 Basic atrous convolution lightweight method

Modern convolutional neural networks such as VGG, Reset, DLA, etc., use two convolutional layers in series to form convolutional groups for feature extraction. Then, the size of the convolution kernel of the first layer is $M_1 \times M_1 \times C1\_in \times D_{1\_out}$, M_1 is the size of the convolution kernel, $C1\_in$ is the number of input channels of the convolution kernel and $D_{1\_in} = D_{in}$, D_{1_out} is the number of output channels of the convolution kernel, and the second layer of convolution the kernel size is $M_2 \times M_2 \times D_{2\_in} \times D_{2\_out}$, M_2 is the size of the convolution kernel, D_{2_in} is the number of input channels of the convolution kernel, D_{2_out} is the number of output channels of the convolution kernel, and D_{out} is the number of output feature map channels, then $D_{1\_out} = D_{2\_in} = D_{2\_out} = D_{out}$. Let the step size of the convolution operation by 1, and there is no padding operation. Then, the feature map receptive field R1 obtained by the concatenation of two ordinary convolution layers is:

$$S_1 = sf_{in} + M_1 + M_2 - 2 \qquad (6)$$

Among them, sf_{in} represents the receptive field of the input feature map. Let the size of the atrous convolution kernel be $EL \times EL \times D_{in} \times D_{out}$, then its receptive field S_2 is:

$$S_2 = sf_{in} + EL - 1 \qquad (7)$$

Set the whole convolution based on the first layer of ordinary convolution and add holes to it. Then its expansion rate m = (M_2 - 1) /(M_1 - 1) + 1, $S_1 = S_2$, $NT_1 = NT_2$.

Set the primary whole convolution lightweight method. The convolution kernel size EL is expanded from the ordinary convolution of size M_1, and the calculation amount V_2 is:

$$V_2 = H_{in} * X_{in} * M_1 * M_2 * D_{in} * D_{out} \qquad (8)$$

It can be seen from equation (2) that the whole convolution structure has similar interval sampling characteristics.

To verify the performance of atrous convolution for lightweight, this paper embeds the atrous convolution lightweight method into several typical convolutional neural network structures VGG (Shin *et al.* 2021), ResNet and DLA (Wang *et al.* 2021).

3.3 Improved atrous convolution lightweight method

Inspired by the design ideas of lightweight neural networks Mobile Net and Squeeze Net, this paper combines atrous convolution with 1×1 point-by-point convolution, further improves the basic atrous convolution residual block, and proposes an enhanced atrous convolution lightweight method.

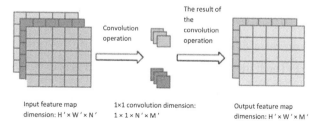

Figure 1. 1×1 convolutional compression feature map process.

Using the channel compression characteristics of 1×1 point-by-point convolution to equalize the receptive field simultaneously, after the point-by-point convolution, a layer of convolutional convolution layers is connected in series to ensure the receptive field of the residual block for further lightweight. The specific structure of the improved atrous convolution lightweight method is shown in Figure 1.

The improved atrous convolution lightweight method combines point-by-point convolution and atrous convolution.

$$V_{impr} = H_{in} * X_{in} * 1 * 1 * D_{in} * M_P + H_{in} * X_{in} * 3 * 3 * D_{out} * M_P \qquad (9)$$

$$Q_{impr} = D_{in} * 1 * 1 * M_P + 3 * 3 * D_{out} * M_P \qquad (10)$$

D_{in}, D_{out}, H_{in}, and X_{in} in equations (16) and (17) are similar to those defined in equation (10).

The same; Lp represents the number of compressed channels, which is 16. It can be seen from the formula calculation that the reduction of the parameter amount and computation amount of the improved atrous convolution lightweight method compared with the primary atrous convolution light method is mainly due to the small convolution kernel of $1 \times$ one point-by-point convolution (Choi and Bajic 2020).

3.4 *Fusion-type atrous convolution lightweight method*

Excessive use of atrous convolution to perform lightweight operations on the network will improve precision attenuation. The main reasons are: (1) From equation (2), it can be seen that the atrous convolution structure has the characteristics of equal interval sampling, and frequent selection will lose each Part of the valuable feature information of the hierarchical feature map; (2) The essence of a deep neural network is to fit an optimal objective function in a solution space, and a deeper network has a more extensive solution space. Fusion-type whole convolution lightweight method as shown in Figure 2.

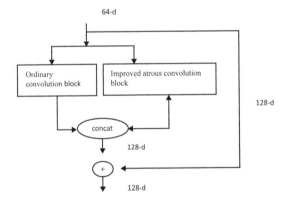

Figure 2.   Fusion-type whole convolution lightweight method.

4   EXPERIMENT AND RESULT ANALYSIS

4.1   *Experimental environment and evaluation criteria*

This paper uses the PyTorch framework for programming and applies CUDA for GPU acceleration. The experiments were completed using the hardware platform configured with Intel Core i5 CPU and NVIDIA GTX 1660TI GPU. Data set and evaluation criteria: In this study, three atrous convolution lightweight methods are embedded into typical network structures to conduct experiments on image classification and object detection tasks (Zhao *et al.* 2021).

4.2 *Experiments on image classification tasks and analysis of the results*

4.2.1 *Experiment and result analysis of primary atrous convolution lightweight method in the image classification task*

This paper uses the CIFAR10 dataset to conduct image classification experiments on the primary atrous convolution lightweight method. Select the direct connection type (VGG) and skip connection type (ResNet, DLA) network structure and apply the primary atrous convolution light method to conduct experiments on the CIFAR10 dataset (Wang *et al.* 2019). Experimental results are shown Table 1 and Figure 3.

Table 1. Experimental results of the basic atrous convolution lightweight method on the CIFAR10 dataset.

Serial	VGG19	ResNet18	ResNet34	DLA18	DLA34
Original network	400	500	1100	600	900
Apply basic atrous convolution lightweight method	200	250	500	260	400

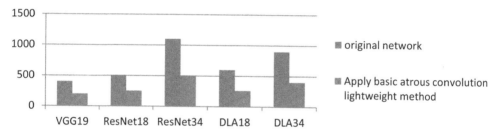

Figure 3. Experimental results of the basic atrous convolution lightweight method on the CIFAR10 dataset.

4.2.2 *Experiments and results analysis of improved and fused atrous convolution lightweight methods in image classification tasks*

Figure 4 show the experimental results of image classification for the CIFAR10 dataset. The closer to the left side of the horizontal axis in the figure, the fewer parameters and computations required by the model, and the lighter the model; the closer to the upper side of the vertical axis, the higher the accuracy of the model and the better the classification effect; the more the model is on the upper left side.

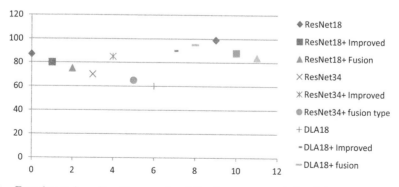

Figure 4. Experimental results of improved and fused atrous convolution lightweight methods in the CIFAR10 dataset.

On CIFAR100, applying the fusion whole convolution method to design the network and the original network does not significantly reduce the Top1 and Top5 accuracy rates, and the reduction is within 0.5%. Therefore, the accuracy of the network structure is higher than that of the original network as shown in Table 2.

Table 2. CIFAR100 dataset improved and fused atrous convolution lightweight.

Network model	Top1/%	Top5/%	Parameters/106	FLOPs/106
ResNet18	76.61	94.05	12.22	557.706
ResNet18+ Improved	67.9	90.85	1.531	50.217
ResNet18+ Fusion	76.11	93.63	5.321	230.948
ResNet34	77.76	94.37	22.328	1162.505
ResNet34+ Improved	69.04	91.55	1.832	84.452
ResNet34+ fusion type	77.84	93.99	9.278	476.135
DLA18	77.94	94.16	13.08	586.895
DLA18+ Improved	75.23	94.04	2.409	97.755
DLA18+ fusion	77.84	94.15	6.189	269.5
DLA34	78.44	94.51	16.32	918.834
DLA34+ Improved	76.79	94.34	2.826	143.995
DLA34+ fusion	78.18	94.55	7.653	420.819

In this paper, several typical lightweight neural networks of Mobile Net (Sun *et al.* 2022), MobileNetv2 (Yao *et al.* 2021), Squeeze Net (Dong *et al.* 2022), Shuffle Net (Tandon *et al.* 2022), and ShuffleNetv2 (Deshmukh *et al.* 2021) are selected to be trained on CIFAR100 using the same training method as the previous experiment. The lightweight characteristics of the improved atrous convolution method and the appropriate macro-network structure design can achieve no significant difference in performance with typical light neural networks as shown in Table 3.

Table 3. Experimental results of typical lightweight neural network and improved atrous convolutional network in CIFAR100.

Network model	Top1/%	Top5/%	Parameters/106	FLOPs/106
MobileNet	66.98	90.44	4.315	48.496
MobileNetv2	69.08	91.98	3.369	68.593
SqueezeNet	70.41	92.64	1.781	55.381
ShuffleNet	71.06	92.65	2.012	45.456
ShuffleNetv2	70.51	92.51	2.361	46.234
ResNet18+ Improved	67.9	90.85	1.531	50.217
DLA18+ Improved	75.23	94.04	2.409	97.755

4.3 *Experiments on target detection tasks and analysis of their results*

The target detection task is divided into single-stage and double-stage. In this paper, a single-stage target detection method without an anchor point, the internet, is used as the target detection task framework, and ResNet34 and DLA34 are used as its backbone network on the KITTI dataset. Conduct object detection task experiments (Gupta 2006). The experimental results are shown in Table 4.

Table 4. DLA34 is the experimental results of the internet backbone network applying the atrous convolution lightweight method on the KITTI dataset.

Backbone network DLA34	AP/%			AOS			BEV AP/%			Parameters/ 106	FLOPs/ 109
	Easy	Mode	Hard	Easy	Mode	Hard	Easy	Mode	Hard		
original network	88.87	78.15	69.15	84.36	74.12	65.16	29.64	23.43	20.01	18.39	71.7
improved	86.02	75.57	66.97	81.97	71.93	63.34	30.36	23.45	21.69	4.89	48.48
Fusion	88.65	78.7	69.71	84.34	74.7	65.71	28.41	25.21	22.48	9.72	56.81

It can be seen from the experimental results that because some other operations occupy the number of parameters and computation on the internet, the improved and fusion-type hollow convolution lightweight methods are not as light as simple image classification tasks but still bring considerable benefits—light effect. Both the original network and the application of the fusion-type atrous convolution lightweight method network have good detection results, and all cars in the picture are detected without missed detection and false detection, compared with the original network application fusion-type atrous convolution lighter weight (Dhiman *et al.* 2021; Limbasiya *et al.* 2018; Qureshi and Gupta 2014). Example of object detection results on the KITTI dataset using the CenterNet as shown in Figure 5.

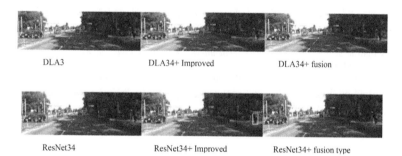

DLA3 DLA34+ Improved DLA34+ fusion

ResNet34 ResNet34+ Improved ResNet34+ fusion type

Figure 5. Example of object detection results on the KITTI dataset using the CenterNet framework using the atrous convolution lightweight method.

5 CONCLUSION

This paper studies the characteristics of atrous convolution applied to lightweight methods. The proposed lightweight methods are based on atrous convolution: primary atrous convolution lightweight method, improved atrous convolution lightweight method and fusion atrous convolution lightweight method quantification method. After theoretical research and experimental analysis, it is found that the light weight of the primary hollow convolution method will reduce the accuracy and the degree of reduction. The lightweight fusion-type whole convolution method can combine speed and accuracy well while getting a better fair result. It can retain the same or even higher accuracy than the original network and has better universality.

REFERENCES

Choi, H. and Bajić, I.V., 2020, November. A lightweight model for deep frame prediction in video coding. In *2020 54th Asilomar Conference on Signals, Systems, and Computers* (pp. 1122–1126). IEEE.
Deshmukh, S., Thirupathi Rao, K. and Shabaz, M., 2021. Collaborative learning based straggler prevention in large-scale distributed computing framework. *Security and communication networks, 2021*: 1–9.

Dhiman, G., Singh, K.K., Soni, M., Nagar, A., Dehghani, M., Slowik, A., Kaur, A., Sharma, A., Houssein, E.H. and Cengiz, K., 2021. MOSOA: A new multi-objective seagull optimization algorithm. *Expert Systems with Applications, 167*: 114150.

Dong, N., Meng, F., Raffik, R., Shabaz, M., Neware, R., Krishnan, S. and Na, K., 2022. Optimization of target acquisition and sorting for object-finding multi-manipulator based on open MV vision. *Nonlinear Engineering, 11*(1): 471–477.

Guo, H., Bai, H. and Qin, W., 2021. ClouDet: A dilated separable CNN-based cloud detection framework for remote sensing imagery. *IEEE Journal of Selected Topics in Applied Earth Observations and Remote Sensing, 14*: 9743–9755.

Gupta, A., 2006. Network management: Current trends and future perspectives. *Journal of Network and Systems Management, 14*(4): 483–491.

Limbasiya, T., Soni, M. and Mishra, S.K., 2018. Advanced formal authentication protocol using smart cards for network applicants. *Computers & Electrical Engineering, 66*: 50–63.

Qureshi, S.R. and Gupta, A., 2014, March. Towards efficient big data and data analytics: a review. In *2014 conference on IT in business, industry and government (CSIBIG)* (pp. 1–6). IEEE.

Risso, M., Burrello, A., Pagliari, D.J., Conti, F., Lamberti, L., Macii, E., Benini, L. and Poncino, M., 2021, December. Pruning in time (PIT): A lightweight network architecture optimizer for temporal convolutional networks. In *2021 58th ACM/IEEE Design Automation Conference (DAC)* (pp. 1015–1020). IEEE.

Shin, Y.G., Sagong, M.C., Yeo, Y.J., Kim, S.W. and Ko, S.J., 2020. Pepsi++: Fast and lightweight network for image inpainting. *IEEE transactions on neural networks and learning systems, 32*(1): 252–265.

Sun, G., Zhou, Y., Pan, H., Wu, B., Hu, Y. and Zhang, Y., 2022. A lightweight NMS-free framework for real-time visual fault detection system of freight trains. *IEEE Transactions on Instrumentation and Measurement, 71*: 1–11.

Tandon, A., Guha, S.K., Rashid, J., Kim, J., Gahlan, M., Shabaz, M. and Anjum, N., 2022. Graph based CNN algorithm to detect spammer activity over social media. *IETE Journal of Research*: 1–11.

Wang, L., Xu, Q., Xiong, Z., Huang, Y. and Yang, L., 2019, October. A multi-level feature fusion network for real-time semantic segmentation. In *2019 11th international conference on wireless communications and signal processing (WCSP)* (pp. 1–6). IEEE.

Wang, S., Pu, Z., Li, Q., Guo, Y. and Li, M., 2021, September. Edge computing-enabled crowd density estimation based on lightweight convolutional neural network. In *2021 IEEE International Smart Cities Conference (ISC2)* (pp. 1–7). IEEE.

Yao, Q., Shabaz, M., Lohani, T.K., Wasim Bhatt, M., Panesar, G.S. and Singh, R.K., 2021. 3D modelling and visualization for vision-based vibration signal processing and measurement. *Journal of Intelligent Systems, 30*(1): 541–553.

Yu, R., Xu, X. and Shen, Y., 2019, July. Rhnet: Lightweight dilated convolutional networks for dense objects counting. In *2019 Chinese Control Conference (CCC)* (pp. 8455–8459). IEEE.

Zhang, B., Zhang, Y. and Wang, S., 2019. A lightweight and discriminative model for remote sensing scene classification with multidilation pooling module. *IEEE Journal of Selected Topics in Applied Earth Observations and Remote Sensing, 12*(8): 2636–2653.

Zhang, X., Zou, Y. and Wang, W., 2018, August. LD-CNN: A lightweight dilated convolutional neural network for environmental sound classification. In *2018 24th international conference on pattern recognition (icpr)* (pp. 373–378). IEEE.

Zhao, Y., Wang, R., Chen, Y., Jia, W., Liu, X. and Gao, W., 2020. Lighter but efficient bit-depth expansion network. *IEEE Transactions on Circuits and Systems for Video Technology, 31*(5): 2063–2069.

Zhu, H., Qiao, Y., Xu, G., Deng, L. and Yu, Y.F., 2019. Dspnet: A lightweight dilated convolution neural networks for spectral deconvolution with self-paced learning. *IEEE Transactions on Industrial Informatics, 16*(12): 7392–7401.

Next Generation Computing and Information Systems – Gupta. (Ed.)
© 2025 The Author(s), ISBN 978-1-032-73865-9

Artificial intelligence based Monte Carlo model for epidemic forecasting for societal aspect

M. Vijayakumar
Department of Computer Technology, NANDHA Arts and Science College, Erode, India

Yana Batla
Department of Electronics and Communication, UIET Chandigarh, India

Mukesh Soni
Dr. D. Y. Patil Vidyapeeth, Pune, Dr. D. Y. Patil School of Science & Technology, Tathawade, Pune, India

Mohan Raparthi
Software Engineer, Alphabet Life Science, Dallas Texas, USA

Aws Zuhair Sameen
College of Medical Techniques, Al-Farahidi University, Baghdad, Iraq

Sheshang Degadwala
Department of Computer Engineering, Sigma University, Vadodara, Gujarat, India

ABSTRACT: After the domestic new crown pneumonia (COVID-19) epidemic has stabilized, the imported people and goods carrying the virus may lead to a slight rebound in the domestic epidemic. Using this model, the second epidemic situation in a city was simulated. The impact of the epidemic's spatial spread and quantitative growth was analyzed using the closure and control measures at different spatial scales (community closure and control, district and county closure and control). The study results show that the first 3 to 5 days after the appearance of an infected person is the best time for containment and control. The effects of the two modes are similar, so they are suitable for dealing with the single-point burst mode. Combined with the analysis results, suggestions are made for formulating closure and control measures after the epidemic rebounds.

1 INTRODUCTION

New coronary pneumonia (COVID-19) is a respiratory infectious disease caused by the new coronavirus SARS-COV-2, which has the characteristics of solid transmissibility and a long incubation period (Cui *et al.* 2021). In December 2019, a large-scale outbreak of COVID-19 occurred in Wuhan, China, and then swept the world, causing massive casualties and economic losses. As the domestic epidemic has stabilized, cities have begun to resume production and life. Currently, the growth forecast of the COVID-19 epidemic and the evaluation and analysis of epidemic prevention policies are mainly conducted through modelling analysis (Tian *et al.* 2015), which are based on group or individual models. Other statistical models or machine learning models to predict the growth of the epidemic and evaluate policies, such as (Claude, Perrin and Ruskin 2009) constructed the SEIR model to predict the turning point of the COVID-19 epidemic in Wuhan, (Huang *et al.* 2007) created a SIQR model to evaluate the prevention and control policies during the COVID-19 epidemic in Shenzhen. Then conducts epidemic prediction and policy evaluation, such as in (Xie *et al.*

2021), using the agent-based simulation modelling method, a multi-scenario simulation experiment was carried out on the policies and measures of the Taiyuan City hospital and government to deal with COVID-19. (Oda *et al.* 2015) established an urban spatial network model to analyze the spread of COVID-19 in the Wuhan situation is simulated. Due to a large amount of calculation in the multi-agent simulation process, the results are uncertain, and the epidemic spread process has Markov characteristics (Chancay-Garcia *et al.* 2018); that is, the future state of the epidemic state is only related to the current state, not the previous state. Therefore, this model uses the Markov chain Monte Carlo method (MCMC) (Nair *et al.* 2022) to obtain the stable distribution of the results and then dynamically calibrate the critical parameters in the model. This paper uses this model to analyze the impact of the scale of urban closure and control on the spread of the epidemic after the epidemic rebounds. It puts forward suggestions to provide a reference for formulating prevention and control measures for a new round of epidemic rebounds that may come.

1.1 *Study area and data*

This paper selects a large city with typical urban characteristics as the research area. The town has a large population, complex personnel, developed traffic, and many crowded regions. Consequently, it is highly vulnerable to the threat of infectious diseases and causes serious consequences. Taking the COVID-19 epidemic in 2020 as an example, the city has experienced four processes: early import and local spread in Hubei, overseas import and spread in spring, gathering epidemic in a market in summer, and multi-point distribution in autumn and winter. This paper collects epidemic data for some of the above processes. At the same time, because the essential geographic information data is required to construct the virtual geographic environment and generate the agent during the simulation process of the spatial agent model, the spread of the epidemic is closely related to the travel activities of the people, this paper collects the primary geographic information data and travel statistics of the city.

2 MODEL DESIGN AND IMPLEMENTATION

Through the preliminary analysis of the existing data, this paper takes the active residents in the city as the agent and designs. It implements the agent model to simulate the spread of the virus in urban population flow and interaction under different policy measures. According to the existing news reports, statistical data and research papers, the model divides the agent into different groups (students, office workers, other groups) according to the travel groups and divides the crowd activity places according to the land type (residential, school, work areas etc.), assign travel trajectories to agents according to travel statistics, and build a virtual geographic environment to simulate the interaction of infected persons with other agents in their places (homes, hospitals, public places, etc.), the spread of viruses, and the health status of infected persons.

2.1 *Framework of the model*

The model is divided into a three-layer overall framework. The first layer is the primary data layer. This layer mainly provides the essential data (shown in Table 1) required by the model. The second layer is the model preparation layer. In this layer, the vector data in the data layer is spatially connected by GIS to obtain the land type of each building, generate environmental entity data (attributes are shown in Table 1), and extract and calculate. Calculate the total residential area and population density of each street, assign the population of each road to each floor of each residential building, generate the same amount of crowd agent data, each agent represents a resident, and design the intelligent agent to travel rules and assigns travel trajectory records to agent data.

Table 1. Environmental entity attributes.

Property field	Illustrate
Gid	building number
Floor	floor number
Level	building type code
X	Longitude
Y	Latitude

2.2 Agent attributes and travel rules

According to the statistical analysis results of the travel purpose of urban residents in the "Annual Report on Traffic Development of new Delhi" (from now on referred to as the annual report), this paper divides the crowd agents into three types of travel groups, namely students, office workers and other groups. Next, it assigns attributes to them. Then, the travel rules of the agent are determined based on the statistical results of the travel time of the three types of travel groups and the step size set by the model. It saves them in the attribute field. In addition, this paper also needs to consider relevant policies for the movement rules of agents and some parameters in the model (travel ratio, infection probability, return to work ratio). Based on the relevant news materials from January to March 2020 and the statistical results of China Unicom's travel ratio, this paper makes the following settings for the management and control measures for the corresponding time of the model.

2.3 Simulation of contact groups of infected persons

The transmission methods of new coronary pneumonia include direct transmission (droplets, breathing, etc.), aerosol transmission, and contact transmission (Soni *et al.* 2022). Therefore, simulating the relationship network of the individuals contacted by the infected person can describe the transmission path of the virus. A complex network is a model that can be used to describe the interaction between different individuals. It can be divided into random networks, small-world networks, self-similar networks, scale-free networks, etc. (Gupta and Awasthi 2011; Gupta *et al.* 2017). When the susceptible person and the untreated infected person are in the same building, according to different building types, a corresponding small-world network is constructed centered on the infected person to represent the infected person and his contacts, and set the probability of infection of the susceptible person after contact as perfect, and determine whether the contact person is infected by comparing the random number with excellent to realize the spread of the virus to the adjacent nodes of the infected person.

2.4 The health state transition rules of infected persons

In the infectious disease model, the population can generally be divided into susceptible (S), latent(E), onset (I), removal (recovery (R) or death (D)), four states (Javed *et al.* 2018). Infections (after the incubation period, they carry the virus, but those who have not yet become ill are recorded) as in (Khan *et al.* 2020), and the diagnosis and treatment of infected persons have a more significant impact on the spread of the epidemic.

2.5 Model realization and parameter calibration

The model realization takes the early growth of the city's epidemic (January 20-March 8) as an example. According to the information released by the city's health and health commission, the model sets two agents in zone D as the initially infected persons (Mazhar *et al.* 2023). The simulation step is 8 hours, and the simulation time is 50 days. After running according to the process in Figure 1, the distribution map of locations where infection events.

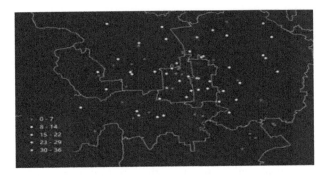

Figure 1.   Example of distribution of infection events in simulated experiment.

The numerical results of the model take the number of people diagnosed with the real epidemic as the observed value and use the MCMC method to conduct large-scale sampling and iteration to obtain a stable distribution of simulated values (Ullah *et al.* 2023). Bayesian fitting explores suitable parameter value ranges that go through. After 1000 MCMC iterations, a linear regression equation is established between the sampling results and the actual values, and the unknown parameter pInfect in the program is determined by measuring the coefficients in the equation:

$$y = \alpha \widehat{y} + \beta \tag{1}$$

In the formula, y is the actual value of the model, and ŷ is the fitted value of the model. When the output result α is close to 1 and β is close to 0, the simulated value of the model is most comparable to the actual value, and the model parameters are the optimal parameters (Yao *et al.* 2021).

At the same time, to verify the rationality of the parameter calibration results, this paper compares the results of the three methods of MCMC sampling, the least-squares method and manual coarse parameter adjustment. It uses the goodness of fit (R^2) and the coefficient of equation (1) to compare; the formula for calculating R^2 is as follows:

$$R^2 = \frac{\sum_{i=1}^{n} \left(\widehat{y}_i - \overline{y} \right)^2}{\sum_{i=1}^{n} \left(y_i - \overline{y} \right)^2} \tag{2}$$

In the formula, $\overline{y}$ is the mean value of the observed value. The best estimates of unknown parameters are shown in Table 2.

Table 2.   Comparison of parameter settings.

Parameter	Parameter Description	MCMC	Least two multiplications	Artificial rough parameter tuning
pInfect1	Phase 1 infection rate	0.236	0.252	0.241
pInfect2	Stage 2 infection rate	0.097	0.101	0.101
pInfect3	Stage 3 infection rate	0.008	0.011	0.012
alpha	Linear equation coefficients after fitting	1.125	1.042	0.67
beta	Linear equation coefficients after fitting	1.015	2.123	93.93
R^2	The goodness of fit of the model	0.995	0.99	0.851

The optimal infection rate MCMC fitting values of the three stages are respectively: the first stage is 0.239, the second stage is 0.099, and the third stage is 0.009. The fitted coefficients $\alpha = 1.126$, $\beta = 1.011$ (Bhatt and Sharma 2023). Since the MCMC sampling results are relatively stable, the goodness of fit is higher than that of the other two methods. The result after sampling and fitting is averaged, and the comparison with the real value is shown in and Figure 2.

As shown in Figure 2, the fitted value of the sampling result is close to the actual value. This indicates that the model can more realistically reflect the evolution of the early epidemic after the parameters are calibrated.

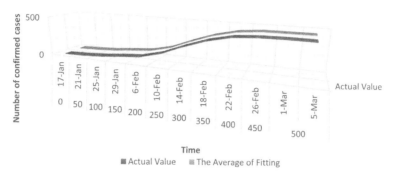

Figure 2.　Comparison of fitting results with true values.

3　SIMULATION EXPERIMENT ANALYSIS OF SEALING AND CONTROL AT DIFFERENT SCALES

With the normalization of epidemic prevention, once the epidemic rebounds, the question of how to carry out closure and control to minimize the impact on urban operations begins to emerge (Rida *et al.* 2016). Set the impact of different containment and control scales on the growth of infected people and the spread of infected areas.

3.1　*Experimental scenario*

The experimental scenario is designed concerning the summer clustered epidemic in the city and the multi-point distribution scenarios in other areas. There are two types of systems: single-point burst and multi-point burst. The detailed settings are as follows:

1) Single-point outbreak: Suppose there are 20 clustered cases in a particular place in District B of the city (the environmental entity number is 171911), and the initial state is set to E.
2) Multi-point sporadic outbreak: Suppose there are 20 sporadic cases in 3 districts of the city, and the locations of occurrence are not fixed in the districts, and the initial state is E.

Among all model parameters, the value of the infection probability (see Section 2.5 for details) is compared with the results of multiple simulations of the city's summer cluster epidemic through the benchmark parameters, and the rate is set to 0.025 (Rida *et al.* 2016). The three scale scenarios are described as follows:

1) No containment and control: After the outbreak began, the government did not issue any containment and control measures and only treated those who were sick promptly.
2) Community closure and control: After the epidemic began, the government closed and controlled the building where the sick person appeared and the patient's address and adjacent structures.
3) District/county closure and control: After the outbreak began, the government launched a closure and power on the districts and counties where the sick person appeared. People in the closure and control areas were not allowed to go out at home, and all close contacts were quarantined.

3.2 Analysis of experimental results

3.2.1 Analysis of the impact of different sealing and control scales on the growth of the number of confirmed diagnoses

For the 25 simulation results, the average, maximum and minimum values were counted in this paper. From the different statistical results of the number of infected people output from the simulation shown in Table 3: in the case of a single-point outbreak, the number of confirmed cases after community lockdown measures was reduced by 19% compared to the number of infections without lockdown measures.

Table 3. Comparison of number of simulated confirmed cases under different scale containment measures.

| Experimental scene | Control measures | Statistical indicators of simulation results | | |
		average value	maximum value	Minimum
single point burst	No sealing control	32	45	22
	Community lockdown	24	37	21
	District and county closures	28	39	24
spread out more	No sealing control	52	78	39
	Community lockdown	43	75	26
	District and county closures	38	47	27

Average the simulation results to draw a growth curve. In the case of the same initial infection, the growth rate of multi-site sporadic is higher than that of the single-site outbreak (Jagota *et al.* 2022). The main growth period of the two epidemic models is 3–10 days, after which the growth rate slows as most infected people are admitted. In the distribution mode, except for no containment and control measures, the growth rate of other experimental groups gradually slowed until stable (Deshmukh *et al.* 2021). However, it is significantly higher than the community closure in the mid-term and lower than the community closure in the later stage (Saini *et al.* 2021).

3.2.2 Analysis of the impact of different containment and control scales on the number of infection incidents

This paper collects the locations of infection events during the simulation experiments of three different containment and control measures. Then, taking the single-point outbreak mode as an example, some experimental results are displayed on the map, as shown in Figure 3.

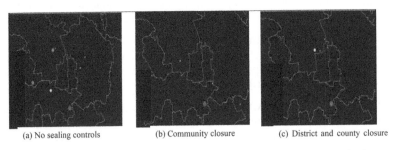

(a) No sealing controls   (b) Community closure   (c) District and county closure

Figure 3. Distribution of infection incident locations under three containment scales in experiment.

As seen from Figure 3, in the single-point mode, infection incidents have occurred in multiple districts if there are no containment and control measures. As a result, the number of incident locations outnumbered the other two cases. Under the two modes, the number of infection incident sites decreased by 50% to 55% under the two modes, and the impact effect was roughly the same. Under the district/county lockdown, the number of infection events in the single-point outbreak mode decreased by 20%, and the number of infection events in the multi-point sporadic manner decreased by 66%. The number of infected areas in the 20-day simulation period increased approximately linearly in the early stage.

3.2.3 *Analysis of the influence of infection event uncertainty on experimental results*

Since the model is a probabilistic model, the number of agents infected by the infected person during each experiment is uncertain, which leads to the uncertainty of the infection location on the one hand and the difference in the number of confirmed diagnoses in different experiments on the other hand. The significant P values of the simulation results are all greater than or equal to 0.2, obeying a normal distribution. The experimental results are 95% set. Furthermore, it calculates the population standard deviation σ of m_n, the sample standard deviation σ_n and the 95% confidence interval c to analyze the convergence and divergence of the mean of the simulation results, where m_n and σ_n are calculated as follows.

$$m_n = \frac{1}{n}\sum\nolimits_{i=1}^{n} y_n \tag{3}$$

$$\sigma_n = \sqrt{\frac{\sum_{i=1}^{n}(m_n - \overline{m_n})^2}{n-1}} \tag{4}$$

After the confidence interval range c is calculated, the convergence interval range is determined to be $\Delta y = c + \sigma = 0.64 + 1.56 = 2.20$ according to the population standard deviation result. Comparing the sample standard deviation of the simulation results, the mean value starts to converge after the 12th time, so the number of simulations can be set to more than 12 times.

4 CONCLUSION

Based on agent modelling and a complex network model, combined with GIS technology, this paper constructs a new crown pneumonia epidemic spread model for a city epidemic. The model uses the MCMC method to determine the critical parameters in the model. Using this model, this paper designs experimental scenarios based on the two modes of the single-point outbreak and multi-point distribution. The study found that the effect of community lockdown is similar in the two epidemic modes on different scales of lockdown measures. From the study, it is recommended to regularly organize a screening in high-risk areas (such as seafood markets, seafood processing plants, cold storage, fishing ports, etc.). Furthermore, more scenarios are designed to help respond to a new round of epidemic rebounds and evaluate other infectious disease prevention measures after the domestic virus-prone areas and foreign epidemics ease.

REFERENCES

Bhatt, M.W. and Sharma, S., 2023. An Object Recognition-Based Neuroscience Engineering: A Study for Future Implementations. *Electrica, 23*(2).

Chancay-Garcia, L., Hernandez-Orallo, E., Manzoni, P., Calafate, C.T. and Cano, J.C., 2018. Evaluating and enhancing information dissemination in urban areas of interest using opportunistic networks. *IEEE Access, 6*: 32514–32531.

Claude, B., Perrin, D. and Ruskin, H.J., 2009, June. Considerations for a social and geographical framework for agent-based epidemics. In *2009 International Conference on Computational Aspects of Social Networks* (pp. 149–154). IEEE.

Cui, C., Huang, C., Zhou, W., Ji, X., Zhang, F., Wang, L., Zhou, Y. and Cui, Q., 2020. AGTR2, one possible novel key gene for the entry of SARS-CoV-2 into human cells. *IEEE/ACM transactions on computational biology and bioinformatics*, 18(4): 1230–1233.

Deshmukh, S., Thirupathi Rao, K. and Shabaz, M., 2021. Collaborative learning based straggler prevention in large-scale distributed computing framework. *Security and communication networks*, 2021: 1–9.

Gupta, A. and Awasthi, L.K., 2011. Peers-for-peers (P4P): an efficient and reliable fault-tolerance strategy for cycle-stealing P2P applications. *International Journal of Communication Networks and Distributed Systems*, 6(2): 202–228.

Gupta, A., Prabhat, P., Gupta, R., Pangotra, S. and Bajaj, S., 2017, December. Message authentication system for mobile messaging applications. In *2017 International Conference on Next Generation Computing and Information Systems (ICNGCIS)* (pp. 147–152). IEEE.

Habib Ullah, K., Sohail, M., Ali, F., Nazir, S., Ghadi, Y.Y. and Ullah, I., 2023. Prioritizing the multi-criterial features based on comparative approaches for enhancing security of IoT devices.

Huang, H.Y., Luo, P.E., Li, M., Li, D., Li, X., Shu, W. and Wu, M.Y., 2007. Performance evaluation of SUVnet with real-time traffic data. *IEEE Transactions on Vehicular Technology*, 56(6): 3381–3396.

Jagota, V., Luthra, M., Bhola, J., Sharma, A. and Shabaz, M., 2022. A secure energy-aware game theory (SEGaT) mechanism for coordination in WSANs. *International Journal of Swarm Intelligence Research (IJSIR)*, 13(2): 1–16.

Javed, F., Khan, S., Khan, A., Javed, A., Tariq, R., Matiullah and Khan, F., 2018. On precise path planning algorithm in wireless sensor network. *International journal of distributed sensor networks*, 14(7),: 1550147718783385.

Khan, F., Khan, A.W., Khan, S., Qasim, I. and Habib, A., 2020. A secure core-assisted multicast routing protocol in mobile ad-hoc network. *Journal of Internet Technology*, 21(2): 375–383.

Mazhar, T., Talpur, D.B., Shloul, T.A., Ghadi, Y.Y., Haq, I., Ullah, I., Ouahada, K. and Hamam, H., 2023. Analysis of IoT Security Challenges and Its Solutions Using Artificial Intelligence. *Brain Sciences*, 13(4): 683.

Nair, R., Soni, M., Bajpai, B., Dhiman, G. and Sagayam, K.M., 2022. Predicting the death rate around the world due to COVID-19 using regression analysis. *International Journal of Swarm Intelligence Research (IJSIR)*, 13(2): 1–13.

Oda, T., Elmazi, D., Spaho, E., Kolici, V. and Barolli, L., 2015, September. A simulation system based on ONE and SUMO simulators: performance evaluation of direct delivery, epidemic and energy aware epidemic DTN protocols. In *2015 18th International Conference on Network-Based Information Systems* (pp. 418–423). IEEE.

Rida, I., Almaadeed, S. and Bouridane, A., 2016. Gait recognition based on modified phase-only correlation. *Signal, Image and Video Processing*, 10: 463–470.

Rida, I., Boubchir, L., Al-Maadeed, N., Al-Maadeed, S. and Bouridane, A., 2016, June. Robust model-free gait recognition by statistical dependency feature selection and globality-locality preserving projections. In *2016 39th International Conference on Telecommunications and Signal Processing (TSP)* (pp. 652–655). IEEE.

Saini, G.K., Chouhan, H., Kori, S., Gupta, A., Shabaz, M., Jagota, V. and Singh, B.K., 2021. Recognition of human sentiment from image using machine learning. *Annals of the Romanian Society for Cell Biology*: 1802–1808.

Soni, M., Singh, A.K., Babu, K.S. and Kumar, S., 2022. Convolutional neural network based CT scan classification method for COVID-19 test validation. *Smart Health*, 25: 100296.

Tian, D., Yang, Y., Xia, H. and Zhang, G., 2015, August. An Evaluation Model for Epidemic Routing in VANETs. In *2015 IEEE 12th Intl Conf on Ubiquitous Intelligence and Computing and 2015 IEEE 12th Intl Conf on Autonomic and Trusted Computing and 2015 IEEE 15th Intl Conf on Scalable Computing and Communications and Its Associated Workshops (UIC-ATC-ScalCom)* (pp. 1661–1664). IEEE.

Xie, Z., Liu, X., Li, Y., Ye, X., Zheng, L. and Xiong, Y., 2021, October. BIM-based optimization of passenger flow organization in urban rail transit stations during the epidemic prevention and control phase. In *2021 China Automation Congress (CAC)* (pp. 4263–4268). IEEE.

Yao, Q., Shabaz, M., Lohani, T.K., Wasim Bhatt, M., Panesar, G.S. and Singh, R.K., 2021. 3D modelling and visualization for vision-based vibration signal processing and measurement. *Journal of Intelligent Systems*, 30(1): 541–553.

Next Generation Computing and Information Systems – Gupta. (Ed.)
© 2025 The Author(s), ISBN 978-1-032-73865-9

Efficient data visualization through novel recurrent neural network-based dimension reduction

Saurabh Sharma, Shafalika Vijyal & Navin Mani Upadhyay
Department of Computer Science & Engineering, Model Institute of Engineering and Technology, Jammu, J&K, India

ABSTRACT: Data visualization can be a challenging task in our everyday lives, especially when dealing with large and complex datasets. Multiple researchers have dedicated their efforts to this topic and have put forth various solutions, such as t-SNE, UMAP, and the Tramp technique. These techniques maintain the overall structure, but they do not accurately depict the local structure in the visualization. Therefore, many of them are not suitable for all types of data sets. Because of these issues, the demand for other novel approaches for data visualization will be high. In this paper, the proposed work is based on a novel Recurrent Neural Network (RNN) based dimension reduction technique to correctly visualize the selected real-world data set. The RNN models can reliably visualize and forecast the numeric values for a general-purpose machine architecture, but for special-purpose machines, the construction of RNN architecture construction has yet to be devised. The proposed model is based on a trial and error-based iteration technique for multiple input parameters. The major challenge for the selection of such input parameters is: that when there are many input factors, it is hard to determine which ones affect model prediction. This problem is reflected as the "curse of dimensionality." The proposed method shows a significant improvement of up to 3% compared to all three compared models (t-SNE, UMAP, & Tramp).

1   INTRODUCTION

Machine learning plays a substantial role in the era of Artificial Intelligence. It varies its models as the nature of data changes (Townes *et al.* 2019). From a large variety of data sets the most complex data sets are categorized as hyper-spectral image data sets which use sometimes an ultraviolet (900 1700 nm) band to represent the images or use more than 3 dimensions. Since the data sets have more than 3 dimensions, it is very difficult to apply direct Linear & nonlinear dimension reduction approaches to visualize such data sets (Daneshfar *et al.* 2021). So, the visualization techniques that were bound with Linear & nonlinear dimension reduction techniques are PCA, MDS, and t-SNE (Mithillesh *et al.* 2022). The second major problem is selecting the correct classifier for the classification of such images. So, based on several selected literature, this paper uses SVM and ELM to classify and grade the selected images. SVM and ELM models of classification identified each hyper-spectral picture pixel to create the prediction map (Mazher *et al.* 2020).

In classification problems, within machine learning the final determination is often based on factors. These factors, known as features can become overwhelming to visualize and work with when there are many of them. Occasionally many of these features are unnecessary duplicates. This is where dimensionality reduction algorithms come into action. Dimensionality reduction involves decreasing the number of variables being examined by identifying a set of variables. It can be divided into two categories; feature selection and feature extraction (Chen Li *et al.* 2023).

DOI: 10.1201/9781003466383-22

1.1 *Feature selection in dimensionality reduction*

Dimensional reduction using PCA and t-SNE has been used with the RNN algorithm for training purposes to improve model prediction (Mithillesh *et al.* 2022). RNN architecture, pre-processing, and task type affected MLP and RNN prediction accuracy. Thus, a scalable method that can handle diverse data is needed. Dimension reduction strategies either retain the pairwise distance pattern across all data samples or preserve local lengths over global distances (Chen Li *et al.* 2023).

Maximum Machine Learning problems involve thousands of features, which leads to two major problems. It makes the training extremely slow, and finding a good solution will be highly difficult. The training models used and compared for this research are based on t-SNE, UMAP, and Tramp techniques.

1.2 *Feature extraction for dimensionality reduction*

The two main approaches introduced for Feature extraction for dimensionality reductions are Projection and Manifold Learning. The RNN algorithm gets input from the output of the PCA algorithm, which has a quite reduced dimensionality (Yaicharoen *et al.* 2023). In this work, the proposed RNN approach is based on two strategies. The first is applying direct RNN modeling without dimensionality reduction. The second is applying PCA with RNN models for dimensionality reduction.

The main contribution of this research is the introduction of a method that uses RNN for reducing dimensionality, in data visualization. This approach aims to address the limitations of techniques and improve the accuracy of visualizing real-world datasets. The study presents a trial-and-error technique for optimizing input parameters in the RNN model. It also discusses how this method can be applied to spectral image datasets, which typically have more, than three dimensions.

2 LITERATURE SURVEY

Numerous research publications on dimension reduction techniques for effective data visualization have been published during the last several years. Existing feature selection approaches are computationally costly when used individually. Applying these algorithms reduces computing costs and improves prediction performance. This paper demonstrates that several dimension reduction strategies may increase machine learning model performance by reducing the feature count (Sinha *et al.* 2021). (Kumari *et al.* 2021) have use one technique i.e., t-SNE. Which divides the data sets into six groups, SVM & ELM prediction sets produced a 100% and 96.35% precise and correct output. (Tingting *et al.* 2021) have used SVM modeling and found six classes for the same selected data sets. Our study shows that the t-SNE visualization technique is optimal for such a data set. (Zaaraoui *et al.*) have proposed an HFRM model to find the effective feature. The HFRM detects and recognizes local features extracted from the Hyper-spectral data sets. The experimental results suggest that the proposed HFRM approach outperforms the current method of (Kumari *et al.* 2021) in terms of rate for prediction and accuracy. Another researcher (Tingting *et al.* 2021), has discovered a classification problem based on weather-related data sets, which rely on both humidity and rainfall and can be collapsed into just one underlying feature. Since both of the aforementioned are correlated to a high degree. Hence, they have reduced the number of features in their selected problem. (Kumari *et al.* 2021) have discovered a 3-D classification problem that includes all 3 dimensions: 2-D, and 1-D image classification problems.

One more important method has been proposed by (Behera *et al.* 2022) called the EBRFE algorithm. The FBRFE algorithm has been used to reduce nonlinear dimensionality. FBRFE visualizes information in a lower dimension with a better mean correlation of 120.17 % with fewer overlapping than standard techniques (Kumari *et al.* 2021).

The existing literature cannot accurately represent the local structure in visualizations. Although these methods are effective in maintaining the data's overall structure, they often face difficulties in accurately representing the local relationships. This is particularly evident when working with complex datasets, when the precise characteristics of data items may not be accurately preserved. As a result, these strategies lose their effectiveness. This makes them unsuitable for certain types of datasets.

3 METHODOLOGY

A detailed methodology, problem formulation, and experimental with Image Data Preparation for Visual Enhancement are shown in this section.

3.1 Problem formulation

As mentioned in earlier section 1, the RNN converts independent stimulation into dependent stimulation by providing all layers having the same weights and biases. It simplifies the variables by adjusting the selected parameters and memorizing each preceding result by feeding them to the subsequent layers. Thus, one needs to combine all these layers into one recurring layer with an identical weight and biases of all the hidden layers. In this case, the data set has still a large dimension. So, let's define the problem in simple mathematical modeling. Suppose, d_t be the current input at time t, the prior cell state c_{t-1}, f, i, o are the selected parameters for the hidden layer at time t, then the function is defined as follows:

$$f, i, o = \sigma((Hy_t + Uh_{t-1}) + b) \tag{1}$$

$$j = \tanh((Hy_t + Uh_{t-1}) + b) \tag{2}$$

$$c_t = f * c_{t-1} + i * j \tag{3}$$

$$h_t = o * tanh(c_t) \tag{4}$$

Where, b is a biased factor, σ Sigmoid function can be defined as eq(5):

$$\sigma(x) = \frac{1}{1 + \exp(-x)} \tag{5}$$

H and U are convolution linear transformation functions applied to the current input and prior hidden state. So, in this paper, we have proposed the RNN followed by PCA for the visualization of c_t, h_t and b. The features are assumed as some symbolized state, and the calculations classify data as normal (N_D) and special (F_D).

3.2 Experimental setup

The system configuration comprised a high-performance cluster, with each node equipped with state-of-the-art hardware. Each node featured a multi-core processor, such as the Intel Xeon E5 series, coupled with a substantial amount of RAM, ranging from 64 to 128 giga-bytes. The processors were selected for their parallel processing capabilities, essential for handling the complex computations involved in training recurrent neural networks. The cluster setup allowed for parallelization of tasks, facilitating faster training and evaluation of the proposed dimension reduction technique. To harness the power of deep learning, we utilized GPUs, specifically the NVIDIA Tesla V100, to expedite matrix operations inherent in neural network computations. The experimental setup was configured to run on a Linux-based environment, leveraging popular deep-learning libraries such as TensorFlow and PyTorch. This comprehensive system configuration ensured the efficient execution of our proposed novel recurrent neural network-based dimension reduction method, paving the way for enhanced data visualization capabilities.

3.3 Data pre-processing

The selected data sets have first pre-processed all the features and removed the noises. A PCA modeling has then been applied to make them into a cluster shown in Figure 1. In Figure 1, the pre-processing and clustering approach employing the PCA algorithm-2 are visually depicted. The data has been transformed into three distinct clusters: cluster-1, characterized by subspace-1; cluster-2, associated with subspace-2; and cluster-3, exhibiting subspace-3.

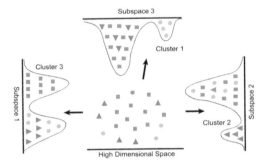

Figure 1.   Pre-processing in high-dimensional space.

Algorithm 1 Principal Component Analysis (PCA)

1: **Input:** Data matrix X, Number of principal components k
2: Subtract mean from each feature in X to center the data
3: Compute the covariance matrix $C = \frac{1}{m}X^T X$
4: Perform eigen decomposition on C to obtain eigenvectors and eigenvalues
5: Select the top k eigenvectors corresponding to the largest eigenvalues
6: Project the data onto the subspace defined by the selected eigenvectors
7: **Output:** Transformed data matrix X_{PCA}

3.4 Recurrent RNN

Recurrent RNNs (Algorithm-3.4 use the previous step's output of the PCA algorithm as input.

Algorithm 2 Recurrent Neural Network (RNN) Algorithm

1: **Input:** Training data (X, Y), learning rate α, number of epochs N
2: **Initialize:** Randomly initialize weights W_{ih} and W_{hh}, biases b_{ih} and b_{hh}.
3: **for** $epoch = 1$ to N **do**
4:    **for** $t = 1$ to T **do**
5:       {Loop through time steps} **Perform forward pass:**
         $h_t = \sigma(W_{ih} \cdot X_t + W_{hh} \cdot h_{t-1} + b_{ih} + b_{hh})$ Compute loss: $L_t = \text{Loss}(h_t, Y_t)$ Perform backward pass to update weights and biases using backpropagation through time (BPTT) Update weights and biases: $W_{ih}, W_{hh}, b_{ih}, b_{hh} \leftarrow \text{UpdateFunction}$
         $\left(W_{ih}, W_{hh}, b_{ih}, b_{hh}, \alpha, \frac{\partial L_t}{\partial W_{ih}}, \frac{\partial L_t}{\partial W_{hh}}, \frac{\partial L_t}{\partial b_{ih}}, \frac{\partial L_t}{\partial b_{hh}}\right)$
9:    **end for**
10: **end for**

In Figure 2, depicting a Recurrent Neural Network (RNN) architecture, three fundamental layers are discernible: the input layer, hidden layer, and output layer. Each layer is interlinked through weights, allowing the network to iteratively, refine its understanding of sequential patterns during training. This recurrent architecture is particularly powerful in tasks involving time series data, natural language processing, and other sequential data domains.

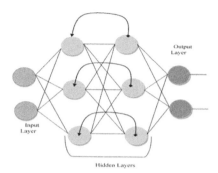

Figure 2.   Recurrent RNN architecture.

3.5   *Principal component analysis*

PCA is a prominent dimensionality reduction method. It gives high-dimensional observations the best linear approximations. PCA finds a subspace with linearity of lower dimensions p ($p < d$) that retains the highest variance for data a with high-dimensionality d. PCA decreases data dimensionality while minimizing variation-related information loss. The data matrix is defined as $A = [a_1 - a_\mu, a_2 - a_\mu, \ldots, a_n - a_\mu] \in \mathbb{R}^{d \times n}$ to make A row-wise zero-mean, as well as the square matrix $M = AA^T \in \mathbb{R}^{d \times d}$. In decreasing sequence, M's d eigenvalues and their associated eigenvectors, dubbed "Principal Components," are $v_1, v_2, \ldots, v_d$. Then, a criterion is set to retain only certain parts of the principal components, such as:

$$1.0 - \frac{\sum_{i=1}^{p} \wedge_i}{\sum_{k=1}^{d} \wedge_k} \leq \sigma \quad \text{with} \quad p \leq d \tag{6}$$

where σ, a modest variable around zero, determines dimensionality reduction accuracy. e.g., if we set $\sigma = 0$, $p = d$ or M has a minimum of one zero eigenvalue, as well as $p < d$. Definition concludes.

$$V = [v_1 \quad v_2 \quad \cdots \quad p_p]_{d \times p} \tag{7}$$

as

$$1.0 - \frac{\sum_{i=1}^{p} \wedge_i}{\sum_{k=1}^{d} \wedge_k} \leq \sigma \quad \text{with} \quad p \leq d \tag{8}$$

where σ, a tiny number around zero influences dimensionality reduction accuracy. e.g., if we set $\sigma = 0, p = d$ or M contains at least a single zero eigen frequency and $p < d$. Definition concludes.

$$V = [v_1 \quad v_2 \quad \cdots \quad v_p]_{d \times p'} \tag{9}$$

and B = VTA, the dimensionally reduced data matrix. The following relation reconstructs the data.

$$A^\wedge d \times n = VB \tag{10}$$

Finally, adding the mean μ to the columns of A yields the approximate data $[a_1, a_2, \ldots, a_n]$. Global linearity limits PCA effectiveness.

Kernel PCA, also known as KPCA is an extension of PCA that incorporates kernel functions commonly used in support vector machines. It allows for the partitioning of data even if they are not linearly separable by leveraging a mapping technique called reproducing kernel Hilbert space. In KPCA observations are transformed into kernel matrices using the method:

$$K = k(x, y) = \phi(x)^{\mathrm{T}}\phi(y) \tag{11}$$

The kernel function, denoted as $k(x, y)$ is used to evaluate the relationship, between observations x and y. The reproducing kernel Hilbert space is where these observations are mapped by the function ϕ. Interestingly due to the kernel trick, there's no need to explicitly compute this mapping function. Instead, we only need to calculate the kernel function.

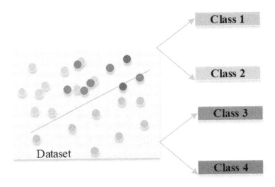

Figure 3. Dataset before dimensionality reduction.

3.6 *Results and discussion*

As mentioned earlier, the pre-processing of the datasets was done before applying the RNN model. The t-SNE also plotted and visualized for multi-dimensional features shown in Figure 4.

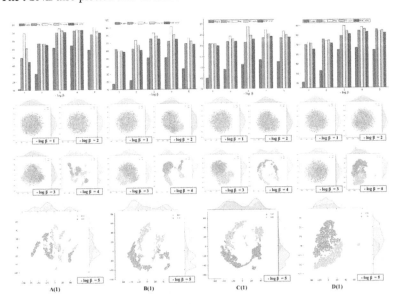

Figure 4. Dataset before dimensionality reduction.

In the above Figure 4, the iteration of 5-dimensional images is shown in the form of cluster and bar graphs. The selected image (Pine Data set) has 5 phases, based on their visibility a log value has been plotted to identify the correct visible area of the pine. The B(1) and D(1) significantly improved visibility from 40% to 47%. which is 3% greater than the t-SNE algorithm (Kumari *et al.* 2021). For validation purposes, Tables 1 and 2 present the best underlying result obtained after applying PCA before applying RNN to the selected data set. In Table 1 the obtained accuracy is presented for the three selected methods (i.e., PCA, t-SNE & RNN). From these tables, it is clear that the RNN achieves a better and more stable result as compared to t-SNE and PCA algorithms. t-SNE tends to yield outcomes suggesting that this approach is more effective, in reducing the dimensionality of data samples. This characteristic is also evident, in the results obtained through Kernel t-SNE.

Table 1. For every iteration up to (20) shown in Figure 4, dimensionality reduction technique for each individual data set.

Method	Pine-12	Pine-13	Pine-14	Pine-16	Pine-17
PCA	0.99	0.93	0.9972	0.92	0.90
t-SNE	0.96	0.97	0.93	0.97	0.91
RNN	0.98	0.97	0.99	0.94	0.96

Table 2. For every cluster up to (20) data-points shown in Figure 4, dimensionality reduction technique for each individual data set.

Method	TF 2012	TF 2013	TF 2014	TF 2016	TF 2017
PCA	0.9965	0.9944	0.9978	0.9912	0.9866
t-SNE	0.9966	0.9942	0.9910	0.9954	0.9926
Kernel t-SNE	0.9857	0.9539	0.9571	0.9852	0.9773

Ultimately, the combination of RNN and PCA surpassed all other approaches in every respect. The calculation speed was excellent, being the second fastest. Additionally, it successfully demonstrated the presence of well-separated clusters, which were anticipated. PCA is used for pre-processing datasets before linear RNN training. Only one model underwent PCA pre-processing. Both RNNs had 25 first-layer nodes and two second-layer nodes. Table 1 highlights the superiority of PCA, with high accuracy scores (e.g., Pine-14: 0.9972). While t-SNE also performed well, it didn't consistently surpass PCA. The efficacy of our RNN models, especially when coupled with PCA preprocessing, is evident in the competitive error differences detailed in Table 1. In Table 2 we can see how PCA consistently achieved accuracy scores ranging from 98.65%, to 99.78% surpassing the performance of both t-SNE and Kernel t- SNE. Although t-SNE and Kernel t-SNE also performed well in the TF 2012 and TF 2014 datasets PCA maintained its superiority. These findings highlight the reliability of PCA in preserving the structure of data during dimensionality reduction. This underscores the computational efficiency and predictive power of our proposed approach.

4 CONCLUSION

PCA may enhance predicted outcomes for linear RNN problems. This study shows that PCA preprocessed data typically yielded more accurate predicted findings. Unfortunately, it isn't always clear whether an RNN has been designed for a linear or nonlinear problem. MLP networks are trained using recorded observations, not deterministic equations like those utilized in this paper. RNN training without PCA preprocessing is possible if the problem's linearity is unknown. If an MLP network trained using PCA-processed data outperforms its quasi-PCA issues, the issue may be linear and PCA may improve prediction accuracy. PCA pre-processing may not increase network model prediction accuracy if the issue is nonlinear. Nonlinear problems need no input dataset preparation. The RNN consistently outperformed other methods and showed improved results. In the Pine 14 dataset, the accuracy achieved with RNN was 99.72% while t-SNE achieved 97% and PCA achieved 93%. These findings highlight the performance of our proposed dimension reduction method based on RNN especially when used in conjunction, with PCA preprocessing. The combination of RNN and PCA outperformed approaches in terms of both accuracy and computational efficiency.

REFERENCES

(2020). Clustering Visualization and Class Prediction using Flask of Benchmark Dataset for Unsupervised Techniques in Machine Learning. *International Journal of Innovative Technology and Exploring Engineering.* http://dx.doi.org/10.35940/ijitee.G5943.059720

Behera, A.P., Singh, J., Verma, S., & Kumar, M. (2022). Data visualization through nonlinear dimensionality reduction using feature-based Ricci flow embedding. *Multimedia Tools and Applications*, 1–20. https://link.springer.com/article/10.1007

Chen, H. (2022). Machine learning and statistics analysis of socioeconomic and health factors impact on the progress of countries' humanitarian commitments. 2022 *International Conference on Computer Applications Technology (CCAT)*, 49–55. https://doi.org/10.1109/CCAT56798.2022.00016

Daneshfar, F., & Kabudian, S.J. (2021). Speech emotion recognition using deep sparse auto-encoder extreme learning machine with a new weighting scheme and spectro-temporal features along with classical feature selection and a new quantum-inspired dimension reduction method. *ArXiv*, abs/2111.07094. https://doi.org/10.48550/arXiv.2111.07094

Gan, S., & Yu, Z. (2020). Partial dynamic dimension reduction for conditional mean in regression. *Journal of Systems Science and Complexity*, 33, 1585–1601. http://dx.doi.org/10.1007/s11424-020-8329-3

Ghosh, T., & Kirby, M.J. (2020). Supervised dimensionality reduction and visualization using centroid-encoder. *J. Mach. Learn. Res.*, 23, 20:1–20:34. https://doi.org/10.48550/arXiv.2002.11934

Hisano, A., Iwasaki, M., Satake, I., Satoh, M., Nagahara, H., Takemura, N., Nakashima, Y., & Nakano, T. (2022). R&D of the Kek Linac Accelerator Tuning Using Machine Learning.

Kumari, J., Patidar, K.C., Saxena, M.G., & Kushwaha, M.R. (2021). A hybrid enhanced real-time face recognition model using machine learning method with dimension reduction. *Indian Journal of Artificial Intelligence and RNNing.* http://dx.doi.org/10.35940/ijainn.B1027.061321

Lafuente, D., Cohen, B., Fiorini, G., Garcia, A., Bringas, M., MorzÃ¡n, E., & Onna, D. (2021). Introduction to machine learning for chemists: an undergraduate course using python notebooks for visualization, data processing, data analysis, and data modeling. *ChemRxiv.* https://doi.org/10.1021/acs.jchemed.1c00142

Li, Y.-C. and Zhan, J. (2023). Effect of dimensionality reduction on uncertainty quantification in trustworthy machine learning. In *Proceedings of the 2023 International Conference on Machine Learning and Cybernetics (ICMLC)*, Adelaide, Australia, pp. 326–332. doi: 10.1109/ICMLC58545.2023.10327953.

Liles, C.A. (2014). RNN Machine Learning and Dimension Reduction for Data Visualization. http://dx.doi.org/10.7763/IJMLC.2015.V5.545

Mazher, A. (2020). Visualization framework for high-dimensional spatio-temporal hydrological gridded datasets using machine-learning techniques. *Water.* http://dx.doi.org/10.3390/w12020590

P, M.K., & Supriya, M. (2022). Throughput analysis with effect of dimensionality reduction on 5g dataset using machine learning and deep learning models. 2022 *International Conference on Industry 4.0 Technology (I4Tech)*, 1–7. https://doi.org/10.1109/I4Tech55392.2022.9952579

Sinha, V., Dash, S., Naskar, M., & Hossain, S.M. (2021). A study of feature selection and extraction algorithms for cancer subtype prediction. 2022 *International Conference for Advancement in Technology (ICONAT)*, 1–6. https://doi.org/10.48550/arXiv.2109.14648

Tingting, F., Jiangming, L., Zhilong, K., Xinhuan, N., & Qingshuang, M. (2021). Nondestructive detection of keemun black tea grade based on hyperspectral imaging technique. https://doi.org.10.13386/j.issn1002-0306.2020090211

Townes, F.W., Hicks, S.C., Aryee, M.J., & Irizarry, R.A. (2019). Feature selection and dimension reduction for single-cell RNA-Seq based on a multinomial model. *Genome Biology*, 20. http://dx.doi.org/10.1101/574574

Yaicharoen, A., Hashikura, K., Kamal, M. A. S., Murakami, I., and Yamada, K. (2023). Effects of dimensionality reduction on classifier training time and quality. In *Proceedings of the 2023 Third International Symposium on Instrumentation, Control, Artificial Intelligence, and Robotics (ICA-SYMP)*, Bangkok, Thailand, pp. 53–56. doi: 10.1109/ICA-SYMP56348.2023.10044946.

Next Generation Computing and Information Systems – Gupta. (Ed.)
© 2025 The Author(s), ISBN 978-1-032-73865-9

An optimized CNN-based model for pneumonia detection

Samridhi Sharma, Suditi Seth, Faisal Rasheed Lone & Azra Nazir
School of Computer Science & Engineering, VIT Bhopal University, Madhya Pradesh, India

ABSTRACT: Convolutional Neural Networks (CNNs) are a breakthrough paradigm shift in medical imaging and diagnostic approaches that can be used to detect pneumonia. CNNs have shown to be incredibly effective at automatically extracting complex and subtle picture features from CT scans and chest X-rays, making them an excellent choice for this task. With the help of deep learning, this initiative aims to improve pneumonia diagnosis accuracy and productivity. Creating an automated system to detect pneumonia is crucial for prompt treatment, especially in remote regions. Given the effectiveness of deep learning, Convolutional Neural Networks (CNNs) are widely recognised for accurately classifying diseases from medical images. The result is a diagnostic tool that may provide quick and precise assessments, giving medical practitioners the power to make educated decisions. This paper employs Transfer Learning and a CNN Model to ascertain the presence of pneumonia in individuals through chest X-ray analysis. The objective is to gain a comprehensive understanding of accurate detection. We built a CNN model from scratch and applied transfer learning to it. Compared to the transfer learning models we used, the CNN model we constructed yielded better results and was more accurate.

1 INTRODUCTION

Children under five and elderly individuals with weakened immune systems face susceptibility to pneumonia. In 2018, pneumonia caused over a million child deaths globally and remains life-threatening without early detection. Streptococcus pneumonia is the primary cause of the sickness in children, caused by viruses, bacteria, and fungus. The second most prevalent cause of bacterial pneumonia is Haemophilus influenza type b (Hib).

Respiratory syncytial virus is the predominant viral cause of pneumonia. In HIV-infected infants, Pneumocystis jiroveci is a prevalent contributor, responsible for a substantial portion, approximately one-quarter, of all pneumonia-related deaths in this demographic (*Pneumonia*, n.d.). The common diagnostic methods for pneumonia include radiography, CT scan, or MRI. Medical professionals assess chest radiographs to identify pneumonia; medical history and laboratory results are additional diagnostic approaches (*WO2003070102A3 – Lung Nodule Detection and Classification – Google Patents*, n.d.). Pneumonia affects 7% of the global population (450 million people) annually, causing four million deaths. It is an acute lower respiratory disease often caused by microorganisms, bacteria, or viruses (Nazir *et al.* 2022). The World Health Organization (WHO) identifies pneumonia as the leading cause of child mortality, responsible for approximately 1.2 million deaths in children under five years of age. Pneumonia is one of the deadliest infections that can strike children under the age of five. India was the only country to record 158,176 deaths in 2016, and the country continues to lead the world in infant pneumonia deaths (Kumar *et al.* 2018). Worldwide, this communicable disease kills people, according to a WHO assessment released. Nowadays, radiography is the most reliable tool that clinicians acknowledge for detecting pneumonia (Pingale and Patil 2017). Nonetheless, some research indicates that variations frequently occur when radiologists interpret chest CT scans. The

DOI: 10.1201/9781003466383-23

disadvantage of human-based observation has spurred technological advancement in this area, where a machine can now distinguish between an aberrant chest X-ray and a normal one (Young and Marrie 1994).

Brighter colours on the lung cavities may indicate many problems, including cancer cells, swollen blood arteries, and heart issues. Chest X-rays are the best approach to confirm the location and size of a lung infection. With these methods, the disease's appearance may be inaccurate or confused with the onset of another condition. As a result, the project is satisfying because it enhances the processing of pneumonia diagnosis in medical circumstances in remote places. Using several classifiers, the researchers could train and evaluate the CNN model and distinguish between chest X-rays that showed no disease and those that showed disease. In recent years, computer-aided design (CAD) has emerged as the most significant area of artificial intelligence and machine learning research. The basic focus of this project is to binary classify chest X-ray images, i.e., divide them into two groups: those with pneumonia and those without. The core emphasis of this paper centres around the pervasive application of neural networks for pneumonia detection within chest X-ray datasets. The images presented visually delineate the distinction between pneumonia-affected and healthy lungs. Our paper aims for precision in detection, and the diagnostic inference will inform patient assessment based on the obtained results. In particular, this study identifies pneumonia instances by using deep learning models to classify chest X-ray pictures precisely. The fundamental method uses deep learning algorithms to extract features from these photos to achieve accurate categorisation.

A custom convolutional neural network (CNN) architecture is carefully designed for this. This CNN is trained on a dataset that includes normal X-ray images and X-ray images with pneumonia, providing the groundwork for further classification tasks. Using transfer learning techniques, a comparison analysis is carried out to assess the efficacy of the bespoke CNN model. Interestingly, the well-known pre-trained models are used to help with classification. This study evaluates the accuracy of this transfer learning model compared to the custom CNN. The paper has the following structure: The project is introduced in Section 1, which also presents the main concept. Related work is explored in Section 2. The approach, max pooling, normalisation, and in-depth process are covered in the following parts. An acknowledgements section and a references section round out the paper.

2 LITERATURE REVIEW

In clinical practice, radiologists scrutinise X-ray images for specific indicators, notably "infiltrates, characterised by prominent white areas within the lungs, which indicate infection [9]. Recognising that pneumonia-like patterns can also appear in tuberculosis and severe bronchitis cases (*WO2003070102A3 – Lung Nodule Detection and Classification – Google Patents*, n.d.). Additional diagnostic procedures, such as Complete Blood Count (CBC), Chest Computed Tomography (CT), and sputum tests, are typically conducted to arrive at a definitive diagnosis. Our endeavour, however, is focused solely on distinguishing whether a chest X-ray image indicates pneumonia or belongs to a category of normal patient cases. This determination is based on identifying cloudy patterns in the X-ray imagery. It is essential to underscore that conclusive diagnosis hinges on the results of pathological tests. In the contemporary landscape of medical diagnostics, Artificial Intelligence (AI) has revolutionised disease detection, with notable successes in breast cancer and brain tumour identification. Convolutional Neural Networks (CNNs), a subset of AI-driven solutions, have exhibited exceptional promise in classification tasks, commanding the trust of medical practitioners worldwide. When considering cost-effective and user-friendly imaging modalities, deep learning and machine learning methods have garnered significant favour in chest X-ray examination. The wealth of data for training diverse machine learning models further amplifies their appeal. Remarkably, in our survey of existing literature, we observed that achieved the highest reported accuracy of 98% (Sharma *et al.* 2017).

Consequently, our choice of deploying a Convolutional Neural Network (CNN) to analyse our dataset was grounded in achieving superior accuracy compared to other deep learning methodologies. These days, artificial intelligence-based solutions are used to diagnose various disorders. Brain tumours, breast cancer, and other illnesses are among them (Sharma *et al.* 2017). Doctors worldwide accept this artificial intelligence (AI) based classification since this Convolutional Neural Network-based detection has proven to be very effective and promising (Sirazitdinov *et al.* 2019). Deep learning and machine learning techniques are becoming increasingly popular when analysing chest X-rays because of their low cost and ease of usage. Furthermore, much data is available for training different machine learning models. Out of all the articles we looked at, it had the highest accuracy, reaching 98% (Sharma *et al.* 2017).

3 CONVOLUTIONAL NEURAL NETWORK (CNN)

These days, artificial intelligence-based solutions are used to diagnose various disorders. Brain tumours, breast cancer, and other illnesses are among them (Bailer *et al.* 2018; *Chest X-Ray Images (Pneumonia) | Kaggle*, n.d.). Doctors worldwide accept this artificial intelligence (AI) based classification since this Convolutional Neural Network-based detection has proven to be very effective and promising. Deep learning and machine learning techniques are becoming increasingly popular when analysing chest X-rays because of their low cost and ease of usage. Furthermore, much data is available for training different machine learning models. Out of all the articles we looked at, had the highest accuracy, coming in at 98% (Bailer *et al.* 2018).

There are three basic building blocks in CNN to preserve the spatial structure:

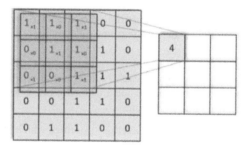

Figure 1. Generic view of a CNN.

3.1 *Normalization*

Normalisation is performed to standardise features, ensuring they have a similar scale. This procedure enhances both the performance and training stability of the model. Normalising a dataset minimises redundancy and ensures that only relevant data is stored. In deep learning research, normalisation has been a dynamic area of study. Let's underscore a few advantages of incorporating normalisation, including significantly reducing training time (*Batch Normalization: Theory and How to Use It with Tensorflow | by Federico Peccia | Towards Data Science*, n.d.).

Occasionally, a feature's value may be significantly higher than that of its surrounding features. In this procedure, we normalise every feature to ensure its preservation. We are creating objectivity in our network by doing this. Normalisation reduces the shift in the internal covariate. Covariate shift refers to the variation in the activation network's distribution caused by network parameter changes throughout the model's training process. To improve training, we must reduce the Internal Covariate Shift. Normalisation preserves the features but eliminates the sharp feature. Normalisation accelerates optimisation by preventing weights from exploding and limiting them to a certain range. One unintentional benefit of normalisation is that it aids CNN in regularisation, albeit marginally not significantly.

3.2 Batch normalization

One of the prevalent methods for normalising activations within a neural network across a mini-batch of a specified size is batch normalisation. This technique involves computing the mini-batch's mean and variance for each feature. Subsequently, the feature is normalised by subtracting the mean and dividing by the standard deviation of the mini-batch. By re-centring and rescaling the inputs of the layers, batch normalisation also referred to as batch norm—allows artificial neural networks to be trained more quickly and steadily. In 2015, Christian Szegedy and Sergey Ioffe made the suggestion.

3.3 Max pooling

One of the most prevalent characteristics of convolutional neural networks is pooling (Sharma *et al.* 2020). The pooling layer's primary goal is to utilise the normalised feature map's "accumulate of collect" feature, produced by applying a convolutional filter to the image. The main task is to continuously reduce both the spatial size of the featured map and the amount of network computation required. The most popular kind of pooling is max pooling. The process of pooling that chooses the maximum element from the area of the feature map that the filter covers is called max pooling. Therefore, a feature map with the most noticeable features from the prior feature map would be the output following the max-pooling layer. The following are max pooling's main advantages:

- *Dimensionality reduction:* It lowers the input data's spatial dimensions, which aids in managing the number of parameters and computational complexity in the neural network's later layers. This can expedite training and avoid overfitting.
- *Translation Invariance:* Max pooling contributes to a certain level of translation invariance, which allows the model to identify features independent of their precise location in the input.
- *Feature Selection:* Max pooling helps capture unique patterns and reduce noise by highlighting the most significant features in each region and choosing the maximum value. Additionally, max pooling has drawbacks, such as losing fine-grained spatial information because only the maximum value is chosen.

Other options for max pooling include average pooling, which determines the average value in each region, and more sophisticated pooling methods that adjust to the input size, like global average pooling and adaptive pooling. Convolutional neural networks use max pooling as a basic operation for dimension reduction and feature extraction, especially in computer vision tasks like object recognition and image classification. Max pooling and average pooling are represented using eq (1) and, (2) respectively.

$$f(m) = [f1(m).....fk(m).... fK(m)]^T, fk(m) = \max(x), x \in Xk \tag{1}$$

$$(a) = [f1(a)...fk(a)...fK(a)]^T, fk(a) = \frac{1}{(|Xk|)} \sum x, x \in Xk \tag{2}$$

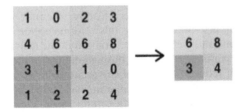

Figure 2.   A representation of Max pooling in CNN.

146

4 PROPOSED MODEL

This study investigates the role of deep learning in computer vision for detecting pneumonia using three convolutional neural network models. The dataset of normal chest x-ray images and diseases infected with pneumonia are obtained. Our analysis allows us to determine which of the three models most effectively detects pneumonia. All the models have done well when identifying pneumonia and normal chest X-rays. We have shown how to categorise X-ray images into positive and negative pneumonia data. Unlike other approaches that heavily rely on transfer learning, we build our model from scratch. We also applied a transfer learning strategy to the problem, and the outcomes were relatively improved compared to other models. Additionally, we employed a transfer learning approach to solve the problem, and the CNN model we developed produced good results and was more accurate than the transfer learning models we employed.

4.1 Dataset

CNN models were instantiated and trained using the Kaggle Chest X-Ray Images (Pneumonia) dataset (*Chest X-Ray Images (Pneumonia) | Kaggle*, n.d.). This dataset, comprising 5856 JPEG images totalling 1.16 GB, was imported from Kaggle and organised into Train, Test, and Validation sets. These sets were further categorised into Pneumonia and Normal, with a distribution of 5216 training images, 624 testing images, and 16 validation images. Employing data augmentation techniques to improve results, the dataset primarily consists of front and back Chest X-ray images obtained from pediatric patients aged one to five years at the Guangzhou Women and Children's Medical Center, Guangzhou.

Table 1. Bifurcation of the dataset into test, train and validation images.

		Number of Images
Testing	Normal	234
	Pneumonia	390
Training	Normal	1341
	Pneumonia	3875
Validation	Normal	8
	Pneumonia	8

4.2 Image augumentation

Deep learning models typically demand substantial datasets for effective training. CNNs, in particular, necessitate thousands of images, posing a practical challenge. To address this, the technique of image augmentation is employed. Image augmentation significantly expands an existing dataset by applying techniques such as standardising features (pixel values), whitening transforms, random rotations, shifts, flipping, rescaling, shearing, zooming, etc. Our methodology implements rescaling, shearing, zooming, and horizontal flipping on training images to enhance the variety and augment the training dataset.

- To prevent models from overfitting.
- The initial training set is too small.
- To improve the model accuracy.
- To Reduce the operational cost of labelling and to clean the raw dataset.

4.2.1 Geometric transformations.

Randomly flip, crop, rotate, stretch, and zoom images. You need to be careful about applying multiple transformations on the same images, as this can reduce model performance.

4.2.2 Color space transformations
Randomly change RGB colour channels, contrast, and brightness.

4.2.3 Kernel filters.
Randomly change the sharpness or blurring of the image.

4.2.4 Random erasing
Delete some part of the initial image.

4.2.5 Mixing images
Blending and mixing multiple images.

4.3 Pooling layer

Pooling layers come after convolutional layers. All four models use max-pooling layers as their type of pooling layer. The 2×2 max-pooling layer chooses the highest pixel intensity values from the image window the kernel currently covers. Images are down-sampled using max-pooling, which lowers the dimensionality and complexity of the image.

4.4 Methodology

Data augmentation is a popular technique for expanding the dataset and improving the model's generalizability. It is the process of using transformations like rotation, flipping, cropping, stretching, and lens correction to increase the size of a dataset artificially. Figure 3 represents the image classification process.

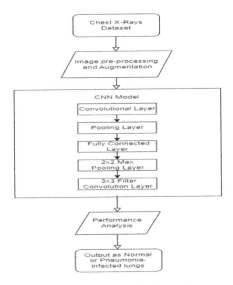

Figure 3. Flowchart representing X-rays image classification using proposed model.

4.4.1 Feature extraction
The network will carry out a number of pooling and convolutional operations in this section to identify the features. This is where the network would identify the zebra's stripes, two ears, and four legs if you had a picture of one.

4.4.2 Classification
These extracted features will be layered onto fully connected layers, acting as a classifier. This section presents the experiment's dataset and the methodology for predicting

pneumonia in X-rays. The pre-trained feature extractor VGG16 is used in this work to extract features.

4.4.3 *Neural networks*

Neural networks are classified as one of the best model architectures for image categorisation. It is unique because it employs a 2x2 filter's max pooling and padding layer and 3x3 filter convolution instead of many hyper-parameters. The max pooling and convolutional layers have a similar structure throughout the system.

5 RESULTS AND DISCUSSION

The model underwent 20 epochs, achieving an accuracy of 91.667% with a loss of 0.24128. The evaluation encompassed three CNN models: VGG16, VGG19, and a custom CNN model. Performance metrics, including accuracy, sensitivity, and specificity, are gauged using True Positive (TP), False Positive (FP), True Negative (TN), and False Negative (FN) values. TP represents correctly classified positive data (pneumonia), TN indicates accurately classified negative data (normal), FN signifies negatively labelled data misclassified, and FP represents positively labelled data inaccurately classified. Evaluation involves a confusion matrix, as depicted in Figure 3.

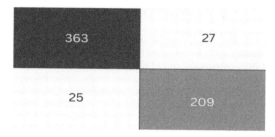

Figure 4.   Confusion matrix obtained by the CNN model.

The model went through a total of 20 epochs and attained an accuracy of 91.667% and a loss of 0.24128. The maximum accuracy achieved was 96.55% at epoch 14, and the minimum loss of 0.0956 was achieved at epoch 15, as presented in Table 2.

Table 2.   Performance of the CNN model at different epochs.

Epoch Stage	Training Loss	Training Accuracy	Validation Loss	Validation Accuracy
1	0.3632	0.8779	18.9482	0.5000
6	0.1321	0.9502	4.2739	0.5032
10	0.1099	0.9605	0.7861	0.6875
15	0.0956	0.9651	1.2604	0.5000
20	0.1008	0.9628	1.2310	0.6250

6 CONCLUSION AND FUTURE WORK

With an accuracy of 91.667%, the CNN 22-layer architecture is shown to have the best accuracy in the results. The model produced an F1 score of 93.3161%, recall of 93.556%, and precision of 93.0769%. Computers' capacity to identify diseases using a range of machine learning and deep learning techniques is very helpful in areas where radiologists are in short supply. especially in countries in South Asia and Africa, where 60–70% of people live in rural areas. Due to their low cost and minimal instrument requirements, these tools are reasonably easy to use in rural areas. These tools will also be very useful in automatically identifying patients who require immediate medical attention versus those who can wait. Our work effectively provides a CNN-based autonomous pneumonia identification technique.

REFERENCES

Bailer, C., De Tewodros, C. B., Habtegebrial, A., Varanasi, K., & Stricker, D. (2018). *Fast Feature Extraction with CNNs with Pooling Layers*. https://arxiv.org/abs/1805.03096v1

Batch normalization: Theory and how to use it with Tensorflow | by Federico Peccia | Towards Data Science. (n.d.). Retrieved January 3, 2024, from https://towardsdatascience.com/batch-normalization-theory-and-how-to-use-it-with-tensorflow-1892ca0173ad

Chest X-Ray Images (Pneumonia) | Kaggle. (n.d.). Retrieved January 3, 2024, from https://www.kaggle.com/datasets/paultimothymooney/chest-xray-pneumonia

Kumar, P., Grewal, M., & Srivastava, M. M. (2018). Boosted Cascaded Convnets for Multilabel Classification of Thoracic Diseases in Chest Radiographs. *Lecture Notes in Computer Science (Including Subseries Lecture Notes in Artificial Intelligence and Lecture Notes in Bioinformatics), 10882 LNCS*, 546–552. https://doi.org/10.1007/978-3-319-93000-8_62/COVER

Nazir, A., Naaz, R., Qureshi, S., & Nazir, N. (2022). HRCT chest analysis for detection of pulmonary arterial hypertension in COVID-19 patients using convolutional neural networks. *2022 IEEE 3rd Global Conference for Advancement in Technology, GCAT 2022*. https://doi.org/10.1109/GCAT55367.2022.9972033

Pingale, T. H., & Patil, H. T. (2017). Analysis of cough sound for pneumonia detection using wavelet transform and statistical parameters. *2017 International Conference on Computing, Communication, Control and Automation, ICCUBEA 2017*. https://doi.org/10.1109/ICCUBEA.2017.8463900

Pneumonia. (n.d.). Retrieved January 3, 2024, from https://www.who.int/health-topics/pneumonia#tab=tab_1

Sharma, A., Raju, D., & Ranjan, S. (2017). Detection of pneumonia clouds in chest X-ray using image processing approach. *2017 Nirma University International Conference on Engineering, NUiCONE 2017, 2018-January*, 1–4. https://doi.org/10.1109/NUICONE.2017.8325607

Sharma, H., Jain, J. S., Bansal, P., & Gupta, S. (2020). Feature extraction and classification of chest X-ray images using CNN to detect pneumonia. *Proceedings of the Confluence 2020 – 10th International Conference on Cloud Computing, Data Science and Engineering*, 227–231. https://doi.org/10.1109/CONFLUENCE47617.2020.9057809

Sirazitdinov, I., Kholiavchenko, M., Mustafaev, T., Yixuan, Y., Kuleev, R., & Ibragimov, B. (2019). Deep neural network ensemble for pneumonia localization from a large-scale chest x-ray database. *Computers & Electrical Engineering, 78*, 388–399. https://doi.org/10.1016/J.COMPELECENG.2019.08.004

WO2003070102A3 – Lung nodule detection and classification – Google Patents. (n.d.). Retrieved January 3, 2024, from https://patents.google.com/patent/WO2003070102A3/en

Young, M., & Marrie, T. J. (1994). Interobserver variability in the interpretation of chest roentgenograms of patients with possible pneumonia. *Archives of Internal Medicine, 154*(23), 2729–2732. https://doi.org/10.1001/ARCHINTE.1994.00420230122014

Next Generation Computing and Information Systems – Gupta. (Ed.)
© 2025 The Author(s), ISBN 978-1-032-73865-9

EEG-based emotion recognition: Leveraging CNNs for precision

Ayushi Kotwal
Department of Computer Science & IT, University of Jammu, Jammu, J&K, India

Vishal Gupta
Model Institute of Engineering and Technology, Jammu, J&K, India

Meetali Verma, Jatinder Manhas & Vinod Sharma
Department of Computer Science & IT, University of Jammu, Jammu, J&K, India

ABSTRACT: Understanding human emotions is essential in recognizing cognitive and emotional functions. A lot of effort has been paid to emotion recognition as a crucial component of psychological study and interaction between humans and computers. Since its non-invasive design and great temporal resolution, electroencephalogram (EEG) signals have become a promising tool for emotion classification. The proposed methodology of emotion classification covers preprocessing of EEG data, extracting significant characteristics, and designing a CNN architecture. CNNs has the tendency to automatically learn key features, perform excellent in tasks identical to the recognition of emotions using EEG signals. They are effectively suited for dealing with the spatial and hierarchical nature of EEG data. The effectiveness and viability of the proposed approach are demonstrated by the experimental findings. The CNN model demonstrates its promise for real-time emotion identification applications by achieving excellent accuracy in its classification. The proposed structure provides fresh perspectives on human affective states and emotion classification. The approach undertaken for the study can be extended with its applications spanning from human-computer interaction systems to mental health evaluation.

1 INTRODUCTION

Affective computing primarily addresses the issue of computer systems helped in correctly processing, identifying, and knowing how humans show their mental and emotional intelligence. This covers the perception (or recognition) of emotions as a significant challenge. Emotion is a complicated psychological state that is connected to both physiological processes and bodily movements [Canon *et al.* 1927]. EEG is a suitable method for obtaining information about human emotions, and it can be used to directly identify emotional states and investigate emotional cognitive processes. From the standpoint of potential applications, EEG-based emotion identification technology has spread to several industries, including safe driving, healthcare, educational institutions, entertainment, retail, and the military [Suhaimi *et al.* 2020]. The recently developed deep learning techniques might offer a way to classify EEG state-of-mind data with excellent accuracy and higher-level domain-invariant representations of features. Various models of deep learning, including recurrent neural networks [Chowdary *et al.* 2022], restricted Boltzmann machines (RBMs) [Gao *et al.* 2015], convolutional neural network models [Yang *et al.* 2019], long short-term memory models [Alhagry *et al.* 2017], deep belief networks [Zheng *et al.*], convolutional recurrent neural network model [Chen *et al.* 2020], transfer learning models [Demir *et al.* 2020], etc. are employed to recognize emotions using EEG data. To use ResNet [Asghar *et al.* 2020] and VGG16

[Cheah *et al.* 2021] for emotion identification based on EEG, these architectures need to adapt to work with sequential data and take into account the unique characteristics of EEG signals. In this paper, we offer a convolution neural network model for detecting emotions utilizing EEG signals on the DEAP dataset that outperforms the current state of the art. The remainder of the paper is organized as follows: The related work is presented in Section 2. Section 3 describes the materials as well as the proposed approach. Section 4 covers the findings and discussion. Finally, Section 5 discusses the conclusion and future scope.

2   RELATED WORK

This section discusses the research that are most important to this work. The following steps are involved in recognizing emotions based on EEG signals: (1) EEG signal acquisition; (2) pre-processing; (3) extraction of features and selection; and (4) classification. Researchers have used a variety of deep learning algorithms to classify emotions using EEG signals.

Li *et al.* 2017 [13] implemented a convolutional recurrent neural network (CRNN) to extract emotions using multiple channels of communication EEG data. The model was trained and tested by the authors using an openly accessible dataset of EEG recordings from thirty-two people. The outcomes showed that the suggested techniques were successful in addressing the affective aspects of arousal and valence. Alhagry *et al.* 2017 [1] implemented a deep learning classifier that uses Long Short Term Memory (LSTM) networks to identify emotions from unprocessed EEG information. Arousal, valence, and liking are utilized to categorize the properties that the LSTM network learns. Using arousal, valence, and liking classes, respectively, the suggested technique when evaluated on the DEAP data set, it achieves an average accuracy of 85.65%, 85.45%, and 87.99%. Chao *et al.* 2019 [4] also introduced a deep learning framework for emotion recognition using multiple channels of electroencephalograph signals that is built on a capsule network (the CapsNet system) and a multiband feature matrix (MFM). Tested on the DEAP dataset, the proposed methodology offers average accuracy with arousal, valence, and dominance classes of 70.3%, 70.2%, and 70.5%, respectively. Pandey *et al.* 2019 [14] provide a deep learning model for classification and variational mode decomposition (VMD) as a feature extraction strategy for subject-independent emotion identification from EEG signals. Using a convolutional neural network (CNN) classifier, the recommended approach is tested on the DEAP dataset, giving an accuracy of 70.89% in identifying four emotions that is valence, arousal, dominance, and liking. Tripathi *et al.* 2017 [16] developed 2 different neural models to characterize user feelings using EEG data obtained from the DEAP database. The deep neural network consists of 4 fully connected layers where the initial neural layer consists of 5000 neurons followed by layers of 500 and 1000 nodes. The second model uses a 2-dimensional Convolutional Neural Network model which is designed to classify pre-processed EEG data presented in the form of 2D effectively. The research also performed the comparison of CNN with deep neural networks and concluded that CNN performed better. The accuracy achieved by the model is 81.41% and 73.35% for 2 class classification and 66.79% and 57.58% for 3 classes on Valence and Arousal and the accuracy of a deep neural network is 75.78% and 73.12% for 2 classes and 58.44% and 55.70% for Valence and Arousal respectively for classification on 3 classes. Perceptual, behavioral, and F. M *et al.* 2015 [9] approach aims to use EEG signals obtained when working through a three-level mental math exercise to differentiate between stress and rest states. The research compared the differences between the tasks (control and stress) using a two-sample t-test with mean p-values of 0.03 for stage one, 0.042 for stage 2, and 0.05 for stage three. It uses a support vector machine classifier for determining mental strain at levels one, two, and three of the math problem, respectively, with accuracy of 94%, 85%, and 80%. According to Kotwal *et al.* 2023 [12] CNN fared better than any other deep learning method when used to evaluate any aspect of the cognitive state of the human brain, but RNN performed poor.

In this study, we used deep learning classifiers for the classification of emotions based on the EEG signals. Arousal, Valence, Liking and Dominance are the emotions that will be used in our study. Convolutional Neural Network model is developed using the programming language Python, the Keras API, and the framework called TensorFlow.

3 MATERIALS AND SUGGESTED APPROACH

This section will include our suggested method of recognizing the emotions as well as the dataset that was used in the study.

3.1 Overview of the dataset

A cutting-edge dataset, DEAP [Koelstra *et al.* 2012], is discussed in this paper, which includes EEG signals that were captured on the scalp using 32 electrodes. The electrical activity of the brain is measured with the help of EEG, which provides information about emotional reactions. Physiological signals including conductance of the skin, respiration, and heart rate, temperature are also included in the dataset. Additional details regarding the participant's emotions may be revealed by these signals. A collection of 40 video clips intended to elicit different emotions, such as fear, anger, sadness, and happiness, were shown to the participants. Following a screening of every movie, the participants self-reportedly rated their emotional experiences. For tasks involving emotion classification and regression, these ratings serve as the ground truth. In the domains of affective computation, recognizing emotions, and brain-computer interfaces, the DEAP dataset is extensively used. A snapshot of the DEAP dataset is given in Figure 1.

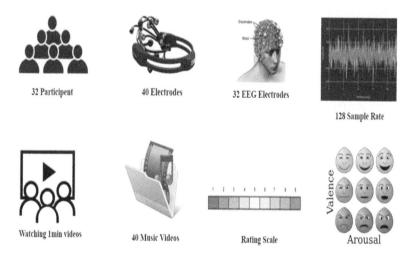

Figure 1. Synopsis of DEAP dataset.

The DEAP dataset's EEG data is 40 × 40 × 8064 in shape as shown in Figure 2. The breakdown of this is as follows:

40: This is generally the total number of individuals or participants in the research.
40: This indicates the total number of electrodes or channels that were utilized to record each participant's EEG signals.
8064: This is the total amount of points of data in the EEG signal for each channel, and it relates to how long the EEG data was captured.

The DEAP dataset's labels are 40 × 4 in shape. Each movie or stimulus has four dimensions, as indicated by the number 4. These categories most likely relate to several emotional characteristics, including:

Valence: A measurement of the emotional reaction's positivity or negativity.
Arousal: is a term used to describe the degree of both physiological and psychological activation.
Dominance: An indicator of one's emotional state of control or dominance.
Liking: An indicator of an individual's degree of liking or disliking an emotional event.

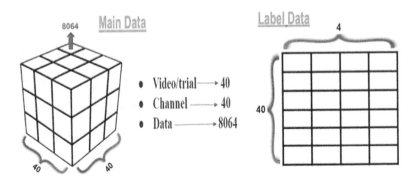

Figure 2.   Data representation.

3.2 *Suggested approach*

Several crucial actions are involved in applying EEG information to identify emotions. First, individuals are given emotional stimuli while electrodes are applied to their scalps to record EEG data. Data quality is improved by pre-processing methods like filtering and artifact removal. Essential components include statistical characteristics and spectral analysis for feature extraction. EEG data is given an emotional state using emotion labeling. After feature extraction, the process of building and training a neural network is usually the following phase of the classification process using a deep learning model. Popular choices for EEG data in this situation are convolutional neural networks (CNNs) and recurrent neural networks (RNNs). To increase the model's performance, hyperparameter tuning—such as learning rates, batch sizes, and regularization strategies is important. Figure 3 displays the process flow diagram for the proposed procedure.

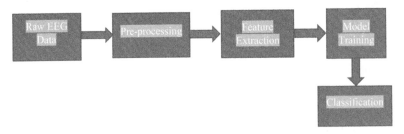

Figure 3.   Proposed method.

3.2.1 *Feature extraction*
One of the techniques for feature extraction is the Fast Fourier Transform. It provides information about the various frequency components contained in the signal by converting the EEG

data from the time domain into the frequency domain. It helps in the extraction of relevant frequency-based features that can be transformed into images and processed further by CNNs for applications such as classification or EEG-based emotion recognition. EEG signals contain a range of frequencies. that correspond to different types of brain activity. The EEG signal can be broken down into its frequency components using FFT. Better resolution of frequencies in the FFT is made possible by a larger window size (256 in this case). This implies that we will be able to record more precise data regarding the frequency components of our EEG signal. This may be crucial for precisely describing the broad range of content of signals from the EEG associated with various emotions. The amount that the windows overlap is determined by the step size (16 in this case). When windows overlap, brief variations in the EEG signal can be recorded that might otherwise go unnoticed. When the step size is 16, it indicates a large overlap between succeeding windows. Fast Fourier transform technique is used for extracting the features, decreasing them from (40,40,8064) to the final dimensions of (58560,70). This led to faster training and improved accuracy. Figure 4 represents the feature extraction technique.

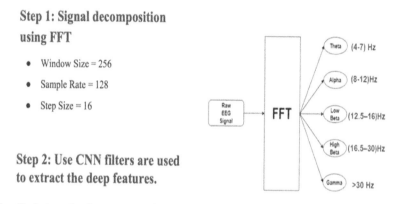

Step 1: Signal decomposition using FFT

- Window Size = 256
- Sample Rate = 128
- Step Size = 16

Step 2: Use CNN filters are used to extract the deep features.

Figure 4.   Technique for feature extraction.

3.2.2 *Emotion classification using CNN model*

Convolutional Neural Network (CNN) models can be used for processing EEG signals to extract meaningful features and perform various tasks, such as classification or feature extraction. The CNN model's approach to emotion classification is shown in Figure 5. Our

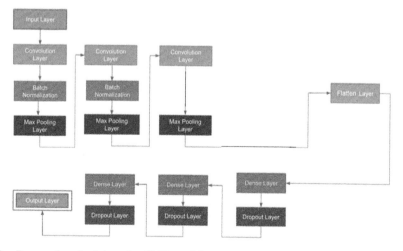

Figure 5.   Proposed methodology for CNN model.

neural network structure was created using the DEAP dataset that incorporates three 1D convolutional layers, followed by three densely connected layers. The final layer utilizes SoftMax activation to classify data into 10 different classes. In the Convolutional neural network model, the main task is to identify which combination of activation functions and optimizers to be used for better results. Softmax is used as the activation function in the final layer, with a Dropout probability of 0.2. After considerable hyperparameter adjustment, the precise number and dimension of filters are determined using Grid Search.

4   RESULTS

There are two phases to the dataset: the first phase includes 80% dataset for training and 20% for testing and 2nd phase includes 75% dataset for training and 25 % for testing purpose to find out how well the suggested method performs. There are 46,848 samples in the training set and 11,712 samples in the test set for an 80-20 ratio while 43,920 data in the training set and 14,640 data in the test set for a 75-25 ratio. For 200 epochs, the proposed CNN model is trained to categorize emotions based on arousal, valence, dominance, and liking. The weights are updated iteratively through the use of the Adam optimization approach during training. Our CNN model performed exceptionally well at generalizing data, obtaining an accuracy of over 80% in each emotion classification. The training and testing accuracy scores for our model are displayed in Figures 6 and 7 with data split in 80-20 ratio. Table 1 shows the performance of our CNN model on each emotion with a 75-25 ratio and 80-20 ratio.

Table 1.   CNN model performance.

Emotion	75-25 ratio	80-20 ratio
Arousal	85.67%	85.94%
Valence	80.73%	82.38%
Dominance	85.23%	87.13%
Liking	88.12%	88.33%

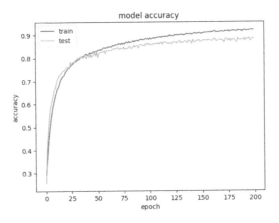

Figure 6.   Training and testing accuracy.

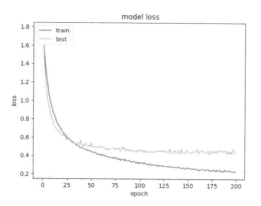

Figure 7. Training and testing loss.

5 CONCLUSION AND FUTURE SCOPE

In this study, we employ the CNN deep learning model to classify emotions based on EEG signals from 32 subjects in a public dataset. This paper makes use of a public dataset known as DEAP. An architecture model for a convolutional neural network with 80-20 train test ratio yields the greatest accuracy of 88.33% and with 75-25 train test ratio gives an accuracy of 88.12% on liking emotion. The model does rather well in generalizing findings for both splits, therefore the change in the train-test split little affects it. Figure 6 depicts the accuracy of training and validation, while Figure 7 displays the loss of training and validation. It is evident that our proposed method has produced superior results than other recent approaches. Deep learning models that are combined or concatenated may improve classification outcomes. A better result has been attained with our recommended methodology. Nevertheless, this proposed approach only addresses one deep learning model. In our upcoming research, hybrid CNN and LSTM models can collaborate on their respective strengths to improve the emotion recognition system's accuracy. We will need to improve its performance in our next future work.

REFERENCES

Alhagry, S., Fahmy, A., & El-Khoribi, R. (2017). Emotion recognition based on EEG using LSTM recurrent neural network. *International Journal of Advanced Computer Science and Applications*, 8. https://doi.org/10.14569/IJACSA.2017.081046.

Asghar, M., F., Khan, M., Amin, Y., & Akram, A. (2020). EEG-based emotion recognition for multi channel fast empirical mode decomposition using VGG-16. *2020 International Conference on Engineering and Emerging Technologies(ICEET), 1–7.* https://doi.org/10.1109/ICEET48479.2020.9048217.

Cannon, W. B. (1927). The James-Lange theory of emotions: a critical examination and an alternative theory. *Am. J. Psychol. 39*, 106–124. doi: 10.2307/1415404

Chao, H., Dong, L., Liu, Y., and Lu, B. (2019). Emotion recognition from multiband eeg signals using capsnet. *Sensors 19*:2212. doi: 10.3390/s19092212.

Cheah, K., Nisar, H., Yap, V., Lee, C., & Sinha, G. (2021). Optimizing residual networks and VGG for classification of EEG signals: Identifying ideal channels for emotion recognition. *Journal of Healthcare Engineering, 2021.* https://doi.org/10.1155/2021/5599615.

Chen, J., Jiang, D., Zhang, Y., & Zhang, P. (2020). Emotion recognition from spatiotemporal EEG representations with hybrid convolutional recurrent neural networks via wearable multi-channel headset. *Comput. Commun., 154,* 58–65. https://doi.org/10.1016/j.comcom.2020.02.051

Chowdary, M., Anitha, J., & Hemanth, D. (2022). Emotion recognition from EEG Signals using recurrent neural networks. *Electronics.* https://doi.org/10.3390/electronics11152387.

Demir, F., Sobahi, N., Siuly, S., & Şengur, A. (2020). Exploring deep learning features for automatic classification of human emotion using EEG rhythms. *IEEE Sensors Journal, 21*, 14923–14930. https://doi.org/10.1109/JSEN.2021.3070373.

F. M. Al-shargie, T. B. Tang, N. Badruddin and M. Kiguchi. (2015) "Mental stress quantification using EEG signals," in *International Conference for Innovation in Biomedical Engineering and Life Sciences*, vol. 56, pp. 15–19, doi: 10.1007/978-981-10-0266-3_4.

Gao, Y., Lee, H., & Mehmood, R. (2015). Deep learning of EEG signals for emotion recognition. *2015 IEEE International Conference on Multimedia & Expo Workshops (ICMEW), 1–5.* https://doi.org/10.1109/ICMEW.2015.7169796.

Koelstra, S., Muhl, C., Soleymani, M., Jong-Seok Lee, , Yazdani, A., Ebrahimi, T., *et al.* (2012). DEAP: a database for emotion analysis; using physiological signals. *IEEE Trans. Affect. Comput. 3*, 18–31. doi: 10.1109/T-AFFC.2011.15.

Kotwal, A., Sharma, V., Manhas, J. (2023). Deep neural based learning of EEG features using spatial, temporal and spectral dimensions across different cognitive workload of human brain: Dimensions, Methodologies, Research Challenges and Future Scope. In: Rathore, V.S., Piuri, V., Babo, R., Ferreira, M.C. (eds) *Emerging Trends in Expert Applications and Security. ICETEAS 2023. Lecture Notes in Networks and Systems*, vol 682. Springer, Singapore. https://doi.org/10.1007/978-981-99-1946-8_7.

Li, X., Song, D., Zhang, P., Yu, Guangliang, Hou, Yuexian, and Hu, Bin (2017). Emotion recognition from multi-channel EEG data through convolutional recurrent neural network. in *IEEE international conference on bioinformatics and biomedicine (BIBM). pp. 352–359. IEEE.*

Pandey, P., and Seeja, K. R. (2019). Subject independent emotion recognition from EEG using VMD and deep learning. *J. King Saud Univ. – Comput. Inf. Sci. 34*, 1730–1738. doi:10.1016/j.jksuci.2019.11.003

Suhaimi, N. S., Mountstephens, J., and Teo, J. (2020). EEG-based emotion recognition: a state-of-the-art review of current trends and opportunities. *Comput. Intell. Neurosci. 2020*, 1–19. doi: 10.1155/2020/8875426

Tripathi, S., Acharya, S., Sharma, R., Mittal, S., and Bhattacharya, S. (2017). "Using deep and convolutional neural networks for accurate emotion classification on DEAP dataset," in *Twenty-ninth IAAI conference.*

Yang, H., Han, J., & Min, K. (2019). A multi-column CNN model for emotion recognition from EEG signals. *Sensors (Basel, Switzerland), 19*. https://doi.org/10.3390/s19214736

Zheng, W, L and Lu, B, L. (2015) "Investigating critical frequency bands and channels for EEG-based emotion recognition with deep neural networks," in *IEEE Transactions on Autonomous Mental Development, vol. 7, no. 3, pp. 162–175, Sept.* doi: 10.1109/TAMD.2015.2431497.

A seven layer DNN approach for social-media multilevel image-based text classification

Veena Tripathi, Navin Mani Upadhyay & Saurabh Sharma
Model Institute of Engineering and Technology, Jammu, J&K, India

ABSTRACT: In recent years, social media platforms have played a significant role in disaster management. To extract valuable insights and understand the meaning behind social media text content, using conventional machine learning methods, text mining solutions have been created. Through these techniques, messages are intended to be categorized into several themes, such counsel and caution. Nevertheless, these methods frequently focus on a single event and have trouble generalizing to cross events classifications. The effectiveness of traditional models trained on historical data for classifying social media messages from future events remains unclear. Our research focuses on the efficacy of a CNN model in classifying Twitter topics across diverse events. Three geotagged datasets from twitter gathered during the COVID-19 pandemic's management are the subject of this study. Two conventional machine learning techniques (LR and SVM) are contrasted with the CNN model's performance. Our experiments conclusively show that the CNN model significantly surpasses the performance of both SVM and LR models in terms of accuracy, regardless of whether the evaluation is based on individual events or across different events. This implies that the CNN model may efficiently categorize messages for impending events and pre-train on Twitter data from past occurrences, improving situational awareness. In conclusion, this study demonstrates the CNN model's capability compared to traditional machine learning techniques for disaster management.

1 INTRODUCTION

The field of natural language processing has made significant progress in the realm of multi-label text categorization. Various practical domains heavily rely on multi-label text categorization to handle the diverse and complex nature of textual data. However, existing approaches that solely depend on word vector representation fall short in fully expressing the semantics of the text, leading to subpar accuracy in classification tasks. This study introduces a technique that enables the classification and search for visually similar texts and images within a dataset based on factors such as form, color, business sector, semantics, general qualities, or a combination thereof, as specified by the user (Dao *et al.* 2023). The proposed approach aims to enhance the efficiency of finding visually similar texts by integrating the results of multiple multilevel deep neural networks, each trained to focus on specific characteristics for conducting similarity searches. However, existing methods often overlook the correlation between labels and the semantic context of each label. To address this limitation, this paper presents a neural network-based approach for text multi-label learning that effectively models the correlation between texts and label vectors (Wang *et al.* 2023) . By capturing the semantic correlation between text and labels, the proposed method surpasses existing approaches, particularly when dealing with large datasets, as evidenced by a series of experiments. The Framework has shown in Figure 1.

DOI: 10.1201/9781003466383-25

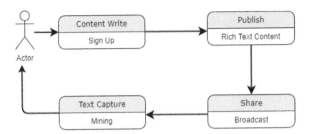

Figure 1.   Framework for identification of multiple labels in social media text.

2 LITERATURE SURVEY

Numerous research publications on the identification of multiple labels in social media have been published during the last several years. Most of the proposed work is based on a hybrid neural network model that combines one or more number of algorithms (Dao *et al.* 2023; Gao *et al.* 2019; Hou *et al.* 2020; Tao *et al.* 2018; Nasersharif *et al.* 2021; Yang *et al.* 2022). Another researcher (Khan *et al.* 2020 and Yuan *et al.* 2021) has worked on DNN based on twitter and abnormal text activities. There findings are also good but not user friendly based on scalability of systems or using the application by more number of users. (Yang *et al.* 2022) has used a regression based text extraction from logo and (Saeedi *et al.* 2013) has improved their concept by using a CNN based regression algorithm.

Strong demand exists in the DNN area for high-performing multi-label classification methods for many purposes, including DNN-based text categorization and detection on social media (Miao *et al.* 2016). The major challenges are based on only a few human annotated textual datasets (Jones *et al.* 2017), and another one is based on lack of high-quality manually labeled data (Takruri *et al.* 2016).

To tackle the aforementioned issues, in this work, we initially introduce the dataset, taken from kaggle[1]. It contains 7K real time cases that were annotated and divided into input features and outcomes. Deep learning and machine learning are two broad categories into which text classification techniques fall. A branch of machine learning called deep learning does away with some data pre-processing. Two deep learning methods that are useful for text categorization are CNN and RNN. To analyze the text content, the study suggests the utilization of multi-channel CNN and BiLSTM for feature fusion. Traditional CNN and RNN models often lose important feature information, leading to suboptimal classification results.

In contrast, the multi-channel CNN (Nasersharif *et al.* 2021) leverages different convolution kernels to extract local semantic features. The proposed text vector representation is obtained by combining these two components and sigmoid activation is used for classification. However, there are still difficulties in investigating object connection, temporal information, and low-level visual notions.

To tackle these issues, the article proposes correlative binary relevance (CBR) classification approach based on DNN and CNN.

3 PROBLEM DEFINATION

Now a days the dataset on different social media platform has increasing day by day which creates problem in identifying the older image text or the real time new image with text (Dao *et al.* 2023). As the size of the dataset expands, so does the level of detail in the architectures. This eventually causes the system accuracy to become saturated, at which point it diminishes to a certain point (say 'E'). The term for this issue is the degradation problem (D_E). The degradation problem may be mathematically represented as follows:

$$E = (y_1, y_2, y_3, \ldots, y_k)^{\mathrm{T}},$$

where E is the degraded final end point, y = {Y_1 U Y_2 U Y_3 ... U Y_k).

$$D_E = x : ||x - \mu|| = \left[\sum_{i=1}^{n} (x_n - \mu)^2 \right]^{\frac{1}{2}} = 1$$

Where D_E is the Degraded end point of the feature selection. x_n is the mean of all the features ($y_1, y_2, y_3, \ldots, y_k$). μ is the mean of all all-selected features having distance of 1.

So, the main work is to reduce the distance vector from the above equation D_E, So that the selection of text from a pre-uploaded image can be verified easily.

4   METHODOLOGY

In this section a detailed methodology has explained in three subsections followed by Experimental setup, proposed algorithm and result and discussion.

4.1   *Experimental setups*

For all the experiments performs in this research work is based on GPUs (NVIDIA GeForce GT 1650RTX). The 1650 RTX uses 16 streaming processors with a frequency of 2.8GHz. It has 1024 MBs of cache and a approx. 128-bit L3 channel. The data transfer rate for the selected GPU is approx. 1000 GB/s. This is fully enabled with parallel programming environment.

4.2   *Dataset selection*

A huge dataset has detected on different repository but for this research a recent dataset with a dimension of 7000, 3462. These datasets contain public comments on social media from the last three-month period specifically tagged with the keywords "disaster" "natural disaster", "god curse", "calamity" etc. The literature has numerous speech, noise, and noisy speech training datasets. The datasets are reviewed in Table 1. The major three selected datasets which were used for this work are: Voice Bank corpus, RassiaSpeech corpus, and 253_Asian corpus. We employed a variety of noise environments, including the Noise corpus and 15 samples from the NOISEX-92 corpus, for everything from pre-processing to model selection. Figure 2 shows a spectrogram of noise added to the selected dataset.

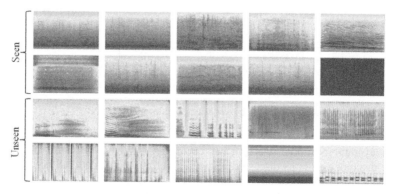

Figure 2.   Spectrogram of noise added to the selected dataset based on seen and unseen the social media post.

Table 1. Selected dataset corpus and their description.

Corpus	Description
Voice Bank	sampled at 48 features
RassiaSpeech	sampled at 16 features
253_Asian corpus	sampled at 48 features
	Noisy Social-Media Datasets
AMI	One hundred hours of real meeting recordings, captured using individual and room-view video cameras as well as close-talking and far-field microphones in three distinct rooms with varying acoustic characteristics [90]
Reverberant Voice bank	A text created artificially for the Voice Bank clean corpus The noisy variant of the corpus is created by artificially adding real noise to the text.
	Noise Datasets
NOISEX-92 UrbanSound8K	noises such as military, pink, white, HF channel, manufacturing, and babble The following noises were reported by 8732,10 urban commenters: air conditioning, car horn, kids playing, barking dogs, drilling, idle motor, gunshot, jackhammer, siren, and street music.

4.3 Training setup

A non-linear CNN has used to train the model. All the dataset has captures with a commonly co-related features having minimum 8 common features. All the images are corrupted with a 0dB noise insertion technique to create a noisy training module. Initially we have selected the 256 samples for verify the level of the selected dataset for the selected model CNN and DNN. The samples are almost 56% overlapped (shown in the Figure 2). The CNN architecture has shown in Figure 3. The features are adjusted based on their magnitude and dimensions for the best fit of model selection.

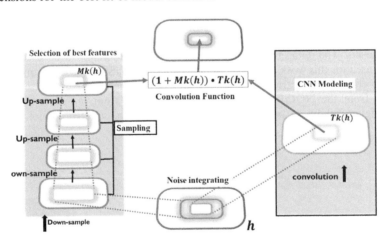

Figure 3. CNN architecture and layering over selected dataset.

A seven-layer DNN model selection has been also applied to measure and predict the correct the distance vector as discussed earlier in section-3. The detailed layer is shown in Figure 4. In Figure 4 a series of convolution layer has shown to invoke the deep neural network. Each layer filters and refine the input images taken from the previous layer. Here the CNN takes only 2D images to extract the text and characters set from the selected image

(Cao *et al.* 2023). This function is done by a series of kernels associated with the GPUs. The last 3 layers are connected to each other only based on a partially fixed overlapped input. This has done only to improve the training speed of the system. All the layers are cleverly depending on the selected sequence number, weight, and measured biases.

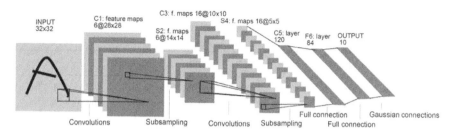

Figure 4.   Seven-layer DNN architecture for character recognition.

By using the last 3 layer of the architecture this work receives an improved evolution metrics which are specific to the selected dataset. The evaluation metrics shows the accuracy of the selected models for classification. All these accuracies signify has calculated by using the following well-known formula:

$$Accuracy\_Model = \frac{TruePositive + TrueNegative}{TruePositive + FalsePositive + FalseNegative + TrueNegative}$$

A batch of 128 epochs were used for 14 times to finalize the minimum mean square error based on only 10 % of testing datasets. The error ratio was only 1.99 % whereas the success rate for each epoch is approx. 97.4%

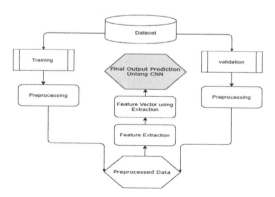

Figure 5.   Proposed framework for data preprocessing, validation, training, model selection architecture.

5   RESULT AND DISCUSSION

In this paper we have performed three types of experiments on all the selected datasets. First-one is on the basis of Objective Evaluation of 2D text based images, second is based on Subjective Evaluation based on 2D text images and last-one is based on hybrid comparison (selection of both objective and subjective). As we have discussed in the previous section all three experiments has divided into 2 parts based on DNN and CNN.

5.1 Objective evaluation method

The result shown in Table 2 is based on three standard measurement techniques: perceptual evaluation technique (PET), short-time objective evaluation technique (STOE) and long-time distortion technique (LTDT). The corresponding bar-graph has shown in Figure 6.

Table 2. PET, STOE, and LTDT, results at high systematic noise integration levels: ~15 dB, ~10 dB, and 20 dB.

Metric		Noisy	A1	A2	A3	A4	A5	A6	A7
PET	high	1.62	1.34	2.14	2.88	2.06	2.92	1.32	1.84
	low	1.18	1.95	1.69	2.18	2.42	2.64	2.86	1.22
	avg	2.19	2.17	2.05	2.53	2.40	1.85	1.39	1.43
STOE	high	1.71	0.80	0.84	0.86	0.83	1.36	1.59	0.68
	low	1.65	0.71	0.98	1.51	0.73	0.69	0.04	0.67
	avg	0.23	0.76	0.34	1.05	0.85	1.02	0.51	0.92
LTDT	high	1.43	2.15	1.14	2.36	2.27	2.98	2.65	2.48
	low	2.93	1.48	1.22	0.76	0.59	1.35	1.66	2.65
	avg	1.02	2.21	2.53	1.50	2.47	1.21	0.55	0.52

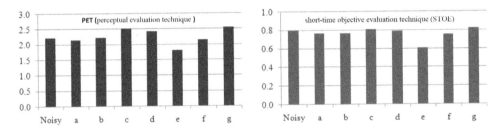

Figure 6. Average PET and STOE results for the seven DNNs levels.

5.2 Subjective evaluation method

There are three types of experiments conducted on all the datasets and found a significant result shown in Figure 7. The performance varies for the experiments in the following manner: 0.1, 0.001, and 0.0001. The performance comparison between all the techniques have also shown in this figure. The DNNs layers has setups according to the highest accuracy. Table 3 shows a comparison between other selected techniques (LR, SVM and CNN model) based on Precision and Recall comparison. Table 3 presents the results of the test, which tallied the SVM, CNN and CSVM algorithms that were applied to the test set. There are a total of 20,000 texts included in the evaluation, with each class representing 2,000 texts' worth of data. According to Table 2 the CNN algorithm and the CSVM algorithm for text classification are quite similar. This is because, after a convolutional neural network has extracted the morphological features, the nonlinear problem is tackled using an SVM multi-classifier, and the dimensional disaster is avoided. When it comes to the categorization of single events, CNN performs better than SVM and LR. The complexity of Twitter data is primarily due to its intricate structure, as tweets often include errors, abbreviations, omissions, and phonetic representations of words. The production of the training set involves a human tagging procedure, which may include varying degrees of uncertainty. CNN, on the other hand, is able to derive high-level abstraction from the tweets, in contrast to SVM, LR, and other machine learning algorithms which are heavily dependent on the validity of the training set. CNN's skill of cross-event topic categorization suggests that the company is able to categorize Twitter topics in the early hours of a new event using an existing knowledgebase developed from previous hurricane events. This knowledgebase was built from previous hurricane occurrences. In case of cross-event classification, CNN outperforms SVM and LR because it can provide a general solution that infers patterns of similarity and dissimilarity across different occurrences. However, because CNN requires more parameters to train than SVM and LR, it takes longer to complete. Due to the constant streaming and rapid accumulation of social media data, using CNN for online learning becomes challenging.

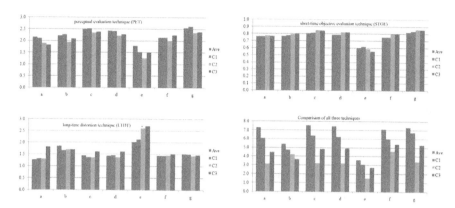

Figure 7.   PET, STOE, LTDT, and comparison of three techniques with average results.

Table 3.   Precision and recall comparison for LR, SVM and CNN model.

Dataset	LR Precision	Recall	SVM Precision	Recall	CNN Precision	Recall	CSVM Precision	Recall
Science and Technology	0.95	0.92	0.93	0.92	0.96	0.95	0.97	0.98
Education	0.75	0.91	0.88	0.90	0.87	0.93	0.94	0.98
Fashion	0.70	0.88	0.78	0.85	0.95	0.96	0.99	0.98
Entertainment	0.94	0.92	0.95	0.92	0.98	0.96	0.99	0.97
Finance	0.92	0.87	0.94	0.90	0.98	0.97	0.98	0.99
Home Furnishing	0.80	0.78	0.82	0.88	0.94	0.94	0.96	0.98
Game	0.96	0.94	0.95	0.94	0.98	0.96	0.96	0.99
House Property	0.94	0.92	0.96	0.94	0.99	0.98	0.98	0.99
Politics	0.77	0.83	0.83	0.91	0.93	0.95	0.95	0.97

6   CONCLUSION

The aim of this work is to examine the classification performance of a CNN-based deep learning model for social media topics in the context of COVID-19 management. Based on the study's specification of the CNN model, we utilized the classifier on three Twitter datasets that were manually labeled. These datasets are categorized into five groups: Aid and Donation, Resources and Infrastructure, Damage and Casualties, Information Sources, and Caution and Advice. Despite the fact that the CNN employed in the tests is a simple model that has not been further adjusted to the job at hand, the findings demonstrate a much superior performance when compared directly to traditional techniques on the dataset that was taken into consideration. In the experiment related to single-event, the CNN classifier consistently outperforms two popular classification techniques, SVM and LR, with precision ranging from 0.63 to 0.98 and recall ranging from 0.60 to 0.99 across various categories. The CNN model demonstrates superior performance compared to SVM and showcasing its effectiveness in handling diverse social media topics.

REFERENCES

Cao, M., Bai, Y., Cao, Z., Nie, L., Zhang, M. (2023). "Efficient image-text retrieval via keyword-guided pre-screening." In *IEEE Transactions on Circuits and Systems for Video Technology.* doi: 10.1109/TCSVT.2023.3339489.

Dao, S. D., Huynh, D., Zhao, H., Phung, D., Cai, J. (2023). "Open-vocabulary multi-label image classification with pretrained vision-language model." In *Proceedings of the 2023 IEEE International Conference on Multimedia and Expo (ICME)*, Brisbane, Australia, pp. 2135–2140. doi: 10.1109/ICME55011.2023.00365.

Gao, Q., & Lim, S. (2019). A probabilistic fusion of a support vector machine and a joint sparsity model for hyperspectral imagery classification. *GIScience & Remote Sensing, 56*, 1129–1147

Hou, Q., Han, M., Qu, F., & He, J.S. (2020). Understanding social media beyond text: a reliable practice on Twitter. *Computational Social Networks*, 8, 1–20.

Jones, P., Sharma, S., Moon, C., & Samatova, N.F. (2017). A network-fusion guided dashboard interface for task-centric document curation. *Proceedings of the 22nd International Conference on Intelligent User Interfaces.*

Khan, M.A., Sarfraz, M.S., Alhaisoni, M.M., Albesher, A.A., Wang, S., & Ashraf, I. (2020). StomachNet: Optimal deep learning features fusion for stomach abnormalities classification. *IEEE Access, 8*, 197969–197981.

Miao, J., Luo, Z., Wang, Y., & Li, G. (2016). Comparison and data fusion of an electronic nose and near-infrared reflectance spectroscopy for the discrimination of ginsengs. *Analytical Methods, 8*, 1265–1273.

Nasersharif, B., & Yazdani, M. (2021). Evolutionary fusion of classifiers trained on linear prediction based features for replay attack detection. *Expert Systems, 38*.

Saeedi, J., & Faez, K. (2013). A classification and fuzzy-based approach for digital multi-focus image fusion. *Pattern Analysis and Applications, 16*, 365–379.

Takruri, M., Rashad, M.W., & Attia, H. (2016). Multi-classifier decision fusion for enhancing melanoma recognition accuracy. *2016 5th International Conference on Electronic Devices, Systems and Applications (ICEDSA)*, 1–5.

Tao, Y., Cui, Z., & Wenjun, Z. (2018). A multi-label text classification method based on labels vector fusion. *2018 International Conference on Promising Electronic Technologies (ICPET)*, 80–85

Wang, L., Chen, X., Ye, Z., Jiang, S. (2023). "Multimodal multi-label text classification based on bidirectional Transformer." In *Proceedings of the 2023 5th International Conference on Robotics and Computer Vision (ICRCV)*, Nanjing, China, pp. 119–123. doi: 10.1109/ICRCV59470.2023.10329110.

Yang, F., & Bansal, M. (2015). Feature Fusion by Similarity Regression for Logo Retrieval. *2015 IEEE Winter Conference on Applications of Computer Vision*, 959–959.

Yang, Y., Hsu, J., Löfgren, K., & Cho, W. (2022). Correction to: Cross-platform comparison of framed topics in Twitter and Weibo: machine learning approaches to social media text mining. *Social Network Analysis and Mining*, 12

Yuan, H., Su, H., Liu, H., Wang, Y., Yang, H., & Hou, J. (2021). PQA-Net: Deep no reference point cloud quality assessment via multi-view projection. *IEEE Transactions on Circuits and Systems for Video Technology, 31*, 4645–4660

Next Generation Computing and Information Systems – Gupta. (Ed.)
© 2025 The Author(s), ISBN 978-1-032-73865-9

CNN based self attention mechanism for cross model receipt generation for food industry

Priya Goyal
Computer Science and Engineering, Manipal University Jaipur, India

Sachin Gupta
Professor, Department of CSE, Maharaja Agrasen Institute of Technology, India

Renato R. Maaliw III
College of Engineering, Southern Luzon State University, Lucban, Quezon, Philippines

Mukesh Soni
Dr. D. Y. Patil Vidyapeeth, Pune, Dr. D. Y. Patil School of Science & Technology, Tathawade, Pune, India

Alikulov Azamat Tuygunovich
The Department of Accounting and Audit, Karshi engineering economics institute, Karshi, Uzbekistan

Aws Zuhair Sameen
College of Medical Techniques, Al-Farahidi University, Baghdad, Iraq

ABSTRACT: Diet management requires keeping track of what you eat. The researchers presented a recipe retrieval technique based on food photos that retrieves the related recipes from the taken images and creates nutritional information accordingly, making recording more convenient. The retrieval of recipes is an example of a cross-modal retrieval challenge. Still, as compared to other challenges, the main challenge is that recipes explain a succession of modifications from raw ingredients to completed goods rather than immediately apparent characteristics. As a result, the model must have a thorough understanding of the raw materials processing process. Additionally, the model improves upon the attention mechanisms of prior techniques to mine the semantics of recipes more effectively. The approach enhances the recall rate of the recipe retrieval task by 22% over the baseline strategy, according to experimental data.

1 INTRODUCTION

People's daily eating habits have a significant impact on people's health. With the continuous improvement of living standards, people pay more and more attention to a healthy diet. As a result, the record of daily dietary intake, as a critical part of dietary management, has received more and more attention (Lokhande *et al.* 2021). Furthermore, for some patients with chronic diseases (such as chronic kidney disease), monitoring nutritional intake is even more crucial due to their need for disease treatment. The basic process of traditional dietary intake recording tools is to predict food categories and raw material types through user food pictures and further estimate their nutrient content. However, due to the variety of styles and appearances of people's homemade foods, the actual results are insufficient to support the inference of nutritional content from food pictures (Salvador *et al.* 2017; Yang

DOI: 10.1201/9781003466383-26

et al. 2021). Semantics and achieve higher accuracy (Ye *et al.* 2020). In contrast to pictures of cooked food, recipes explain the transformation of raw ingredients into finished products rather than explicitly expressing the qualities of the finished product (Lohani *et al.* 2021).

2   DIFFERENT MODAL RETRIEVAL

2.1   *Cross-modal retrieval Model*

The main difficulty of the cross-modal retrieval problem is how to measure the similarity between data of different modalities. Compare. A classic approach in the field of cross-modal retrieval is canonical correlation analysis (CCA). Then CCA has developed many variants, such as kernel CCA (kernel CCA) (Wu *et al.* 2021). Later, with the development of deep neural networks, many works also began to use deep neural networks for latent space mapping (Li *et al.* 2022).

2.2   *Food-related research Model*

Food-related research has always been a concern and valued by researchers. Common research directions include food identification (Yang and Yang 2020), raw material identification (Saini *et al.* 2021), and nutrient composition estimation. Author took the lead in introducing the classic latent space alignment idea of cross-modal retrieval into recipe retrieval, and their model framework. Such a structure enables retrieval not to be limited by fixed food attributes and requires less manual extraction of labels during training.

3   CROSS-MODEL FRAMEWORK

3.1   *Different parts of recipe text*

The data of the two modalities of text and pictures respectively obtain corresponding feature representations through their respective encoding modules. Then the feature vectors of the two modalities enter the joint embedding module to learn the embedding representation in a shared latent space so that the similarity between the matched text and the image is as high as possible. The entire model is trained end-to-end. Next, the different modules in the model will be introduced separately.

3.2   *Different parts of cross-model*

3.2.1   *Transformer*

The Transformer model is a model proposed by Google in 2017 (Phasinam *et al.* 2022). Its original purpose was to replace the recurrent neural network to solve the task of seq2seq in natural language processing. Compared with the recurrent neural network, the Transformer solves two problems: one is to get rid of the serial calculation order and improve its parallel ability; the other is to solve the problem of the recurrent neural network processing long text because the text is too long "Forgetting" phenomenon, which makes it challenging to deal with the issue of long-distance dependence. Model framework has been shown in Figure 1.

The ability of the Transformer model to deal fig with long-distance dependencies mainly depends on the self-attention mechanism it adopts (Li *et al.* 2019). Unlike the recurrent neural network (RNN), which can only learn the previous information through the hidden layer of information transmitted in the last time slice, with the help of the self-attention mechanism, the Transformer model can "globally browse" the input data and find a higher correlation with it.

Chocolate chip
cookies recipe

Lasagna recipe

Figure 1. Model framework.

3.2.2 *Primary cooking method*

Each recipe contains a title. The title of a recipe is usually a high-level summary of the recipe. It usually includes the primary raw materials used in the recipe and the primary cooking method (such as scrambled eggs with tomatoes), and some also include flavors or cuisines (such as spicy shredded potatoes, Sichuan-style spareribs) Wait). Each vector can obtain an n-dimensional hidden layer representation ht:

$$y_u = \text{word2vec}(x_u), u \in [1, U] \tag{1}$$

Then this paper adopts the attention mechanism to calculate the encoding of the entire title. The principle of the attention mechanism is to calculate the weights for different hidden layer representations and obtain the final feature representation by weighted summation, to reflect the importance of other text parts (Tharewal *et al.* 2022).

$$v_u = \tanh\left(X_{att}i_u + c_{att}\right) \tag{2}$$

$$\alpha_u = \frac{\exp\left(v_u^S v_d\right)}{\displaystyle\sum_u \exp\left(v_u^S v_d\right)} \tag{3}$$

Among them, X_{att} and c_{att} are the parameters of the perceptron layer, which can be trained by back propagation. The final weight, α_u obtained by the attention mechanism, is measured by the similarity between v_u and v_d. v_d is called the context vector and represents the preference for the hidden layer vector. The selection method of the environment vector is detailed in section 3.2.5.

Finally, the encoded subtitle of the title (64 dimensions) is obtained by the weighted average of the hidden layer representations of each word in the title:

$$emb_{title} = \sum_t \alpha_u i_u \tag{4}$$

3.2.3 *Material encoding*

Recipes usually have a section listing the various raw materials used, visible (e.g., tomatoes, potatoes, etc.) and invisible (e.g., salt, sugar, etc.).

The raw material encoding emb_{ingr} is the same way as the title encoding. First, the hidden layer representation is obtained by formula (2) and formula (3). Then, the weight is calculated by the formula (4) and formula (5), and finally, the final encoded emb_{ingr} (64 dimensions) is obtained by procedure (6).

3.2.4 *Different operation steps*

The operation steps describe the food preparation process in detail and are the most semantically informative part of the recipe but also the most complex part. This paper does not use the hierarchical encoding method but uses a single Transformer model to get the final general representation from the word vector directly.

3.2.5 *Environment vector*

A similar attention mechanism is used in this paper to improve the v_d(environment vector) selection method of the attention mechanism. Here, v_d is randomly initialized, continuously adjusted during training, and shared among all recipes.

3.2.6 *Encoding process*

The overall encoding of the recipe text is obtained by splicing three parts of the encoding:

$$emb_{recipe} = [emb_{title}, emb_{ingr}, emb_{inst}] \tag{5}$$

3.3 *Processing pictures*

In this paper, the 50-layer residual network version ResNet50 is used as the image encoding module, and the network parameters are initialized with the parameters pre-trained on the ImageNet dataset. After the picture is input into the residual neural network, this paper selects the penultimate layer of the network (i.e., removes the last layer of the SoftMax classification layer) as the encoding image of the picture data, with a dimension of 2 048 dimensions (Wawale *et al.* 2022).

3.4 *Neural network encoding*

In this paper, the corresponding information in the text and the picture is mapped to the same latent space through a layer of fully connected neural network encoding, which are denoted as ϕR and ϕv, respectively. The latent space dimension is 1 024 dimensions.

$$\varnothing_x = \tanh\left(X_x emb_{image} + c_x\right) \tag{6}$$

3.5 *Different loss function*

This paper uses the same loss function as the (Mehbodniya *et al.* 2022), including the retrieval loss M_{retr} and the semantic loss M_{sem}. The total loss function is:

$$M = M_{retr} + \lambda M_{sem} \tag{7}$$

Among them, λ is a hyper parameter that controls the relative size of retrieval loss and semantic loss.

The input to the retrieval loss contains two triplets ϕ_x, ϕ_S+, ϕ_S- and $(\phi_S, \phi_x+, \phi_x)$. The first element of the triplet is the feature representation of the image (ϕ_x) or text (ϕ_S), and the sample is called the anchor sample; the last two elements are the matching sample (positive example) in another modality and the non- Representation of matched samples (negative examples).

3.6 Learning the parameters

Since the overall model is large and has many parameters, learning the parameters of the two encoding modules of text and pictures simultaneously may cause shocks in the training results. Then, during back propagation, only the parameters of the text encoding module and the joint representation learning module are updated until the accuracy converges on the validation set. In the second stage, the parameters of the text encoding module and the joint representation learning module are fixed.

4 DIFFERENT STAGES OF TRAINING

4.1 Domestic eating habits

Each recipe in the LCR dataset includes three parts: title, raw materials, and operation steps. According to statistics, the average title length of each recipe is 6.83 words, and each formula contains an average of 6.89 raw materials and 7.34 operating steps. At the same time, the method of (Saini *et al.* 2021) in this paper, through binary/triple matching.

4.2 Different model

4.2.1 Pytorch framework

The code in this article is implemented using the Pytorch framework. The model was trained using the Adam optimizer with a learning rate of 10-4, an interval α of 0.3, λ of 0.1, and a batch size of 50. The above hyper parameters are selected on the validation set. When computing the loss function, we select positive and negative examples for anchor samples in each batch.

4.2.2 Median rank

In this experiment, two tasks of using images to retrieve recipes (im2recipe) and using recipes to retrieve images (recipe2im) were selected to measure the model's accuracy. In addition, they were tested on three alternative sets of 1 000/5 000/10 000 sizes.

4.3 Different algorithms

This paper selects three models proposed in recipe retrieval in recent years as the benchmark algorithm and compares the accuracy with the proposed algorithm. They are: (1) JNE (joint neural embedding) (Rida, Almaadeed and Almaadeed 2019); (2) ATTEN (Rida *et al.* 2016); (3) Adamine (Khan *et al.* 2020). The experimental results are shown in Figure 2. It can be seen from Table 1 that the accuracy of the model in this paper is significantly better than the three benchmark algorithms under different tasks and different candidate set sizes.

Table 1. Comparison between two attention methods.

Size	Method	im2recipe				recipe2im			
		MedR	R@1	R@5	R@10	MedR	R@1	R@5	R@10
1000	Uc_glob	2	0.4	0.733	0.833	2	0.41	0.734	0.836
	Uc_title	2	0.407	0.733	0.828	2	0.413	0.739	0.834
5000	Uc_glob	6.8	0.193	0.46	0.592	6.4	0.202	0.469	0.597
	Uc_title	6.5	0.194	0.462	0.598	6.1	0.203	0.475	0.606
10000	Uc_glob	12.2	0.132	0.345	0.468	12	0.138	0.354	0.476
	Uc_title	12	0.133	0.344	0.471	11.5	0.139	0.358	0.483

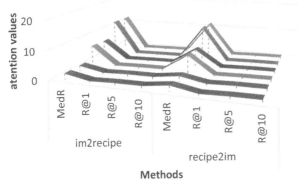

Figure 2. Comparison between two attention methods.

4.4 Different mechanism in the model

While comparing the accuracy, this paper also conducts some control experiments to verify the effectiveness of each module in this model by changing or deleting some modules in the model and comparing the accuracy.

4.4.1 Environment vector

This paper adopts a different mechanism from the (Khan *et al.* 2021; Kapoor *et al.* 2015; Soni and Jain, 2018) to select the environment vector in the attention mechanism (see Section 3.2.5 for details). This paper compares the two selection methods. Among them, Uc_glob is the selection method of the environment vector and Uc_title is the selection method in the model of this paper. (Bhatt and Sharma 2023) uses the same end-to-end learned parameter vectors as the environment vector for all recipes.

4.4.2 Training model

Figure 3 with Table 2 shows the results of training the model on the data of each part of the recipe and its combination (reporting only the accuracy of the im2recipe task on the 10,000-sized candidate set), which can reflect the importance of each part in the recipe.

Table 2. Experiments results of using different recipe parts and their combinations.

Size	Method	R@1	R@5	R@10
1000	JNE	0.239	0.546	0.674
	ATTEN	0.253	0.555	0.69
	Adamine	0.352	0.672	0.771
	Proposed	0.407	0.733	0.828
5000	JNE	0.097	0.277	0.393
	ATTEN	0.105	0.3	0.408
	Adamine	0.162	0.405	0.534
	Proposed	0.194	0.462	0.598
10000	JNE	0.064	0.194	0.286
	ATTEN	0.068	0.202	0.304
	Adamine	0.109	0.295	0.412
	Proposed	0.133	0.344	0.471

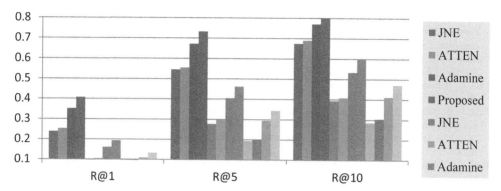

Figure 3. Experiments results of using different recipe parts and their combinations.

5 CONCLUSION

If you would want to keep track of what you eat and how much you consume, this research presents a cross-modal recipe retrieval model that uses a self-attention mechanism to automatically retrieve the recipes that are most relevant to the input food images. The model in this paper uses the Transformer model's self-attention mechanism to capture better the long-distance dependencies appearing in recipes and improves the semantic understanding ability of the model. This research also improves the attention mechanism that is utilized in the classic recipe retrieval model, which better captures the uniqueness across various recipes and further boosts the model's accuracy. Therefore, the current method only takes targeted consideration in the recipe text encoding module, while a more commonly used image processing model is used in the picture encoding module. In the future, the image encoding module will be adjusted and optimized for the recipe retrieval problem.

REFERENCES

Bhatt, M.W. and Sharma, S., 2023. An object recognition-based neuroscience engineering: A study for future implementations. *Electrica*, *23*(2).

Kapoor, L., Bawa, S. and Gupta, A., 2015. Peer clouds: a P2P-based resource discovery mechanism for the Intercloud. *International Journal of Next-Generation Computing*, *6*(3): 153–164.

Khan, F., Khan, A.W., Khan, S., Qasim, I. and Habib, A., 2020. A secure core-assisted multicast routing protocol in mobile ad-hoc network. *Journal of Internet Technology*, *21*(2): 375–383.

Khan, S.U., Khan, A.W., Khan, F., Khan, M.A. and Whangbo, T.K., 2021. Critical success factors of component-based software outsourcing development from vendors' perspective: A systematic literature review. *IEEE Access*, *10*: 1650–1658.

Li, C., Niu, H., Shabaz, M. and Kajal, K., 2022. Design and implementation of intelligent monitoring system for platform security gate based on wireless communication technology using ML. *International Journal of System Assurance Engineering and Management*: 1–7.

Li, T., Guo, Y. and Ju, A., 2019, December. A self-attention-based approach for named entity recognition in cybersecurity. In *2019 15th International Conference on Computational Intelligence and Security (CIS)* (pp. 147–150). IEEE.

Lohani, T.K., Ayana, M.T., Mohammed, A.K., Shabaz, M., Dhiman, G. and Jagota, V., 2023. A comprehensive approach of hydrological issues related to ground water using GIS in the Hindu holy city of Gaya, India. *World Journal of Engineering*, *20*(2): 283–288.

Lokhande, M.P., Patil, D.D., Patil, L.V. and Shabaz, M., 2021. Machine-to-machine communication for device identification and classification in secure telerobotics surgery. *Security and Communication Networks*, *2021*: 1–16.

Mehbodniya, A., Webber, J.L., Neware, R., Arslan, F., Pamba, R.V. and Shabaz, M., 2022. Modified Lamport Merkle Digital Signature blockchain framework for authentication of internet of things healthcare data. *Expert Systems, 39*(10): 12978.

Phasinam, K., Kassanuk, T. and Shabaz, M., 2022. Applicability of internet of things in smart farming. *Journal of Food Quality, 2022*: 1–7.

Rida, I., Almaadeed, N. and Almaadeed, S., 2019. Robust gait recognition: a comprehensive survey. *IET Biometrics, 8*(1): 14–28.

Rida, I., Boubchir, L., Al-Maadeed, N., Al-Maadeed, S. and Bouridane, A., 2016, June. Robust model-free gait recognition by statistical dependency feature selection and globality-locality preserving projections. In *2016 39th International Conference on Telecommunications and Signal Processing (TSP)* (pp. 652–655). IEEE.

Saini, G.K., Chouhan, H., Kori, S., Gupta, A., Shabaz, M., Jagota, V. and Singh, B.K., 2021. Recognition of human sentiment from image using machine learning. *Annals of the Romanian Society for Cell Biology,* pp.1802–1808.

Salvador, A., Hynes, N., Aytar, Y., Marin, J., Ofli, F., Weber, I. and Torralba, A., 2017. Learning cross-modal embeddings for cooking recipes and food images. In *Proceedings of the IEEE conference on computer vision and pattern recognition* (pp. 3020–3028).

Soni, M. and Jain, A., 2018, February. Secure communication and implementation technique for sybil attack in vehicular Ad-Hoc networks. In *2018 Second International Conference on Computing Methodologies and Communication (ICCMC)* (pp. 539–543). IEEE.

Tharewal, S., Ashfaque, M.W., Banu, S.S., Uma, P., Hassen, S.M. and Shabaz, M., 2022. Intrusion detection system for industrial Internet of Things based on deep reinforcement learning. *Wireless Communications and Mobile Computing, 2022*: 1–8.

Wawale, S.G., Shabaz, M., Mehbodniya, A., Soni, M., Deb, N., Elashiri, M.A. and Naved, M., 2022. Biomedical Waste Management Using IoT Tracked and Fuzzy Classified Integrated Technique. *Human-centric Computing and Information Sciences, 12*(32).

Wu, C., Lu, P., Xu, F., Duan, J., Hua, X. and Shabaz, M., 2021. The prediction models of anaphylactic disease. *Informatics in Medicine Unlocked, 24*: 100535.

Yang, J. and Yang, J., 2020, October. Aspect based sentiment analysis with self-attention and gated convolutional networks. In *2020 IEEE 11th International Conference on Software Engineering and Service Science (ICSESS)* (pp. 146–149). IEEE.

Yang, M., Kumar, P., Bhola, J. and Shabaz, M., 2021. Development of image recognition software based on artificial intelligence algorithm for the efficient sorting of apple fruit. *International Journal of System Assurance Engineering and Management*: 1–9.

Ye, R., Wang, W., Ren, Y. and Zhang, K., 2020, October. Bearing fault detection based on convolutional self-attention mechanism. In *2020 IEEE 2nd International Conference on Civil Aviation Safety and Information Technology (ICCASIT* (pp. 869–873). IEEE.

Next Generation Computing and Information Systems – Gupta. (Ed.)
© 2025 The Author(s), ISBN 978-1-032-73865-9

Re-weighting X-Ray images classification using multi-class deep learning CNN models

Saurabh Dhanik
School of Computing, Graphic Era Hill University Bhimtal Campus, Uttarakhand, India

Bhargavi Posinasetty
Masters in Public Health, The University of Southern Mississippi, Hattiesburg, USA

Mohammed Wasim Bhatt
Model Institute of Engineering and Technology, Jammu, J&K, India

Renato R. Maaliw III
College of Engineering, Southern Luzon State University, Lucban, Quezon, Philippines

Haewon Byeon
Department of Digital Anti-Aging Healthcare, Inje University, Gimhae, Republic of Korea

Aws Zuhair Sameen
College of Medical Techniques, Al-Farahidi University, Baghdad, Iraq

ABSTRACT: The problem of long-tailed distribution of samples in the data set would considerably affect the detection model's performance in medical picture identification due to the imbalance in the number of samples of each kind in the data set. Their goal was to study the phenomenon of over-fitting. When trained on multi-class unbalanced data sets, re-weighting the network improves the initial loss function. To emphasize the image's intrinsic intricacies, the CLAHE approach is employed to preprocess it. The number ResNext50 has been chosen. As the experimental data set for a feature extraction network, utilize the COVID-chest x-ray data set. Experiments were used to evaluate the model's accuracy, precision, recall, F1 value, and efficiency.

1 INTRODUCTION

Deep learning and neural networks have been widely used in remote sensing image recognition, video surveillance, medical image detection and other fields. However, since most medical image datasets have small sample sizes, high sample acquisition costs, manual labeling costs, and severe category inequality, deep convolution neural network models require many labeled training samples to ensure their final performance (Gudmundsson, El-Kwae and Kabuka 1998). Therefore, when using deep learning methods for medical image detection, it is necessary to process the data itself or the neural network itself to a certain extent to make up for the problems caused by the insufficiency of medical image data itself. (Nikolaev *et al.* 2021) proposed the ChestXNet fine-grained diagnostic network, which optimizes the neural network by introducing an attention mechanism module and a bidirectional parallel global maximum average pooling structure between channels (Islam and Mondal 2019). A deep separable dense network DWSDenseNet is proposed using 2905 COVID-19 chest X-ray images as an experimental dataset for binary detection experiments (Singh and Dutta 2014). When dealing with real-world application problems, the datasets that can be collected often have class imbalances (Ye *et al.* 2010).

2 RELATED RESEARCH

2.1 Reweighting

The loss function computation is versatile and easy, allowing it to better respond to the changing trend of picture characteristics in various data sets. As a result, reweighting is one of the most often used solutions to the long-tail issue in most research and experiments (Soni et al. 2022; Soni and Kumar 2020; Bharti et al. 2021). The entire loss function's balance of the head classification loss function may be decreased so that the network can classify the tail during the training phase, avoiding the overfitting issue of head categories caused by the neural network's long-tailed distribution while learning the dataset (Ajaz et al. 2022).

2.2 ResNext network

ResNext (Mehbodniya et al. 2021) is a particular residual network which belongs to the combination of ResNet network (Rida et al. 2020) and the Inception network (Ayvaz et al. 2022) etwork block structure is composed of the simplified Inception structure block plus the skip layer shortcut in ResNet (short-cut), which can reduce the hyper parameters of the neural network while ensuring the network's performance. The web design breaks away from the fixed mindset of improving network performance by deepening and widening the network layers, increasing the number of paths with the same topology, and utilizing the split-transform-merge strategy for group convolution in a simple and scalable way.

3 PROPOSED METHODS

To adapt to the disease detection task of multi-class chest X-ray images, this paper uses ResNext50 as the feature extraction network, uses the re-weighting method to improve the cross-entropy loss function, and uses the CLAHE algorithm for image preprocessing.

3.1 Improvement of loss function by reweighting method

This study recommends adding reweighting to improve the original loss function to alleviate the overfitting issue caused by the data's long-tailed distribution (Laila et al. 2022). The two weights are calculated using two distinct weight calculation algorithms, then linearly blended, and the resultant weight value is used to weight the loss function. The following is the general formula for reweighting the cross-entropy loss function:

$$N_{ce} = weight\left(-In\frac{\exp(A_j)}{\sum_{i=1}^{O}\exp(A_i)}\right) \tag{1}$$

Assume the neural network's input is the sample x, where O is the length of the vector generated by the web's final layer, i.e., the number of categories in the sample, A j and I are the jth and i values of the output vector, and weight is the loss function. As a result, based on the re-weighting concept, this study provides the first weight-setting technique as follows:

$$b_i = In\frac{\sum_{j=1}^{O}OA_i}{O_j} \tag{2}$$

Ni is the number of samples in the dataset from the i-th class, ai is the weight value of the i-th kind of samples, n is the number of courses in the model, and a = [a1, a2, an]. When the base n is bigger than 1, the logn X function is a monotonically growing function due to its monotonic character. When the value of X is greater, on the other hand, the process trend is milder. As a result, the weight value is calculated using the ln X function. The calculation method of the second weight is obtained through the definition of the adequate sample size, and the (Tufail *et al.* 2021) has its specific inference process. The formula for calculating this weight is:

$$c_i = \frac{1 - \Omega}{1 - \Omega^{n_i}} \tag{3}$$

Among them, bi is the weight value of the i-th type of samples, ni is the number of samples of the i-th type of samples, and set $\Omega = 0.99$ according to the parameters of the (Ahmad *et al.* 2022). The formula for applying this weight to the loss function is:

$$M_2 = b \left(-In \frac{\exp(A_j)}{\sum_{i=1}^{O} \exp(A_i)} \right) \tag{4}$$

Formula (2) is used to compute the weight by dividing the number of samples of each kind by the total number of pieces. A weight that makes the model more stable. Both systems of weighing have merits and cons. As a result, the two weights a and b are blended linearly. The following is the enhanced cross-entropy loss function:

3.2 *Using ResNet50 as a feature extraction network*

Wang et al. demonstrated the feasibility of neural network application in X-ray chest disease detection and proposed the COVID-Net network. The network model adds a depth wise convolution layer (depth wise convolution, DWConv) to the ordinary neural network model.

Next, this paper conducts experiments on the accuracy of the models of COVID-Net and ResNext50, respectively, detecting X-ray chest images and classifying images into three categories as the target in the literature accuracy, as shown in Figure 1. ResNext50 can have a similar performance to COVID-Net while reducing the network's number of parameters and computation. To sum up, ResNext50 is chosen as the feature extraction network for this model.

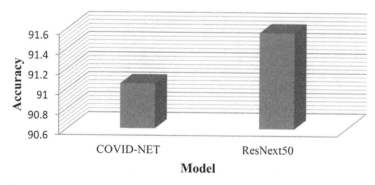

Figure 1.   Comparative accuracy of network under same task.

3.3 *CLAHE algorithm for X-ray image preprocessing*

The CLAHE algorithm, also known as contrast limited adaptive histogram equalization (contrast limited adaptive histogram equalization) (Gupta and Awasthi 2009), uses this method to process X-ray images, highlighting the image's internal details and suppressing noise, and better preserve X-ray images Lesion information in (Islam and Mondal 2019). The implementation steps of the algorithm are as follows:

1) Divide the image into three layers, R, G, and B, convert them into three single-channel images, and divide them into multiple image squares of equal size and non-overlapping.
2) Calculate the grayscale histogram of each image square:

$$AVG(M) = \frac{M_x - M_y}{M_g} \tag{5}$$

Among them, AVG(N) represents the average number of pixels of the grey level, M_x and M_y represent the number of pixels in the X-axis and Y-axis directions, respectively, and M_g is the number of grey levels in the image block.

3) Set a clipping coefficient O of the number of grey-scale pixels, and calculate the clipping threshold O_k
4) Equalize the histogram of each image square.

Two image data are randomly selected from the covid-chest x-ray-dataset dataset and preprocessed using the CLAHE algorithm. The results are shown in Figure 2. The left side is the original image of the chest X-ray, and the right side is the result of the image preprocessing through the CLAHE algorithm. It can be seen that the internal light-dark contrast is more apparent, and the details are more prominent.

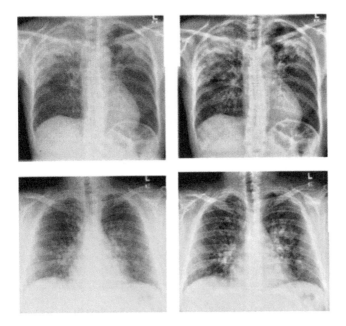

Figure 2.   Enhancement of chest X-ray images with CLAHE algorithm.

4   EXPERIMENTS SETUP & RESULT ANALYSIS

4.1   *Experimental platform*

This experiment is carried out in the Linux system. The Ubuntu version number of the system is 18.04, the code is written in the python3.6 environment built in Anaconda, and the algorithm model is uniformly constructed using the PyTorch framework.

4.2 Algorithm model metrics

The data set itself has an unbalanced distribution of class samples in the training set and the test set. Therefore, this paper uses four metrics to assess the performance of the proposed model, namely Accuracy, Precision rate, Recall rate, and F1-measure. As shown in Table 1, it is a confusion matrix.

Table 1.   Confusion matrix.

	Actual positive	Actual negative
Predict positive examples	TP, (Positive class is judged as the positive class)	FP, (Negative class is judged as the positive class)
Predict Negatives	FN, (Positive class is judged as negative class)	TN, (Negative class is judged as the negative class)

4.3 Data preparation

The dataset used in this article is the open-source dataset covid chest xray dataset on GitHub. The sample distribution of this dataset has the characteristics of a typical long-tailed distribution. Before this experiment, for the consideration of dividing the training set and the test set, the categories with samples less than 5 in the original data set were firstly eliminated. Secondly, to better construct a long-tailed distribution data set so that the data distribution conforms to the "28 law", 30 samples of the standard category were randomly divided into the training set from the RSNA pneumonia data set Kaggle competition.

4.4 Results analysis

In this paper, eight experiments are designed to evaluate the model's performance and the effectiveness of the reweighting method applied to imbalanced medical image datasets by comparing the experimental results.

The experimental results are shown in Table 2. Among them, "+a" indicates that the detection model uses the reweighted loss function in formula (3), "+b" suggests that the detection model uses the loss function in formula (5), and "+a+b" indicates that the detection model the model uses the loss function in Equation (6). This table shows that the reweighted classification detection network has a certain degree of improvement in its four classification evaluation indicators.

Table 2.   Accuracy, precision, recall and F1 value of each experiment.

Experiment	Accuracy	Precision	Recall	F1
① ResNext50	71.03	46.89	43.47	43.93
② ResNext50+CLAHE	72.9	48.91	43.3	44.66
③ ResNext50+CLAHE+a	75.7	60.18	51.72	53.77
④ ResNext50+CLAHE+b	74.77	59.02	47.11	51.5
⑤ ResNext50+CLAHE+focal loss	71.96	55.79	50.63	52.21
⑥ ResNext50+CLAHE+ LDAM loss	70.84	47.25	52.95	46.53
⑦ ResNext50+CLAHE+Seesaw loss	42.59	7.68	10.59	8.16
⑧ ResNext50+CLAHE+a+b	77.57	61.79	53.36	56.35

Figure 3 with Table 3 shows the accuracy changes in the last 100 epochs in the test set for Experiment ③, Experiment ④, and Experiment 8, and Figure 4 with Table 4 shows the accuracy changes in all ages. It can be seen that although the test accuracy rate of experiment

③ is higher than that of experiment ④, the stability is weaker, and the accuracy rate fluctuates seriously. The test accuracy trend of experiment 8 combines the advantages of the first two. When the accuracy reaches a higher value, the convergence of the accuracy is relatively gentle, and the stability is increased.

Table 3. Accuracy changes of experiments ③, ④ and ⑧ in the last 100 epochs.

Serial	Experiment ③	Experiment ④	Experiment ⑧
410	0.75	0.86	0.79
420	0.74	0.84	0.78
430	0.76	0.85	0.782
440	0.73	0.83	0.79
450	0.76	0.85	0.789
460	0.73	0.86	0.77
470	0.77	0.88	0.772
480	0.77	0.88	0.772

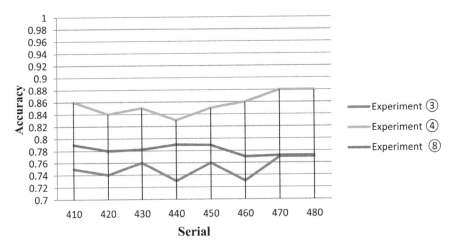

Figure 3.   Accuracy changes of experiments ③, ④ and ⑧ in the last 100 epochs.

Table 4.   Accuracy variation of experiments ③, ④ and ⑧ in all epochs.

Serial	Experiment ③	Experiment ④	Experiment ⑧
410	0.65	0.76	0.69
420	0.64	0.74	0.68
430	0.66	0.75	0.682
440	0.63	0.73	0.69
450	0.66	0.75	0.689
460	0.63	0.76	0.67
470	0.67	0.78	0.672
480	0.67	0.78	0.672

At the same time, to further prove the advanced nature of this method, this paper compares it with the focal loss algorithm (Nair *et al.* 2022), LDAM loss algorithm (Gomathi *et al.* 2020) and Seesaw loss algorithm, in the same environment. The experimental results correspond to Experiment ⑤, Experiment ⑥ and Experiment ⑦. These four experimental

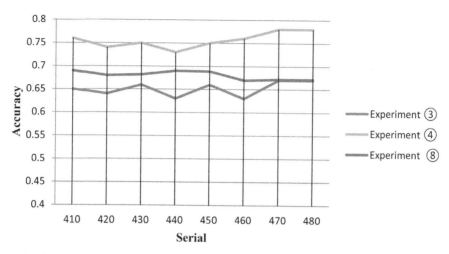

Figure 4.   Accuracy variation of experiments ③, ④ and ⑧ in all epochs.

evaluations are compared by conducting multi-class detection experiments on the covid-chest x-ray-dataset dataset. The comparison of experiment ⑤ and experiment 8 can confirm that this paper's reweighting method is more suitable for the multi-class X-ray chest detection task than the focus loss algorithm proposed. In experiment ⑤, the precision, recall and F1 value of the model after using the focal loss algorithm were improved by 6.88 percentage points, 7.33 percentage points and 7.55 percentage points, respectively, compared with the data in experiment 2, which proves that the focal loss can be effective to a certain extent. Please solve the problem of over fitting caused by long-tailed distribution, but its accuracy is not high, and there is even a downward trend. Combined with the analysis of (Yao *et al.* 2021) and (Bhatt and Sharma 2023), the use of a focal loss algorithm in the literature can effectively improve the model's performance.

5   CONCLUSION

This paper uses deep learning methods to detect disease on multi-class imbalanced X-ray image datasets. Aiming at the problem of the long-tailed distribution of sample data in the dataset, the model uses a reweighting the loss function to counteract the impact of the unbalanced long-tailed distribution of data classes on the detection model. The experimental model uses ResNext50 as the detection network and uses the CLAHE algorithm to pre-process the dataset images. The comparative illustration of 8 experiments confirmed that the reweighting proposed in this paper is feasible and effective for dealing with long-tail problems, and the effect is better than other loss functions. The paper's experimental results are limited by the number of samples in the dataset. To further improve the evaluation values and make the model better adapted to the multi-category X-ray chest disease detection task, it is still necessary to invest more data in the neural network training stage. Improving model performance in limited data is the focus of future research.

REFERENCES

Ahmad, S., Ullah, T., Ahmad, I., Al-Sharabi, A., Ullah, K., Khan, R.A., Rasheed, S., Ullah, I., Uddin, M.N. and Ali, M.S., 2022. A novel hybrid deep learning model for metastatic cancer detection. *Computational Intelligence and Neuroscience, 2022.*

Ajaz, F., Naseem, M., Sharma, S., Shabaz, M. and Dhiman, G., 2022. COVID-19: Challenges and its technological solutions using IoT. *Current Medical Imaging*, *18*(2): 113–123.

Ayvaz, U., Gürüler, H., Khan, F., Ahmed, N., Whangbo, T. and Bobomirzaevich, A., 2022. Automatic speaker recognition using mel-frequency cepstral coefficients through machine learning. *CMC-Computers Materials & Continua*, *71*(3).

Bharti, R., Khamparia, A., Shabaz, M., Dhiman, G., Pande, S. and Singh, P., 2021. Prediction of heart disease using a combination of machine learning and deep learning. *Computational intelligence and neuroscience, 2021*.

Bhatt, M.W. and Sharma, S., 2023. An object recognition-based neuroscience engineering: A study for future implementations. *Electrica*, *23*(2).

Gomathi, S., Kohli, R., Soni, M., Dhiman, G. and Nair, R., 2020. Pattern analysis: Predicting COVID-19 pandemic in India using AutoML. *World Journal of Engineering*, *19*(1): 21–28.

Gudmundsson, M., El-Kwae, E.A. and Kabuka, M.R., 1998. Edge detection in medical images using a genetic algorithm. *IEEE transactions on medical imaging*, *17*(3): 469–474.

Gupta, A. and Awasthi, L.K., 2009, December. Peer enterprises: A viable alternative to Cloud computing?. In *2009 IEEE International Conference on Internet Multimedia Services Architecture and Applications (IMSAA)* (pp. 1–6). IEEE.

Islam, S.M. and Mondal, H.S., 2019, July. Image enhancement based medical image analysis. In *2019 10th International Conference on Computing, Communication and Networking Technologies (ICCCNT)* (pp. 1–5). IEEE.

Laila, U.E., Mahboob, K., Khan, A.W., Khan, F. and Taekeun, W., 2022. An ensemble approach to predict early-stage diabetes risk using machine learning: An empirical study. *Sensors*, *22*(14): 5247.

Mehbodniya, A., Alam, I., Pande, S., Neware, R., Rane, K.P., Shabaz, M. and Madhavan, M.V., 2021. Financial fraud detection in healthcare using machine learning and deep learning techniques. *Security and Communication Networks, 2021*: pp.1–8.

Nair, R., Vishwakarma, S., Soni, M., Patel, T. and Joshi, S., 2022. Detection of COVID-19 cases through X-ray images using hybrid deep neural network. *World Journal of Engineering*, *19*(1): 33–39.

Nikolaev, A.V., De Jong, L., Weijers, G., Groenhuis, V., Mann, R.M., Siepel, F.J., Maris, B.M., Stramigioli, S., Hansen, H.H. and De Korte, C.L., 2021. Quantitative evaluation of an automated cone-based breast ultrasound scanner for MRI–3D US image fusion. *IEEE transactions on medical imaging*, *40*(4): 1229–1239.

Rida, I., Al-Maadeed, N., Al-Maadeed, S. and Bakshi, S., 2020. A comprehensive overview of feature representation for biometric recognition. *Multimedia Tools and Applications*, *79*: 4867–4890.

Singh, A. and Dutta, M.K., 2014, November. A blind & fragile watermarking scheme for tamper detection of medical images preserving ROI. In *2014 International Conference on Medical Imaging, m-Health and Emerging Communication Systems (MedCom)* (pp. 230–234). IEEE.

Soni, M. and Kumar, D., 2020, September. Wavelet based digital watermarking scheme for medical images. In *2020 12th international conference on computational intelligence and communication networks (CICN)* (pp. 403–407). IEEE.

Soni, M., Gomathi, S., Kumar, P., Churi, P.P., Mohammed, M.A. and Salman, A.O., 2022. Hybridizing convolutional neural network for classification of lung diseases. *International Journal of Swarm Intelligence Research (IJSIR)*, *13*(2): 1–15.

Tufail, A.B., Ma, Y.K., Kaabar, M.K., Martínez, F., Junejo, A.R., Ullah, I. and Khan, R., 2021. Deep learning in cancer diagnosis and prognosis prediction: a minireview on challenges, recent trends, and future directions. *Computational and Mathematical Methods in Medicine, 2021*.

Yao, Q., Shabaz, M., Lohani, T.K., Wasim Bhatt, M., Panesar, G.S. and Singh, R.K., 2021. 3D modelling and visualization for vision-based vibration signal processing and measurement. *Journal of Intelligent Systems*, *30*(1): 541–553.

Ye, S., Zheng, S. and Hao, W., 2010, October. Medical image edge detection method based on adaptive facet model. In *2010 international conference on computer application and system modeling (ICCASM 2010)* (Vol. 3, pp. V3-574). IEEE.

Next Generation Computing and Information Systems – Gupta. (Ed.)
© 2025 The Author(s), ISBN 978-1-032-73865-9

Machine learning-based approaches for the diagnosis and detection of brain stroke

Trupti Panchal & Sugayaalini Ramasubramanian
Department of Mechanical Engineering, Pandit Deendayal Energy University, Gandhinagar, Gujarat, India

Yogesh Kumar
Department of Computer Science and Engineering, Pandit Deendayal Energy University, Gandhinagar, Gujarat, India

ABSTRACT: This paper presents an in-depth analysis on brain stroke prediction using machine learning based approaches for prediction and classification. We take on a wide range of data sources, including demographic data, medical records, lifestyle conditions, and clinical measures for a prediction model. We managed to determine significant patterns and risk factors related to stroke owing to the dataset utilised in this study. Feature selection techniques are utilised to identify the most relevant predictors, enhancing the efficiency and interpretability of the model. In addition, we study alternative data preparation methods, such as managing missing values, outlier identification, and feature scaling, to ensure data quality and model resilience.

1 INTRODUCTION

A blockage in the blood circulation or a breakage of blood vessels leading to leakage particularly in the brain is known as a brain stroke. It is a life-threatening condition as once occurred there are a lot of chances to have another. Blood and oxygen are unable to penetrate the brain due to the blood clots that have developed. Usually, these blood clots originate in arteries filled with lesions or constricted by fatty deposits. People who have high blood pressure (hypertension), high cholesterol (hyperlipidaemia), type 2 diabetes, a history of heart disease, or abnormal heart rhythms such as atrial fibrillation have a greater probability to have a stroke. Artificial intelligence techniques used in healthcare for disease prediction. Artificial intelligence strives to emulate cognitive processes in human beings.(Shah et al. 2022) A standard evolution in healthcare has been resulted by the increasing accessibility of health-related data and a rapid growth in analytics tools. For processing unstructured data, favoured AI compromises of natural languages and for structured data machine learning techniques has utilized. The three primary disease categories that utilize AI approaches are neuroscience, tumors, and cardiovascular. (Jiang et al. 2017)

The artificial intelligence-based learning approaches are used for prediction and classifications. These methods are applied to the diagnosis and prognosis of a wide range of illnesses, especially those whose diagnosis depends on signal analysis or imaging. AI can also be used to identify people or environments that are more likely to contract illnesses or carry out dangerous tasks. Medical visual report analysis using machine learning (ML) techniques has been improved as a result of advanced algorithms that make it possible for the automated acquisition of increased features. (Ghaffar Nia et al. 2023)

DOI: 10.1201/9781003466383-28

1.1 *About the paper*

We have used the dataset "Brain_Stroke Data" which has a usability of 1.76. It's a cleaned dataset with no duplicate values and inconsistencies. In this article, Various algorithms based on machine learning have been employed for predicting the disease caused by certain medical problems such as brain stroke, hypertension, and heart diseases and by smoking. We have used two alternative scaling strategies, Minmax and Standard, to enhance the prediction and to obtain the right outcomes. Numerous performance criteria have been used to examine the conclusions, including accuracy, loss, precision, recall and F1-score.

1.2 *Organization of the paper*

The remaining part of this paper is laid out below. The research methodology is explained in the second section. The method used to choose the literature is covered in this section. Cancer consequences and clinical Applications are outlined in the third section. The reported research in the fourth section includes the importance and role of deep learning in the field of tumors and malignancies. The comparison analysis is further explored in this part, along with the difficulties of the ongoing task and performance evaluation using several other metrics. All these investigations are covered in the fifth section with a detailed reasoning. The last part, or the sixth section, enlightens us on future directions.

2 LITERATURE SURVEY

This section discusses the work done by different researchers in the field of healthcare and disease prediction using machine and deep learning-based approaches.

(Bentley *et al.* 2014) examined the hypothesis that patients who would develop SICH could possibly be identified using machine learning regarding CT scans rather than demonstrate clinical improvement with no bleeding after taking tPA.. The scientists have concluded that CT images can be used for automated and improved prediction of SICH following thrombolysis by the application of machine learning to acute stroke. They have proposed that such techniques should be evaluated with larger cohorts and the inclusion of sophisticated imaging.

Having a solid customize for long-term stroke risk prediction, (Dritsas and Trigka 2022) have examined a few models that were developed and evaluated using machine learning. This research's most important contribution is a staking approach that works well and is backed by several metrics, such as accuracy, precision, recall, AUC, and F-measures. The results of the investigation show that stake classification outperforms other methods, with an accuracy, precision, recall, and AUC of 99%, as well as F-measures of 98.9%. (Sailasya & Aruna Kumari, n.d.) have implemented five distinct models to predict the possibility of stroke in individuals using a variety of psychological parameters for prediction and detection. Naive Bayes, with an accuracy of almost 82%, was the method with the highest accuracy for this classification.

For the accurate stroke prediction, (Dev *et al.* 2022) thoroughly exclaimed numerous components in electrical health records. They determined the variables that are essential for stroke prediction used by a variety of statistical methods and principal components evaluation. They concluded that the most crucial cause for stroke prediction was heart disease, age, hypertension, and average glucose level. In addition, when compared to using every input characteristic and other techniques, the perception neural level used by the four parameters possessed the greatest accuracy and lowest miss rates.

In their research, the (Emon *et al.* 2020) recommended implementing a number of machine learning techniques to the early detection of stroke issues exhibiting symptoms related to age, smoking, body mass index, average blood sugar, heart disease, hypertension, and average glucose levels. These effective characteristics have been used to develop ten different classifiers for hypertension and for prediction the stroke used XGBoost classifier. The output of the basic classifiers was combined using a weighted voting approach to get the highest accuracy. Furthermore, the weighted voting classifier in this study strikes the base classifiers and achieves a 97% accuracy rate.

In this paper, (Kaur *et al.* 2022) managed time series-based prediction and arrived at a helpful conclusion. This investigation proposed time series-based techniques including FFNN, LSTM, GRU and biLSTM. Although all the computational algorithms utilised in this paper efficiently detect the limitations associated with initial stroke identification, GRU achieved the highest accuracy of 95.6%, subsequent to FFNN with 83%, LSTM with 87%, then biLSTM with 91%. The result of the experiment allowed for the measurement of neural activity to identify stroke symptoms. The researchers concluded that it can undoubtedly help identify stroke early on and save patient's lives.

In this study, (Chang *et al.* 2019) employed a method for anticipating results determined by symptoms of hypertension found during physical examinations. The result prediction for patients was divided into two components in this study. The initial step was identifying the main straits from the patient's extensive physical examination signs.

(AlKaabi *et al.* 2020) intended to build assess prediction models in order to recognize individuals who are most likely to develop hypertension without requiring intrusive clinical testing. The construction and evaluation of prediction models of hypertension involved the use of fivefold cross-validation and three supervised machine learning algorithms.

(Sharma *et al.* 2020)study attempted to develop machine learning models for heart disease utilising appropriate inputs in their study. They worked with a standard dataset from UCI's heart disease prediction for this research, which comprises of 14 distinct heart disease-related parameters. Results indicated that Random-Forest outperformed other ML techniques in terms of prediction accuracy while consuming less time.

3 PROPSED METHODOLOGY

Figure 1 shows the methods used in predicting brain stroke by machine algorithms and data science. It involved analyzing crucial healthcare details and developing a model that can detect potential stroke risks

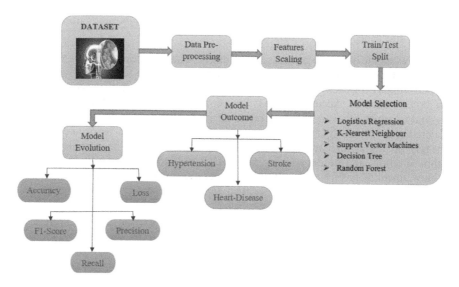

Figure 1.   System design for brain stroke prediction.

Four libraries—Panda, NumPy, Seaborn, and Matplotlib.pyplot—have been utilized. Then, using sklearn preprocessing, we changed all the text values in the subsequent columns to float number values.

The first method consists of Data Processing and Scaling which is cleaning and pre-processing the data to remove any inconsistencies, missing values, or outliers. We utilized Minmax and Standard scalers, which are two different sorts of scalers.

Predictive models can be constructed as machine learning algorithms are able to recognize correlations and pattern recognition in data. These models can determine an individual's probability of having a stroke by accounting for a wide range of parameters, including medical history, lifestyle decisions, demographic data, and genetic predispositions. The second method involves selection of appropriate model algorithms for brain stroke prediction. In this paper, we have selected 5 models namely:

(1) Logistics Regression:
It models the likelihood of a binary result using the logistic function, also known as the sigmoid function. It's a particular kind of generalized linear model (GLM), which can be expanded to solve classification issues involving several classes.

(2) Support-Vector Machines:
SVMs work especially effectively in applications where the objective is to locate the optimum hyperplane in a high-dimensional space to divide data points of distinct classes. The data points that are closest to the hyperplane are referred to as "support vectors" since they are essential in establishing the decision boundary.

(3) K-Nearest Neighbour:
K-NN is an instance-based, non-parametric learning algorithm. Based on how similar data points are to one another in a feature space, this straightforward yet efficient method generates predictions. Assist Vendor Devices

(4) Decision Tree:
This is an easy-to-understand model that divides the data into subsets recursively according to the values of input features. The result of these splits is a structure that resembles a tree, with each internal node denoting a decision made on the basis of a feature, each branch representing an option that may be made, and each leaf node representing the final predicted class or value.

(5) Random Forest:
The basic principle underlying Random Forest, which is well-known for its accuracy and resilience, is to train several decision trees and then combine their predictions to get more accurate and dependable forecasts.

We categorized the data into training and testing sets, and the results associated with brain strokes induced by hypertension, strokes, and heart disease were retrieved.

The performance of each model with testing sets was subsequently checked using evaluation criteria such as Accuracy, Loss, Precision, Recall, and F1 Score. However, the quality and quantity of available data make an important part in how successfully these models predict the future. It is vital to keep in mind that these models' expected effectiveness heavily rely on the calibre and quantity of the available data. The parameter selection, algorithm selection, and model training approaches utilised directly influence the way the algorithms' function. To ensure the utility and generalisation of such models, it is essential to periodically validate and enhance them using a variety of representative datasets.

4 RESULTS AND DISCUSSION

This section explores the results and outcomes of utilizing machine learning in brain stroke prediction. It delves into the key aspects of how these models work, the data sources they rely on, and the implications for healthcare and patient outcomes.

Table 1.   Hypertension prediction.

Models	Hypertension prediction				
	Accuracy	Loss	Precision	Recall	F1-score
Logistics Regression	0.899	0.0008	0.90	1.00	0.94
Support-Vector Machines	0.902	0	0.90	0.99	0.94
K-Nearest-Neighbour	0.899	0.00089	0.90	1.00	0.94
Decision-Tree	0.900	0	0.90	1.00	0.94
Random-Forest	0.866	0.05	0.90	0.96	0.93

While predicting Brain Stroke due to hypertension, as shown in Table 1 and Figure 2, the Support Vector Machine model yields the best accuracy, with a precision of 90.2% in Minmax scaling and 90.1% in Standard scaling, followed by Decision Tree model with 90.0% accuracy. The models logistic regression and K-NN show same accuracy with precision of 89.9%. The least accuracy is shown by the Random Forest Model, with 86.6% accuracy.

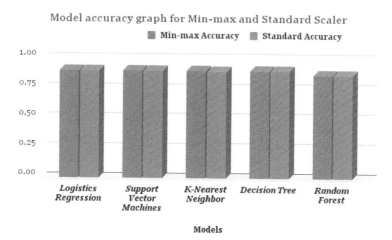

Figure 2.   Model accuracy graph for hypertension prediction.

As shown in Table 2 and Figure 3, the Support Vector Machine model predicts strokes with the highest accuracy, 90.1% in Minmax scaling and 90.0% in Standard scaling. The Decision Tree model comes in second with 90.0% accuracy. With 89.9% precision, both the logistic regression and K-NN models exhibit the same accuracy. With 86.6% accuracy, the Random Forest Model exhibits the lowest accuracy.

Table 2.   Stroke prediction.

Models	Stroke prediction				
	Accuracy	Loss	Precision	Recall	F1-score
Logistics Regression	0.899	0.00089	0.90	1.00	0.94
Support-Vector Machines	0.901	0	0.90	1.00	0.94
K-Nearest-Neighbour	0.899	0.00089	0.90	0.99	0.94
Decision-Tree	0.900	0	0.90	1.00	0.94
Random-Forest	0.866	0.049	0.90	0.96	0.93

As shown in Table 3 and Figure 4, the Support Vector Machine model predicts strokes with the highest accuracy, 90.2% in Minmax scaling and 90.1% in Standard scaling. The Decision Tree model comes in second with 90.1% accuracy. With 89.9% precision, logistic regression is followed by K-NN model with accuracy 89.7%. With 86.5% accuracy, the Random Forest Model exhibits the lowest accuracy. It is evident that, in both Minmax and Standard scaling, the Support Vector Machine model achieves its optimum accuracy.

Table 3. Heart disease prediction.

| Models | Heart Disease prediction | | | | |
	Accuracy	Loss	Precision	Recall	F1-score
Logistics Regression	0.899	0.00089	0.90	1.00	0.94
Support-Vector Machines	0.902	0	0.90	0.99	0.94
K-Nearest-Neighbour	0.897	0.00087	0.90	1.00	0.94
Decision-Tree	0.901	0	0.90	1.00	0.94
Random-Forest	0.865	0.05	0.90	0.96	0.93

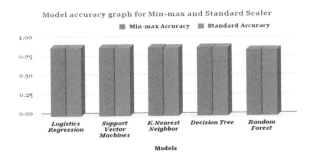

Figure 3. Model accuracy graph for stroke prediction.

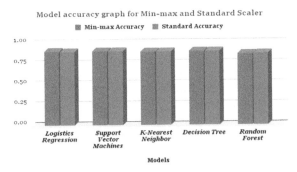

Figure 4. Model accuracy graph for hypertension prediction.

5 CONCLUSION AND FUTURE WORK

In conclusion, the use of data science and machine learning approaches to predict brain strokes has improved early detection and prevention strategies. By analysing large datasets and extracting valuable insights, these techniques have the capacity to upgrade our understanding of the risk factors and warning signs associated with strokes.

The usage of data science and machine learning in stroke prediction offers several benefits. Firstly, it makes it possible to determine high-risk individuals who may benefit from specialised therapies and proactive measures which results in prompt healthcare, lifestyle changes, and pharmaceutical changes that reduce the risk of stroke incidence.

Secondly, these approaches can help healthcare professionals make decisions that are more precise and updated. Machine learning models might provide beneficial insights and risk evaluations through analysing massive amounts of data and taking into consideration multiple variables at once, aiding healthcare professionals in their diagnosis and treatment protocols.

In the end, using data science and machine learning to predict brain strokes has the potential to transform stroke prevention and improve patient outcomes. Prediction of brain stroke using data science and machine learning has the potential to completely change stroke prevention and enhance patient outcomes. These methods might help discover patients who are in danger, providing proactive medications, and providing medical professionals with beneficial decision-making assistance. +Future research may include real-time data streams and powerful deep learning algorithms to increase the precision and timeliness of stroke risk prediction.

REFERENCES

AlKaabi, L. A., Ahmed, L. S., Al Attiyah, M. F., & Abdel-Rahman, M. E. (2020). Predicting hypertension using machine learning: Findings from Qatar Biobank Study. *PLoS ONE*, *15*(10 October 2020). https://doi.org/10.1371/journal.pone.0240370

Bentley, P., Ganesalingam, J., Carlton Jones, A. L., Mahady, K., Epton, S., Rinne, P., Sharma, P., Halse, O., Mehta, A., & Rueckert, D. (2014). Prediction of stroke thrombolysis outcome using CT brain machine learning. *NeuroImage: Clinical*, *4*, 635–640. https://doi.org/10.1016/j.nicl.2014.02.003

Chang, W., Liu, Y., Xiao, Y., Yuan, X., Xu, X., Zhang, S., & Zhou, S. (2019). A machine-learning-based prediction method for hypertension outcomes based on medical data. *Diagnostics*, *9*(4). https://doi.org/10.3390/diagnostics9040178

Dev, S., Wang, H., Nwosu, C. S., Jain, N., Veeravalli, B., & John, D. (2022). A predictive analytics approach for stroke prediction using machine learning and neural networks. *Healthcare Analytics*, *2*. https://doi.org/10.1016/j.health.2022.100032

Dritsas, E., & Trigka, M. (2022). Stroke Risk Prediction with Machine Learning Techniques. *Sensors*, *22*(13). https://doi.org/10.3390/s22134670

Emon, M. U., Keya, M. S., Meghla, T. I., Rahman, M. M., Mamun, M. S. Al, & Kaiser, M. S. (2020). Performance Analysis of Machine Learning Approaches in Stroke Prediction. *Proceedings of the 4th International Conference on Electronics, Communication and Aerospace Technology*, ICECA 2020, 1464–1469. https://doi.org/10.1109/ICECA49313.2020.9297525

Ghaffar Nia, N., Kaplanoglu, E., & Nasab, A. (2023). Evaluation of artificial intelligence techniques in disease diagnosis and prediction. *Discover Artificial Intelligence*, *3*(1). https://doi.org/10.1007/s44163-023-00049-5

Jiang, F., Jiang, Y., Zhi, H., Dong, Y., Li, H., Ma, S., Wang, Y., Dong, Q., Shen, H., & Wang, Y. (2017). Artificial intelligence in healthcare: Past, present and future. In *Stroke and Vascular Neurology* (Vol. 2, Issue 4, pp. 230–243). BMJ Publishing Group. https://doi.org/10.1136/svn-2017-000101

Kaur, M., Sakhare, S. R., Wanjale, K., & Akter, F. (2022). Early Stroke Prediction Methods for Prevention of Strokes. *Behavioural Neurology*, 2022. https://doi.org/10.1155/2022/7725597

Sailasya, G., & Aruna Kumari, G. L. (n.d.). Analyzing the Performance of Stroke Prediction using ML Classification Algorithms. In *IJACSA) International Journal of Advanced Computer Science and Applications* (Vol. 12, Issue 6). www.ijacsa.thesai.org

Shah, H. A., Saeed, F., Yun, S., Park, J. H., Paul, A., & Kang, J. M. (2022). A Robust Approach for Brain Tumor Detection in Magnetic Resonance Images Using Finetuned EfficientNet. *IEEE Access*, *10*, 65426–65438. https://doi.org/10.1109/ACCESS.2022.3184113

Sharma, V., Yadav, S., & Gupta, M. (2020). Heart Disease Prediction using Machine Learning Techniques. *Proceedings – IEEE 2020 2nd International Conference on Advances in Computing, Communication Control and Networking*, ICACCCN 2020, 177–181. https://doi.org/10.1109/ICACCCN51052.2020.9362842

Next Generation Computing and Information Systems – Gupta. (Ed.)
© 2025 The Author(s), ISBN 978-1-032-73865-9

Predicting asthma control test score using machine learning regression models

Krishna Modi
Department of CSE, Indus Institute of Technology and Engineering, Indus University, Ahmedabad, India

Ishbir Singh
Department of ME, Indus Institute of Technology and Engineering, Indus University, Ahmedabad, India

Yogesh Kumar
Department of CSE, School of Technology, Pandit Deendayal Energy University, Gandhinagar, Gujarat, India

ABSTRACT: Asthma, a chronic respiratory condition, affects millions of individuals worldwide. Accurate prediction of Asthma Control Test (ACT) scores holds immense value for personalized healthcare interventions. This study embarks on the challenge of forecasting ACT scores by harnessing weather and demography features. Employing advanced feature engineering and rigorous data preprocessing, we investigate eight regression models: K-Nearest Neighbors, Logistic Regression, Decision Tree, Support Vector Machine (SVM), Random Forest, Deep Neural Network (DNN), ADA Boost, and XGBoost. Our primary objective is twofold: to evaluate the predictive ability of these models in the context of asthma and to conduct a comprehensive comparative analysis. The results unveiled each model's efficacy and limitations in estimating ACT scores, paving the way for their real-world applicability. Among our findings, we observed noteworthy variations in model performance, with Random Forest and XGBoost emerging as the top-performing models due to their adeptness in minimizing prediction errors. Furthermore, our investigation illuminates the profound correlations between weather and demography features and ACT scores, underscoring the potential of environmental data in optimizing asthma management and control. This research offers valuable insights into the prediction of asthma control test scores based on weather and lifestyle features. Through comprehensive experimentation, we vary key hyperparameters such as the number of neighbours in KNN, the maximum tree depth in DT, the number of estimators in RF, kernel functions in SVM, and the number of iterations in DNN, among others. Our findings, coupled with the proposed method for future research, offer promising prospects for enhancing asthma management and personalized healthcare.

1 INTRODUCTION

Asthma is a considerable health and economic burden globally. Due to increasing cases of Asthma, there is a growing demand for effective strategies to predict asthma control, enabling tailored interventions and improved patient outcomes. In this pursuit, machine learning is proven to be a promising tool for the early diagnosis and management of lifestyle diseases.

The primary goal of this study is to determine the efficiency of ML models in the diagnosis of asthma, and, to conduct a comprehensive comparison of the performances of several ML models. Specifically, our focus lies in forecasting ACT scores, a pivotal measure of asthma control, by leveraging a combination of weather and lifestyle features. This approach not

DOI: 10.1201/9781003466383-29

only contributes to the field of asthma management but also provides a foundation for the broader application of ML in healthcare diagnostics.

Asthma diagnosis and management have historically relied on conventional clinical assessments and expert judgment. However, the limitations of these methods, including subjectivity and the inability to capture nuanced patterns, necessitate a paradigm shift towards data-driven approaches. ML, with its capacity to discern complex relationships within vast datasets, offers a promising avenue to enhance asthma care.

Our research employs a diverse set of ML models, ranging from traditional regression techniques to advanced ensemble methods, each integrated with feature engineering and data preprocessing. This approach not only evaluates the predictive capabilities of these models but also uncovers insights into their relative strengths and limitations in estimating ACT scores. Our evaluation criteria include Mean Absolute Error (MAE), Mean Squared Error (MSE), Mean Absolute Percentage Error (MAPE), Root Mean Squared Error (RMSE), and Variance, allowing for a comprehensive assessment.

2 BACKGROUND

Asthma represents a global healthcare challenge, impacting the lives of approximately 262 million people in 2019. (Vos, T. *et al.* 2020) Tragically, this condition also contributed to nearly 4.5 million deaths during the same year (Vos, T. *et al.* 2020), highlighting the critical need for effective asthma management strategies.

The World Health Organization (WHO) has emphasized the significance of avoiding asthma triggers as a fundamental approach to reducing asthma symptoms and enhancing overall control. This underscores the importance of proactive asthma management. In response to this imperative, our study seeks to predict ACT scores early, employing lifestyle and weather conditions as predictive factors. The overarching aim is to empower individuals with the knowledge and tools necessary to preemptively avoid asthma triggers, ultimately reducing the incidence of asthma cases.

The Asthma Control Test (ACT) is a well-established and widely recognized tool in the field of asthma management. The ACT is designed to assess and quantify the level of asthma control in individuals with asthma. Development of the ACT and its validation as a reliable measure of asthma control is discussed by Nathan, R. A *et al.* (Nathan *et al.* 2004) ACT scores typically range from 5 to 25, with higher value indicating improved asthma control. These score help categorize individuals into well-controlled, partially controlled, or uncontrolled asthma categories.

By forecasting ACT scores based on lifestyle choices and weather conditions, our research aligns with global efforts to mitigate the burden of asthma. It aspires to provide individuals with actionable insights that can inform their decisions and enable them to make lifestyle adjustments to avoid known triggers. In this way, our study not only contributes to the advancement of asthma management but also carries the potential to impact public health positively.

In recent years, researchers have been increasingly taking advantage of the power of machine learning (ML) to advance our healthcare system.(Amaral JLM *et al.*) Many studies have been proposed for detecting asthma using EHR data.(Bose S. *et al.*) Katsuyuki *et al.* have explored deep learning models for adult Asthma diagnosis, concluding that Deep Neural Networks (DNN) outperform classical Machine Learning.(Tomita K *et al.*) They have claimed to achieve an accuracy of 0.98. Radiah H. and Sin-Ban utilized DNN model to predict asthma exacerbation. (Haque *et al.* 2021) Their approach is recognizing the significant impact of environmental factors on asthma control. They were able to achieve 0.94 accuracy with ADAM optimizer. Notably, the dataset employed in our study is derived from their research efforts, serving as the cornerstone of our investigation.

3 METHODOLOGY

In this section, we discussed an overview of the experiments conducted to predict ACT scores using machine learning regression models. Figure 1 provides a visual representation of the comprehensive workflow process we followed throughout our experiments. This workflow encapsulates the key stages of our research, including data preprocessing, model training, and evaluation. Each step in this process is crucial to predict ACT scores accurately. By visualizing our workflow, we aim to offer a clear and insightful glimpse into the systematic approach for our research.

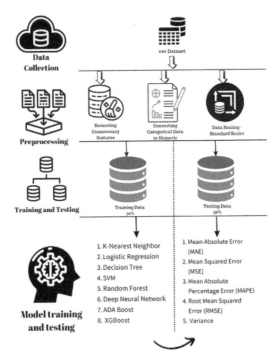

Figure 1. Workflow of the study.

3.1 *Data collection and visualization*

The dataset used for this experiment contains 1010 samples and initially comprises 14 attributes. This dataset was collected as part of the research conducted by Radiah H. and Sin-Ban. (Haque *et al.* 2021) The dataset's attributes include the following: User Number, User ID, Location, Age, Gender, Outdoor Job, Outdoor Activities, Smoking Habit, Humidity, Pressure, Temperature, UV Index, Wind Speed and ACT Score. Figure 2 depicts the frequency distribution of each attribute within the dataset. That will give a better understanding of each attribute's characteristics.

3.2 *Data preprocessing*

We have removed the "UserNo" and "UserID" attributes, as they do not contribute to our predictive model. Several attributes, including "Age", "Gender", "OutdoorJob", "OutdoorActivities", "SmokingHabit" and "UVIndex" were categorical. To convert these attributes into numeric form, we applied label encoding, ensuring that the models could effectively utilize this information.

In our analysis, we employed the SelectKBest method from scikit-learn's feature_selection module. This technique enabled us to assess the significance of each attribute concerning the target variable. The SelectKBest method employs the chi-squared (χ^2) scoring function, to rank attributes based on their relevance to the target variable. This method supports in the selection of the top 'k' attributes that demonstrate the strongest relationships with the target variable.

From Figure 3, it is concluded that attributes such as humidity, windspeed, age, and gender exhibit a strong influence on the predictive accuracy of our models. In contrast, location, outdoor job, and pressure established minimal relevance in the context of ACT score prediction. As a result, we made an informed decision to eliminate these less influential attributes from our dataset. Consequently, our final dataset consists of eight key attributes: humidity, windspeed, age, gender, outdoor activities, temperature, smoking habit, and UV index. This refined dataset allows our machine learning models to focus on the most relevant features, thereby enhancing the overall predictive performance and interpretability of our models.

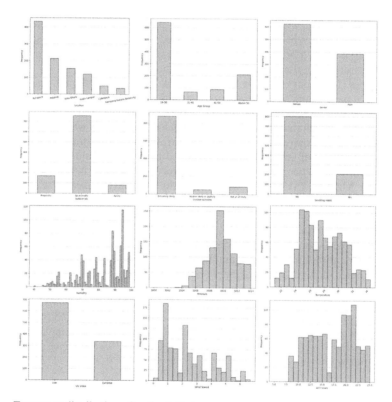

Figure 2.   Frequency distribution of each attribute.

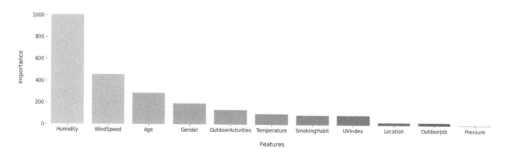

Figure 3.   Importance of each feature.

3.3 Data splitting

We divided the dataset into training and testing sets using a 70–30% split, where 70% of the data was allocated for training the models and 30% for testing their performance. The splitting of data into these subsets allowed us to evaluate the models' ability to generalize to new, unseen data. Out of total 1010 records, 707 were given to train the model and the remaining 303 were used for testing.

3.4 Attribute scaling

To speed up the calculation of the algorithm and to scale the data in a particular range, Standard Scaling is performed. After splitting data, we standardized the numerical attributes using StandardScaler from Scikit-Learn.

3.5 Machine learning regression models for training

Subsequently, we applied a comprehensive set of regression models to predict Asthma Control Test (ACT) scores based on the preprocessed dataset. These model-specific experiments allow us to provide a broad understanding of each model's behavior and the impact of hyperparameter tuning on their predictive performance.

The K-Nearest Neighbors (KNN) algorithm was examined with a range of neighbor values, making it particularly valuable for applications where nearby data points significantly impact the prediction. Logistic Regression is a widely-used linear model. It offeres simplicity and interpretability, which make it an excellent choice for scenarios where the model's inner workings need to be transparent. Decision Trees, through the adjustment of tree depths and the application of the Gini impurity method, proved to be robust and intuitive, showcasing suitability in contexts where decision-making processes should be easily interpretable. Support Vector Machine (SVM) Regression gives its flexibility in kernel selection, is well-suited for complex, non-linear relationships between attributes. Random Forest (RF) emerged as a robust choice for mitigating overfitting, and is ideal when the prediction requires resistance to noise and variability. Deep Neural Network (DNN) provides valuable insights into complex, data-driven patterns, making it a compelling option for applications where intricate relationships in the data need to be captured. ADA Boost's ensemble approach combines multiple weak learners and form a strong predictor. That is beneficial when predictive accuracy is utmost. Finally, XGBoost, known for its efficiency, offered a compelling solution for cases where model accuracy is crucial, combining both predictive power and computational speed. This thorough examination allowed us to showcase strengths of each model, enabling more informed choices in real-world applications.

3.6 Model evaluation

To evaluate the performance of these machine learning regression models, we used a range of evaluation metrics. Each metric provides predictive capabilities of the models, and we used them to compare and evaluate the performance of each algorithm. Table 1 represents various performance metrics we have considered for this research. In the formulas, n represents the total number of samples, Y_i represents the actual value of the i-th sample, and $\widehat{Y}_i$ represents the predicted value for the i-th sample.

4 RESULTS

In this section, we present the outcomes of our experiments in predicting Asthma Control Test (ACT) scores using a range of machine learning regression models. These results are illustrated through Tables 2 to 10, that highlight the performance of the models. Our research aims to provide a detailed analysis of the model's predictive capabilities and their effectiveness in estimating ACT scores.

Tables 2 through 9 display the performance of the aforementioned algorithms across various hyperparameters, while Table 10 provides a comparative analysis of the best results achieved by each learning algorithm.

Table 1.   Evaluation metrics.

Metric	Formula
Mean Absolute Error (MAE)	$\frac{1}{n}\sum_{i=1}^{n} \lvert Y_i - \widehat{Y}_i \rvert$
Mean Squared Error (MSE)	$\frac{1}{n}\sum_{i=1}^{n} (Y_i - \widehat{Y}_i)^2$
Mean Absolute Percentage Error (MAPE)	$\frac{1}{n}\sum_{i=1}^{n} \left\lvert \frac{Y_i - \widehat{Y}_i}{Y_i} \right\rvert \times 100$
Root Mean Squared Error (RMSE)	$\sqrt{\frac{1}{n}\sum_{i=1}^{n} (Y_i - \widehat{Y}_i)^2}$
Variance	$\frac{1}{n}\sum_{i=1}^{n} (Y_i - \bar{Y})^2$

Table 2.   Performance analysis of kNN with different values of k.

Value of k	MAE	MAPE	MSE	RMSE	Variance
2	0.83	0.05	2.66	1.63	0.88
3	0.90	0.06	2.65	1.63	0.88
4	0.91	0.06	2.37	1.54	0.89
5	0.93	0.06	2.36	1.54	0.89
6	1.05	0.07	2.58	1.61	0.88
7	1.13	0.08	2.78	1.67	0.87

Table 3.   Performance analysis of logistic regression.

Data	MAE	MAPE	MSE	RMSE	Variance
Training Data	1.95	0.14	8.17	2.86	0.63
Testing Data	1.96	0.15	7.56	2.75	0.66

Table 4.   Performance analysis of decision tree for different maximum depth.

Max_depth	MAE	MAPE	MSE	RMSE	Variance
3	1.90	0.13	6.80	2.61	0.69
4	1.60	0.11	5.20	2.28	0.76
5	1.49	0.10	4.58	2.14	0.79
6	1.23	0.08	3.72	1.93	0.83

Table 5.   Performance analysis of random forest for different number of estimators.

N_estimators	MAE	MAPE	MSE	RMSE	Variance
100	0.74	0.05	1.62	1.27	0.93
125	0.74	0.05	1.57	1.25	0.93
150	0.73	0.05	1.51	1.23	0.93
175	0.73	0.05	1.54	1.24	0.93

Table 6. Performance analysis of support vector regressor for different kernel.

Kernel	MAE	MAPE	MSE	RMSE	Variance
Linear	1.95	0.15	7.02	2.65	0.69
RBF	1.65	0.13	5.27	2.30	0.77
Poly (degree = 3)	2.08	0.16	8.30	2.88	0.62

Table 7. Performance analysis of DNN for various number of iterations.

iteration	MAE	MAPE	MSE	RMSE	Variance
100	2.38	0.17	8.93	2.99	0.59
150	2.21	0.16	7.61	2.76	0.65
200	2.11	0.15	7.05	2.66	0.68
250	2.03	0.15	6.66	2.58	0.69

Table 8. Performance analysis of ADABoost for different number of estimators.

N_estimators	MAE	MAPE	MSE	RMSE	Variance
50	1.72	0.12	4.20	2.05	0.81
75	1.68	0.12	4.02	2.01	0.81
100	1.70	0.12	4.10	2.02	0.81
125	1.70	0.12	4.13	2.03	0.81

Table 9. Performance analysis of XGBoost for different number of estimators.

N_estimators	MAE	MAPE	MSE	RMSE	Variance
100	1.17	0.08	2.45	1.57	0.89
125	1.13	0.08	2.33	1.52	0.89
150	1.10	0.07	2.26	1.50	0.90
175	1.06	0.07	2.17	1.47	0.90

Table 10. Comparative analysis of different models.

Results	MAE	MAPE	MSE	RMSE	Variance
kNN	0.83	0.05	2.66	1.63	0.88
LR	1.96	0.15	7.56	2.75	0.66
DT	1.23	0.08	3.72	1.93	0.83
RF	0.73	0.05	1.51	1.23	0.93
SVM	1.65	0.13	5.27	2.30	0.77
DNN	2.03	0.15	6.66	2.58	0.69
ADABOOST	1.68	0.12	4.02	2.01	0.81
XGBOOST	1.06	0.07	2.17	1.47	0.90

Table 10 clearly illustrates that errors, as measured by various metrics, are minimized in kNN, Random Forest and XGBoost models. kNN approach allowed the model to discern local relationships, particularly useful when certain demographic or environmental factors exhibit localized effects on asthma control. Ensemble methods (Random Forest and XGBoost) are capturing complex patterns within the dataset. XGBoost and ADABoost are combining weak learners to form strong predictors. By iteratively adjusting the weights of misclassified samples, XGBoost and ADABoost contributed to refining the overall predictive accuracy of our models. These models demonstrated their ability to provide accurate and reliable estimates of ACT scores.

5 CONCLUSION

Our research offers valuable insights into the prediction of ACT scores based on weather and lifestyle features. The exceptional performance of the Random Forest and XGBoost models underscores the potential of ensemble methods in capturing complex patterns within the data. The findings presented in this study contribute to the ongoing efforts to optimize asthma management and personalized healthcare. The implications extend beyond the limitations of asthma control and provide a template to use the potential of machine learning in the management of lifestyle diseases. As we navigate the ever-evolving landscape of healthcare and data-driven solutions, the research presented here assists in improving patient outcomes and enhancing the quality of healthcare delivery. We hope that our work inspires further research and innovation in this field, ultimately leading to more effective asthma management, reduced healthcare costs, and improved patient well-being.

REFERENCES

Amaral JLM, Lopes AJ, Veiga J, Faria ACD, Melo PL (2017) High-accuracy detection of airway obstruction in asthma using machine learning algorithms and forced oscillation measurements. *Comput Methods Programs Biomed* 144:113–125.

Bansal, K., Bathla, R. K., Kumar, Y. (2022). Deep transfer learning techniques with hybrid optimization in early prediction and diagnosis of different types of oral cancer. *Soft Computing* 26(21), 11153–11184.

Bhardwaj, P., Bhandari, G., Kumar, Y., Gupta, S. (2022). An Investigational Approach for the Prediction of Gastric Cancer Using Artificial Intelligence Techniques: A Systematic Review. *Archives of Computational Methods in Engineering* 1–22.

Bose S, Kenyon CC, Masino AJ (2021) Personalized prediction of early childhood asthma persistence: a machine learning approach. *PLoS ONE* 16(3):e0247784.

Haque, R., Ho, S. B., Chai, I., Abdullah, A. (2021). Optimised deep neural network model to predict asthma exacerbation based on personalised weather triggers. *F1000Research*. 10:911.

Kaur, I., Sandhu, A. K., Kumar, Y. (2022). Artificial Intelligence Techniques for Predictive Modeling of Vector-Borne Diseases and its Pathogens: A Systematic Review. *Archives of Computational Methods in Engineering* 1–31.

Kaur, K., Singh, C., Kumar, Y. (2023). Diagnosis and Detection of Congenital Diseases in New-Borns or Fetuses Using Artificial Intelligence Techniques: A Systematic Review. *Archives of Computational Methods in Engineering*.

Koul, A., Bawa, R. K., Kumar, Y. (2022). Artificial intelligence techniques to predict the airway disorders illness: a systematic review. *Archives of Computational Methods in Engineering* 1–34.

Kumar, Y., Kaur, I., Mishra, S. (2023). Foodborne Disease Symptoms, Diagnostics, and Predictions Using Artificial Intelligence-Based Learning Approaches: A Systematic Review. *Archives of Computational Methods in Engineering* 0123456789.

Modi, K., Singh, I., Kumar, Y. (2023). A Comprehensive Analysis of Artificial Intelligence Techniques for the Prediction and Prognosis of Lifestyle Diseases. *Archives of Computational Methods in Engineering* 0123456789.

Nathan, R. A., Sorkness, C. A., Kosinski, M., Schatz, M., Li, J. T., Marcus, P., Murray, J. J., Pendergraft, T. B. (2004). Development of the asthma control test: a survey for assessing asthma control. *The Journal of allergy and clinical immunology* 113(1), 59–65.

Singh, J., Sandhu, J. K., Kumar, Y. (2023). An Analysis of Detection and Diagnosis of Different Classes of Skin Diseases Using Artificial Intelligence-Based Learning Approaches with Hyper Parameters. *Archives of Computational Methods in Engineering* 0123456789.

Sisodia, P. S., Ameta, G. K., Kumar, Y., Chaplot, N. (2023). A Review of Deep Transfer Learning Approaches for Class-Wise Prediction of Alzheimer's Disease Using MRI Images. *Archives of Computational Methods in Engineering* 1–21.

Tomita K, Nagao R, Touge H, Ikeuchi T, Sano H, Yamasaki A, Tohda Y (2019). Deep learning facilitates the diagnosis of adult asthma. *Allergol Int* 68(4):456–461.

Vos, T. (2020) Global burden of 369 diseases and injuries in 204 countries and territories, 1990–2019: a systematic analysis for the Global Burden of Disease Study 2019. *Lancet*. 396(10258):1204–22.

Next Generation Computing and Information Systems – Gupta. (Ed.)
© 2025 The Author(s), ISBN 978-1-032-73865-9

An approach for early prediction of multi-disease using ML techniques

Vani Malagar, Mekhla Sharma & Navin Mani Upadhyay
Model Institute of Engineering and Technology, Jammu, J&K, India

ABSTRACT: Leveraging the explosion of medical data including patient histories, electronic health records & medical imaging, the integration of machine learning techniques with healthcare for multi-disease prediction has experienced tremendous growth. Several algorithms have already been proposed by many researchers but they cannot fit for all the solutions at once. So, this research paper uses Machine Learning (ML) techniques to investigate the complexity involved in multi-disease prognosis in the context of convergence. It explores three well-knows ML- algorithms (Decision Tree, Random Forest classifier and Gradient Boosting) to investigate and compare the best way to identify and diagnose the prediction problems. Further a detailed comparison with existing algorithms shows a better result from 70%–77%. The accuracy of used algorithm has shown 78.57% for Decision Tree, 77.92% for Random classifier and for Gradient Boosting 95.38%. Similarly, the F1-score is also improved up to 3%. The paper also redefines several machine learning algorithms used by several researchers to solve the prediction problem, by emphasizing the importance of gradient boosting as a successful strategy. This work also validates the used models based on a statistical analysis "ANOVA and Chi-square".

1 INTRODUCTION

Multi-disease classification and identification in healthcare and medical sciences focus on using advanced artificial intelligence-based technology to diagnose, detect and categorize multiple dis- eases or medical conditions simultaneously (Khurana *et al.* 2019 and Sagaro *et al.* 2020). This approach aims to improve the accuracy and efficiency of detection and classification of disease using advanced AI-based machine learning algorithms mainly in the cases where patients may express symptoms or risk factors associated with multiple health conditions. It is an interdisciplinary field involving various areas like medicine, data science, and technology to improve healthcare and enhance patient outcomes (Shirsath *et al.* 2018 and Ardabili *et al.* 2020) and some key aspects of multi-disease identification and classification are as follows:

- Collection and analysis of different types of medical or health record data.
- Application of advanced machine learning algorithms to collected data for developing models that can identify and classify multiple diseases.
- Automating diagnostic process that reduces workload of healthcare professionals.
- Improving speed and accuracy of disease prognosis.
- Early detection of diseases for better treatment outcomes and minimizing healthcare costs.
- Prioritizing and managing chronic diseases helping healthcare providers.
- Providing insights into disease patterns, patient population and many more for research and drug development.

Traditional disease identification and classification methods are often based on established medical guidelines, manual interpretations and mainly focus on the experience of healthcare professionals (Polsterl *et al.* 2016).

DOI: 10.1201/9781003466383-30

Other challenges involve variability in diagnosis, time constraints, compromised accuracy, extensive manual review, and challenges in handling large patient data. On the contrary, machine learning based techniques provide the probability for increased accuracy, efficiency, scalability, and adaptability as compared to traditional methods of disease identification and classification (Anitha *et al.* 2016 and Alotaibi et al. 2019)

1.1 *Steps in multi-disease identification and classification*

Figure 1 illustrates the various steps involved in the identification and classification of diseases including some of the machine learning algorithms which are explained below:

- Collection of data including medical records containing patient data, diagnostic information and many more.
- Pre-processing involving cleaning and normalizing data for analysis.
- Splitting data including training, validation, and test sets for evaluation of model performance.
- Feature engineering using domain-specific knowledge for extracting features for distinguishing between diseases.
- Selection of model for multiple disease identification and classification.
- Evaluation of model using suitable evaluation metrics.
- Testing model for real-world performance.
- Interpret ability for ensuring models are interpretable and make correct predictions.
- Deployment of model after satisfactory performance in the healthcare environment.
- Monitoring healthcare model for ensuring accuracy and reliability.
- Feedback loop to get continuous input on model performance and future scope for improvements.

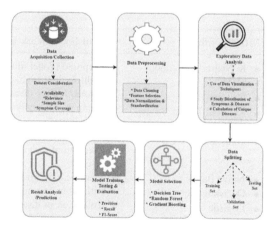

Figure 1.   Steps in multi-disease identification/classification.

Several research papers have been published that provide details regarding the work done in integrating machine learning techniques for identifying and classifying diseases. Table 1 gives a list of some of the research articles that explains the work in this area.

1.2 *Challenges*

Challenges involved in multiple disease identification and classification can adversely affect the correctness and reliability of the suitable model used in healthcare diagnosis. Some of the challenges are:

- Poor quality of data including limited or imbalanced dataset, missing values, errors/outliers, or inconsistencies.

- Interoperability issues, variability in data formats and standards.
- Concerns related to interpretability of machine learning models.
- Privacy and security concerns maintaining patient sensitive information.
- Incorrect or mislabelled data affecting model training
- Complexity in extraction of relevant features from medical data.
- Complexity issues involved in data integration from different sources.
- Issues involved in handling categorical data.
- Complex and computationally expensive pre-processing techniques.

1.3 *Contribution*

In the domain of healthcare for multiple disease identification & classification, detecting diseases based on symptoms is extremely critical for the enhancement of patient care and maintaining healthcare resource allocation. Study proposed in this paper uses the various machine learning techniques in this realm and focus on the following contributions:

- Application of suitable machine learning algorithm for prediction of diseases depending upon the symptoms identified.
- Statistical analysis of dataset keeping in mind the complexity of the data.
- Keeping track of prediction accuracy obtained in order to ensure that it surpasses the traditional prognostic approaches.
- Seamless integration of machine learning algorithms on used medical dataset.

Thus, the paper is a transformative initiative in the field of medical healthcare in prediction and classification of diseases based on symptoms using suitable machine learning algorithms. Section 2 covers the proposed work; Section 3 focuses on the results and finally Section 4 concludes the paper.

Table 1. Existing work.

Research Work	ML Algorithm	Area of Focus/Work Done
Multiple Disease Prediction System using Machine Learning and Stream lit [Gopisetti *et al.* 2023]	• K-Nearest Neighbor • SVM • Decision Tree • Random Forest • Logistic Regression • Gaussian Naive Bayes.	• Proposed model in the paper can forecast multiple diseases like diabetes, heart disease, chronic kidney disease and cancer • Comparative analysis of accuracy of each algorithm • Multiple datasets • Creation of web application for forecasting diseases
Identification and prediction of chronic diseases using ma chine learning approach [Alanazi *et al.* 2022]	• Naïve Bayes • Decision tree • Logistic regression (KNN)	• Identification and prediction of patients with more common chronic illnesses • Collection of disease symptoms & preparation of datasets • Comparative analysis of proposed work
Multiple disease prediction using ML algorithms [Arumugam *et al.* 2021]	• SVM • Naive Bayes • Decision Tree	• Proposed framework worked on forecasting the likelihood of heart disease in diabetes individuals
An Approach to detect multiple diseases using ML algorithm [Mohit *et al.* 2021]	• SVM • Logistic Regression • KNN	• Medical test web application built that makes predictions about various diseases like Breast cancer, Diabetes, and heart diseases.
Disease prediction from various symptoms using machine learning [R *et al.* 2022]	• Weighted KNN	• Disease prediction system based on symptoms, age, gender using multiple ML algorithms

(continued)

Table 1. Continued

Research Work	ML Algorithm	Area of Focus/Work Done
Multiple Disease Prediction Using Different Machine Learning Algorithms Comparatively [Godse *et al.* 2019]	• Naïve Bayes Algorithm • K-Nearest Algorithm • Decision Tree Algorithm • Random Forest Algorithm • SVM	• High accuracy achieved using weighted KNN • Proposed system aimed at bridging the gap between doctors and patients • Support prediction for multiple sickness prediction • Builds doctor's recommendation system • Web/android application deployed for user
A ML Model for Early Prediction of Multiple Diseases to Cure Lives [Kumar *et al.* 2021]	• Decision Tree • Random Forest • Naïve Bayes • KNN algorithms	• Model build predicts the disease in compliance with symptoms entered by patients • GUI based interaction system with users

2 PROPOSED WORK

Methodology used in the paper involves several steps to predict diseases from symptoms using machine learning algorithms. The dataset contains features and corresponding disease labels. The aim is to predict diseases from symptoms using machine learning algorithms applied on the dataset based on two major specifications of binary classification, one on the basis of True Positives [TP] and the other False Negatives [FN].

2.1 *Methodology*

Methodology used in this paper is based on decision tree, random classifier and gradient boosting algorithm followed by several steps including data pre-processing, exploratory data analysis, model selection and evaluation, cross-validation, and model testing. The decision tree algorithm shown in Algorithm-1, correctly classify the dataset discussed in Table 2.

Algorithm 1: Decision-Tree with TP-FN

INPUT: Datasets including features, F {itching, skin-rash, nodal-eruption,
continuous_sneezing, ... , yellow_crust_ooze}
OUTPUT: Chosen Decision Tree D_TP-FN
if F is null **then** return (failure) **end if**
for all i = 1 **to** n **do**
    set F_m = List of features (F_i) in
    DFS order set F_{bm} = { } , //
    Selected Nodes
    call update (F, F_m) set F_i = F_{bm}
end for choosing the tree procedure
for all i = 1 **to** n **do**
    set $D_{Fm} = \sum^n D_{Fm}$ // computer the average distance to select the node.
Distance
F_m = argmin$_i D_{Fm}$ // choose the representative decision tree model D_TP-FN
Return F_m End Procedure

Table 2. Classification report for selected features for the models

Class	Decision Tree			Random Forest Classifier			Gradient Boosting		
	Precision	Recall	F1-Score	Precision	Recall	F1-Score	Precision	Recall	F1-Score
Paroxysmal Positional Vertigo	1	0.95	0.97	1	0.89	0.94	1	0.95	0.97
AIDS	1	0.93	0.96	1	1	1	1	0.83	0.91
Arthritis	1	0.9	0.95	1	1	1	1	0.9	0.95
Diabetes	1	1	1	1	1	1	1	1	1
Tuberculosis	1	1	1	1	1	1	1	1	1
hepatitis A	1	0.02	0.04	0	0	0	1	1	1

From the Algorithm-1 first we choose the feature from the given dataset defined as 'F' and name that as 'F_m' which is a member based on DFS algorithm. If the selected feature is a member of the selected decision group, then make a separate function called 'F_{bm}.' 'Set F_i' to select the next feature from the given dataset. Finally compute the average distance of selected feature from the decision tree nodes and return the F_m. Furthermore, the number of unique diseases and their occurrences are calculated by using the below equation:

$$F_E = -\sum_{i=1}^{n} P_i * log P_i \tag{1}$$

And the Selection of FBM is calculated by using the below equation:

$$F_{BM} = 1 - \sum_{i=1}^{n} P_i^2 \tag{2}$$

Furthermore, for a better understanding of the data, Figure 2 shows the framework with a detailed step involved.

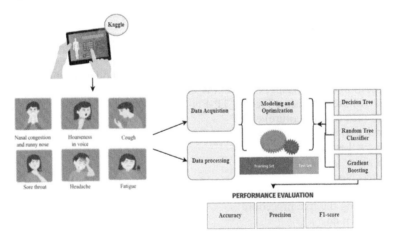

Figure 2. Proposed framework for multi-disease prediction.

In the above Figure 2, the first step is dataset acquisition, then selection of appropriate data features containing comprehensive symptom-based disease information is done. While doing this task the major considerations taken into the accounts are:

- Dataset readily available and accessible from reputable sources.
- Datasets include symptom information associated with various diseases.
- Dataset is sufficiently large having a sample size of 4920 samples and 133 characteristics to provide enough instances for each disease class.

2.2 Dataset description

Dataset[1] used has 4920 samples and 133 characteristics, including a wide spectrum of medical data involving 41 different classifications of diseases. Each sample is distinguished by 133 traits, making them a valuable source of data for disease prediction.

2.3 Data pre-processing

For data pre-processing techniques, missing data handling (imputation, removal, etc.), outlier detection & treatment, feature scaling & normalization and feature selection & dimensionality reduction is followed.

2.4 Training and model selection

For assessing the model performance, data is split into training and validation sets using the train test split () function. Models are trained using training set and evaluated on the validation set using metrics such as accuracy, confusion matrix, and classification report shown in Table 2.

3 RESULTS AND DISCUSSION

The paper conducts a thorough comparative analysis of three potential classifiers, including the decision tree, random forest classifier, and gradient boosting for the binary classification of True Positive (TP) and False Negative (FN) scenarios. Classifiers are evaluated based on their performance and outcome. Performance of the classifiers is thoroughly examined which contains confusion matrices, training and validation accuracy, and classification reports including precision, recall, F1-score, and support values for class of disease.

3.1 Performance analysis of decision tree model

Decision tree model performs admirably when it comes to disease prediction. The model adequately generalizes to new data without exhibiting over-fitting, as seen by the fact that both the training and validation accuracies are 78.57%. Notably, classes like "Diabetes" and "Tuberculosis" excel, scoring perfectly on F1-score tests for precision, memory, and performance. On the other hand, categories like "AIDS" and "Paroxysmal Positional Vertigo" also perform well, with good accuracy and recall. Some classifications like "hepatitis A" however display difficulties with a very low recall, indicating that the model may have trouble detecting these cases. Total weighted average F1-score for all classes comes in at about 0.77 and offers a thorough evaluation of the model's performance. The values for the classification report that includes accuracy, recall, and F1-score for a few classes are listed in Table 1.

3.2 Performance analysis of random forest classifier

Random forest classifier performs well in the classification of diseases when the parameter values (criterion='entropy', max depth=10, and max features='sqrt') are implemented. Training and validation accuracy are both commendable and about 83.37%, showing that the model generalizes well with little or no over-fitting to the training set. When analyzing certain classes, the model excels at diagnosing diseases like "AIDS", "Arthritis", "Diabetes" and "Tuberculosis", obtaining the highest possible accuracy, recall, and F1-score values. This displays the model's incredible accuracy in identifying certain diseases. Model retains great precision and an adequate recall for classes like "Paroxysmal Positional Vertigo", indicating its usefulness in a variety of disease circumstances. Weighted average F1-score taking into account the total model performance is almost 0.82 as shown in Figure 3. Random Forest model shows potential in disease prediction in terms of implications and suggestions, especially for diseases with high performance metrics.

3.3 *Performance analysis of gradient boosting classifier*

The Gradient Boosting Classifier, performs well in the classification of diseases when its parameters are set to (criterion='entropy', max depth=10, and max features='sqrt'). During training, the model shows a constant decrease in training loss. This demonstrates that the model successfully picks up complicated data patterns without over-fitting. 95.38% accuracy for both training and validation shows that it can generalize effectively to newly acquired data. Model excels at detecting diseases like "Diabetes", "Tuberculosis" and "Hepatitis A" obtaining near- perfect accuracy, recall, and F1-score values.Model retains excellent accuracy for some classifications such as "AIDS" but has a lower recall for some of the diseases. The F1-score is approximately 0.96% when the selected feature is based on performance of the entire model as shown in Figure 3.

3.4 *Statistical analysis of selected models*

The statistical test shows a significant improvement in Decision Tree, Random Forest and Gradient Boosting shown in Figure 4. From the Figure 4, it is clear that the Gradient Boosting approach is suitable for applications requiring high accuracy in disease categorization as seen by its better result. ANOVA and chi-square have been used for the evaluation of all three algorithms. Hence, the knowledge of the model's performance by including feature significance analysis and domain-specific insights will be used in future work. Figure 3a below provides the performance analysis of the models depicting the higher and the low accuracy depending upon the algorithms on the FN and TP cases. The number of point fields in the statistics indicates that there are 9 data points or observations in our dataset. In the context of Regression Analysis, the degree of freedom is often referred to as the number of datapoints minus the number of estimated model parameters.

We have eight degrees of freedom in this situation. Residual sum of squares (RSS) calculates the difference between the sum of the squared values of observed and residual values from the regression model. RSS in this instance is 13921.33. For this paper, the Pearson's Correlation Coefficient, commonly abbreviated as "r" is 0.70, indicating a moderately strong positive linear relationship between the variables. R-square measures the amount of variance seen in the dependable variable (C) that is defined by the independent variable in the model. Additionally, the adjusted R-squared value is a modified value form of R- squared.

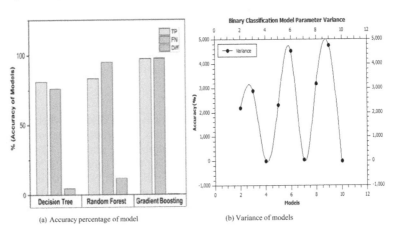

(a) Accuracy percentage of model   (b) Variance of models

Figure 3.   Accuracy percentage and variance.

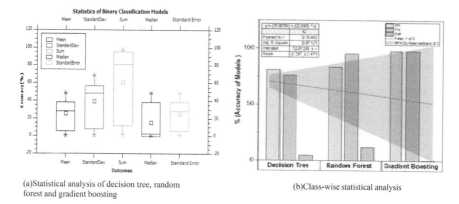

(a)Statistical analysis of decision tree, random forest and gradient boosting

(b)Class-wise statistical analysis

Figure 4.   Statistical analysis.

The adjusted R-squared value is 0.67, which represents the percentage of variance explained after taking model complexity and degrees of freedom into account as shown in Figure 3b, and 4a. As per the statistical analysis shown in Figure 4(a, b), applying the linear fitting on the percentage accuracy of the individual models represented with "C" by considering the parameters such as intercept, slope-value, standard error, t-value as shown in Figure 4b. These statistics offer important insights into the extent to which the dependent variable's variability is explained in section-3.

4   CONCLUSION

The research work focused on predicting disease from symptoms using three classifiers: Decision Tree, Random Forest Classifier and Gradient Boosting. These classifiers were evaluated based on validation accuracy, training accuracy, confusion matrix, and classification report. Overall, all three classifiers demonstrated a reasonable performance in predicting disease. Gradient Boosting achieved the highest training and validation accuracy of 95.38%. Confusion matrices provide the detailed information about the distribution of predicted and actual prognosis across different disease categories. Classification reports presented precision, recall, F1-score, and support metrics for each disease category, further evaluating the classifiers performance. Based on the results, Gradient Boosting emerged as the best model for predicting diseases from symptoms in this research work. Overall, the research work contributes to the field of medical prognosis by showcasing viability and feasibility of using machine learning algorithms based on disease symptoms. Hence, the findings highlight the potential of these classifiers to assist healthcare professionals in making a correct and efficient disease prognosis.

REFERENCES

Alanazi, R. 2022. Identification and prediction of chronic diseases using machine learning approach. *Journal of Healthcare Engineering*, 2022: Article ID 2826127. DOI: 10.1155/2022/2826127.

Alotaibi, F.S. 2019. Implementation of machine learning model to predict heart failure disease. *International Journal of Advanced Computer Science and Applications*, 10(6): 261–268.

Anitha, S., & Sridevi, N. 2019. Heart Disease Prediction Using Data Mining techniques. *Journal of Analysis and Computation*, 13(2): 48–55.

Ardabili, S.F., Mosavi, A., Khamosi, P., Ferdinand, F., Varkonyi-Koczy, A.R., Reuter, U., Rabczuk, T., & Atkinson, P.M. 2020. COVID-19 Outbreak prediction with machine learning. *Journal of Algorithms*, 13 (249): 1–36.

Arumugam, K. *et al.* 2021. Multiple disease prediction using machine learning algorithms. *Materials Today: Proceedings*, 2021.

Battineni, G., Sagaro, G.G., Chinatalapudi, N., & Amenta, F. 2020. Applications of machine learning predictive models in the chronic disease diagnosis. *Journal of Personalized Medicine*, 10(2): 21.

Godse, R.A. *et al.* 2019. Multiple disease prediction using different machine learning algorithms comparatively. *International Journal of Advanced Research in Computer and Communication Engineering*, 8(12): 50–52.

Gopisetti, L.D., Lambavai Kummera, S.K., Pattamsetti, S.R., Kuna, S., & Niha. 2023. 5th International Conference on Smart Systems and Inventive Technology (ICSSIT). *IEEE* DOI: 10.1109/ICSSIT55814.2023.10060903.

Keniya, R. *et al.* 2022. Disease Prediction from Various Symptoms Using Machine Learning.

Khurana, S., Jain, A., Kataria, S., Bhasin, K., Arora, S., & Gupta, A.D. 2019. Disease prediction system. *International Research Journal of Engineering and Technology*, 6(5): 5178–5184.

Kumar A., A., & Pathak B., A. 2021. A machine learning model for early prediction of multiple diseases to cure lives. *Turkish Journal of Computer and Mathematics Education*, 12(6): 4013–4023.

Mohit, I. *et al.* 2021. An approach to detect multiple diseases using machine learning algorithm. *Journal of Physics: Conference Series*, 2089(1).

Polsterl, S., Conjeti, S., Navab, N., & Katouzian, A. 2016. Survival Analysis for high-dimensional, heterogeneous medical data: Exploring feature extraction as an alternative to feature selection. *Artificial Intelligence in Medicine*, 72: 1–11.

Rathee, P.A. 2023. Multiple disease prediction using machine learning. *Iconic Research and Engineering Journals*, 6(12).

Shirsath, S.S., & Patil, S. 2018. Disease prediction using machine learning over big data. *International Journal of Innovative Research in Science and Technology*, 7(6): 6752–6757.

Next Generation Computing and Information Systems – Gupta. (Ed.)
© 2025 The Author(s), ISBN 978-1-032-73865-9

Early prediction of down syndrome using deep transfer learning-based approaches

Nirmit Patel, Tarang Ghetia, Devraj Jhala, Shubh Kapadia & Yogesh Kumar
Department of CSE, School of Technology, Pandit Deendayal Energy University, India

ABSTRACT: Down Syndrome, a genetic disorder caused due to an extra copy of chromosome 21. With around 1 in 700 newborn babies being affected from it, it remains one of the most prevalent chromosomal disease. Children with down syndrome typically have physical and intellectual disabilities, but with early intervention and support, they can live long and fulfilling lives. This research investigates the use of Convolutional Neural Networks (CNNs) and transfer learning for early detection of Down Syndrome. The dataset used in this study is taken from Kaggle and consists of about 3000 facial images of children with and without Down Syndrome. The highest accuracy achieved in this study is over 95% and further shows the efficacy of transfer learning for more reliable detection of down syndrome.

1 INTRODUCTION

John Langdon Down, a doctor from Cornwall, England, is credited with describing the illness for the first time in 1866 (Ropper and Bull 2020). Down Syndrome is one of the most common chromosomal disorder in India, affecting about 1 in every 600 babies born. In the United States, it affects about 1 in every 700 babies born. Down syndrome occurs due to an additional copy of chromosome 21, leading to a genetic condition. Down syndrome cannot be cured, yet there exist treatments and therapies that offer assistance to individuals with the disorder live healthy and fulfilling lives (NIH 2017).

About 80% of people with Down syndrome have a moderate intellectual handicap, while some are severely damaged and others have development that is similar to that of typically developing people (Sideropoulos *et al.* 2023). People with Down syndrome make significant contributions to society in many different fields, including art, music, and business. Down syndrome has intrigued researchers due to its distinctive physical and intellectual characteristics for decades. A recent study has found that about 44.9% of people with down syndrome that they studied, suffered with one of the following skin conditions: Folliculitis, Keratosis pilaris, Acne vulgaris, Hidradenitis suppurativa, Furunculosis (Firsowicz *et al.* 2020). Also, individuals with trisomy 21 (genetic causation of down syndrome) are at heightened susceptibility to experiencing exacerbated symptoms, higher hospitalization rates, increased likelihood of intensive care, secondary bacterial infections, and elevated mortality from SARS-CoV-2 infections compared to the general population (Espinosa 2020).

Some of the common physical features of people diagnosed with down syndrome are flattened face, almond-shaped eyes slanting up, short neck, small ears, hands and feet, poor muscle tone and joints and are shorter in height as adults and children (Centers for Disease Control and Prevention [CDC] 2021). Exploiting these features using deep convolutional

DOI: 10.1201/9781003466383-31

neural network models, an effective model can be created for down syndrome classification task.

This paper sheds some light on the detection of Down Syndrome via the face images of children infected with this disease. The proposed method uses transfer learning to accurately classify the images in the dataset. We have used DenseNet121, Inception, and XceptionNet as our base models, which are then fine-tuned to achieve acceptable performance on this classification task.

In the following sections we delve into the literature proposed earlier. After that, we dive into the details of dataset preparation, the proposed algorithm, the results and discussion along with conclusion.

2 RELATED WORK

There has been extensive research and activity in detection of Down Syndrome using machine learning, deep learning models. Below given are some of the literature works published in recent works:

In (Qin *et al.* 2020), the authors proposed a novel facial recognition method that utilizes deep CNNs to detect down syndrome from facial images. The method employs transfer learning and image preprocessing techniques to train a binary model for down syndrome detection. The DCNN achieved high accuracy, recall, and specificity, with 95.87%, 93.18%, and 97.40% respectively.

In (Thomas & Arjunan 2022), the authors proposed a model that utilizes a SegNet architecture based on the Visual Geometry Group (VGG-16) for the segmentation of the NT region. The segmented NT images are trained using an AlexNet model for classification of Down Syndrome. Model was trained and tested on a data of 100 fetal images. The SegNet model achieved an accuracy rate of 98.18% and a Jaccard index of 0.96 in segmenting the NT region. The classification model based on the AlexNet achieved an overall accuracy rate of 91.7%, a sensitivity rate of 85.7%, and a receiver operating characteristic (ROC) value of 95%.

In (Wang *et al.* 2023), the authors suggested an integrated process (DSD) using the modern transformer technique and transfer learning strategy to forecast cases of Down syndrome from the initial metaphase micrographs. Several models, including faster, mrcnn, retina, defor, spar, and swin, were used in the study. The Swin Transformer model distinguishes out because it constantly exceeds the competition on a number of metrics, demonstrating its reliability and effective operation.

In (Zhao *et al.* 2014), the suggested method uses Gabor wavelet transform and local binary patterns (LBP) to extract morphological and structural variation information. A number of classifiers model, including Support vector machine (SVM), random forest (RF), Linear discriminant analysis (LDA), and K-nearest neighbour, were trained to detect cases of down syndrome after feature fusion and selection. By combining geometric and Gabor jet features with LDA, the combined features had the highest accuracy (0.967) and F1 score (0.956).

In (Mishima *et al.* 2019), the study evaluated the performance of Face2Gene, a facial recognition software, for suggesting candidate syndromes for patients with congenital dysmorphic syndromes in Japan. The software achieved a top 1 sensitivity rate of 42.9% and with top 10 sensitivity of 60.0% in Group 1, which included all cases. When considering only Face2Gene-trained syndromes, the top 1 sensitivity rate increased to 61.2%. In Group 2, which focused on cases with Down syndrome (DS), the software achieved a top 1 sensitivity rate of 100% for both youngest and oldest facial images.

In (Vk & Rajesh 2022), the authors proposed a method for recognizing DS in face images is described, which includes feature extraction, facial fiducial point detection, feature reduction, and classification. The system developed using the Back Propagation algorithm shows a detection rate and accuracy that contribute to the effective diagnosis of DS. The overall efficiency of the system is 93.58%.

In (Setyati *et al.* 2021), the paper proposes two CNN architectures for the identification of Down Syndrome, William Syndrome, and Normal. The first CNN architecture, trained and evaluated six times, achieves an average accuracy of 91%. The second architecture, which includes 15 layers, has an average accuracy of 89%.

In (Pooch *et al.* 2020), the authors proposed two algorithms for automated identification of Down Syndrome through images: one SVM with accuracy of 0.84 and another, a deep CNN having accuracy of 0.94. The results were validated using 10-fold cross validation. The dataset here used is a custom dataset having 170 images where 50% are of healthy children and rest of the images were of children diagnosed with down syndrome.

3 METHODOLOGY

3.1 *Dataset description*

The dataset chosen here is a publicly available dataset on Kaggle (Kaggle, n.d.). The dataset contains two directories containing – images of healthy children and images of children diagnosed with down syndrome. It contains about 1500 images of healthy children and 1500 images of children diagnosed with down syndrome. Generally, the age of children ranges from 0–18, but the most of the images of children has age between 0 to 15. Below Figure 1. shows the proposed framework used on the dataset.

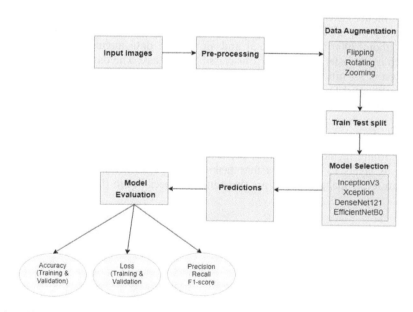

Figure 1.   Flowchart of the proposed framework.

3.2 *Libraries*

In this research, various Python libraries were utilized to facilitate the processing, analysis, and visualization of data, as well as the development and training of neural network models. The os library was utilized for managing image paths, enabling seamless access to image files and directories. The OpenCV was utilized for image loading and preprocessing, allowing for efficient manipulation and processing of image data. NumPy, a fundamental package for numerical computing in Python, was employed for efficient handling and manipulation of image arrays. TensorFlow, along with its high-level API Keras, was utilized for various purposes including creating neural network layers and architectures, implementing optimization algorithms, selecting pre-trained models for transfer learning, employing the ImageDataGenerator for data augmentation. The scikit-learn library was used for evaluating model performance by providing essential metrics, including accuracy, precision, recall, and generating confusion matrixes. Matplotlib and Seaborn were utilized for data visualization, aiding in the clear representation of training and validation loss, as well as training and validation accuracy during model training. Kaggle's GPU-integrated notebooks with a CPU RAM of 13 GB and GPU P100 with a memory of 15.9 GB were employed for training the neural network models, providing the necessary computational resources for efficient model training and experimentation.

3.3 *Data preprocessing*

Prior to feeding the image data into the neural network for classification, a series of preprocessing steps were applied to ensure uniformity and enhance the quality of the data. First, all images were resized to a standardized dimension. Subsequently, pixel values in the images were normalized to fall within the range of [0, 1]. Normalization enhances model convergence and performance by ensuring that each pixel's intensity contributes proportionately to the network's learning. Furthermore, data augmentation techniques, including random rotation, horizontal and vertical shifts, zooming, and horizontal flipping, were applied to the training dataset. Augmentation diversifies the dataset, augmenting the learning process and helping the model generalize better by exposing it to a broader range of image variations. This step can prove to be vital in the cases when you have a small dataset, like in this case we have only about 3000 images in total and along with data augmentation, we have also preferred to use transfer learning for this classification task. The dataset was then split into train and test in 3:1 ratio. The results of different CNN algorithms like Inception, Xception, EfficientNetB0, and DenseNet121 are compared to provide insight as to which Network architecture performs the best on the dataset.

3.4 *Model architecture*

Transfer learning includes utilizing a pre-trained neural network as a starting point and modifying its layers to adapt to a new task. The pre-trained model's early layers retain general features, while later layers are fine-tuned to specialize in the specific features of the

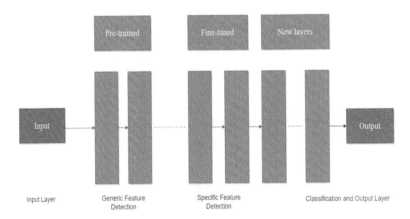

Figure 2.   Transfer learning model architecture

new task. The number of top layers of pre-trained neural network fine-tuned depends on similarity of the task for which pre-trained model was designed for and the current task. This approach speeds up training and enhances performance, especially when the new task has a limited dataset. The Figure 2. shows the transfer learning model architecture used in this paper. Initially, we froze earlier layers to retain pre-trained weights and then selectively unfreezing specific layers for retraining to enhance model learning. Adapting the learning rate, batch size, and optimizer like Adam and Adagrad while considering the dataset size and complexity for fine-tuning. Additionally, the dropout rates, kernel sizes, were experimented and architecture-specific hyperparameters like module configurations and number of layers in models such as Inception or Xception were refined for the model's ability to extract discriminative features from facial images, augmenting its performance in Down syndrome detection through transfer learning.

3.5 *Algorithms*

The architectures mentioned below represents different design philosophies and innovations in convolutional neural network design, covering areas such as depthwise separable convolutions (Xception), inception modules (Inception), efficient scaling (EfficientNet), and densely connected layers (DenseNet). This diversity allows for a comprehensive comparison of various architectural approaches in detecting down syndrome.

3.5.1 *DenseNet121*
DenseNet, short for Densely Connected Convolutional Networks, is a cutting-edge convolutional neural network architecture characterized by a unique connectivity pattern among layers. This dense interconnection facilitates feature reuse, encourages gradient flow, and promotes deep feature propagation. DenseNet's architecture alleviates the vanishing gradient problem and enhances parameter efficiency while maintaining model compactness.

3.5.2 *Inception*
Inception, also known as GoogLeNet, is a seminal deep learning architecture notable for its innovative inception module. This module utilizes multiple parallel convolutional operations with varying filter sizes within a sole layer, allowing the network to capture features at different scales. The architecture integrates these inception modules in a carefully designed and computationally efficient manner, leading to increased model depth and width without an overwhelming rise in computational cost.

3.5.3 *Xception*
Xception, or "Extreme Inception," is an advanced convolutional neural network architecture that deviates from traditional CNN designs by utilizing depthwise separable convolutions. This architecture, each convolutional operation is split into two separate layers: a depthwise convolution responsible for filtering spatially, followed by a pointwise convolution for combining information across channels.

3.5.4 *EfficientNet80*
EfficientNetB0 comprises a stack of convolutional layers, including depthwise separable convolutions, followed by global average pooling and fully connected layers. It employs skip connections, similar to ResNet, to facilitate gradient flow and aid in training deeper networks efficiently.

4 RESULTS AND DISCUSSION

The training and validation loss curve shown in Figure 3., DenseNet and Inception have a relatively smoother learning curve towards the end, suggesting that the model is learning steadily and effectively. This indicates a good balance between model complexity and dataset fit, likely resulting in better generalization and stable performance on unseen data.

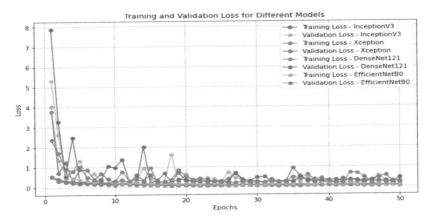

Figure 3. Training and validation loss curve of different models.

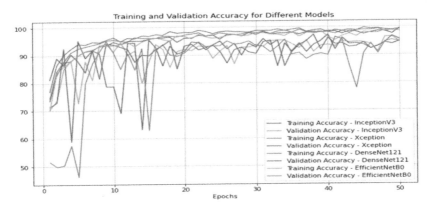

Figure 4. Training and validation accuracy curve of different models.

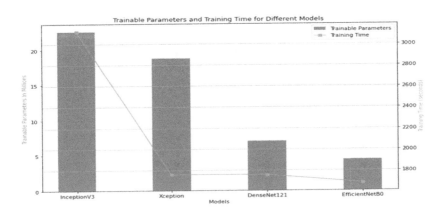

Figure 5. Trainable parameters V/s training time.

On observing Figure 4., the plot of training and validation accuracy curve of the models, it is evident that the validation accuracy curve of Inception and Xception is somewhat smoother as compared to that of DenseNet and EfficientNet indicating that the model is effectively capturing patterns in the training data and is able to generalize those patterns to unseen validation data. As evident from Figure 5., the Xception model, with around 18 million trainable parameters, outperforms in training time at approximately 1760 seconds, closely comparable to DenseNet with about 6.9 million parameters; in contrast, InceptionV3, boasting nearly 22.5 million parameters, exhibits the longest training time at 3116.6 seconds.

In clinical contexts focusing on Down syndrome detection via facial image analysis, precision, recall, and the F1 measure are pivotal. These metrics signify the accuracy and reliability of identifying individuals with Down syndrome. High precision ensures that flagged cases are accurate, reducing unnecessary stress and misdiagnosis. High recall ensures that affected individuals aren't missed, facilitating timely interventions. The F1 measure provides an overall gauge of accuracy, guiding the optimization of detection systems for better clinical outcomes and support for affected individuals and their families.

Table 1. Evaluation metrics of different models as measured on the test dataset.

Model	Class Label	Testing			
		Precision	Recall	F1-Score	Accuracy
Xception	Positive	0.95	0.95	0.95	95%
	Negative	0.94	0.94	0.94	
Inception	Positive	0.93	0.96	0.94	94%
	Negative	0.96	0.92	0.94	
DenseNet121	Positive	0.96	0.91	0.94	93.6%
	Negative	0.91	0.96	0.94	
EffecientNetB0	Positive	0.90	0.96	0.93	92.5%
	Negative	0.96	0.88	0.92	

The above Table 1. shows the overall performance metrics of the models for both the classes with Xception having the highest accuracy of 95%. Xception's use of Depthwise Separable Convolutions helps in capturing intricate features from the images efficiently, which might be crucial in differentiating between the characteristics associated with Down syndrome and those without. It's ability to extract hierarchical representations of features from images can be vital in discerning subtle differences between the groups.

5 CONCLUSION AND FUTURE WORK

In conclusion, this study explored the performance of transfer learning utilizing various neural network architectures—Inception, Xception, DenseNet, and EfficientNet—as base models for the purpose of detecting Down syndrome via image analysis. The research revealed that using pre-trained models greatly improves classification performance, with Xception outperforming other models in terms of Accuracy, Precision, Recall, and F1-score. Transfer learning approaches provide efficient knowledge transfer from pre-trained models, assisting in the extraction of discriminative features and overcoming the hurdles posed by small dataset sizes. Furthermore, the comparative analysis provided useful insights into each architecture's strengths and limitations, including the training time, allowing for informed selection and optimization of the best model for this critical medical imaging application. Future work should focus on further fine-tuning and customization of these architectures to optimize their performance for Down syndrome detection, and potentially adding newer and more complex neural architectures like NasNetLarge or ResNeXt, contributing to more effective early diagnoses and improved patient outcomes.

REFERENCES

Centers for Disease Control and Prevention (CDC) 2021, Facts about Down Syndrome | CDC, online, Available at: https://www.cdc.gov/ncbddd/birthdefects/downsyndrome.html#:~:text=People%20with%20Down%20syndrome%20usually.

Espinosa, J.M., 2020. Down syndrome and COVID-19: a perfect storm?. *Cell Reports Medicine*, *1*(2).

Firsowicz, M., Boyd, M. and Jacks, S.K., 2020. Follicular occlusion disorders in Down syndrome patients. *Pediatric Dermatology*, *37*(1), pp.219–221.

Kaggle. (n.d.). Detection of Down Syndrome in Children. [online] Available at: https://www.kaggle.com/datasets/mervecayli/detection-of-down-syndrome-in-children [Accessed 7 Oct. 2023].

Mishima, H., Suzuki, H., Doi, M., Miyazaki, M., Watanabe, S., Matsumoto, T., Morifuji, K., Moriuchi, H., Yoshiura, K.I., Kondoh, T. and Kosaki, K., 2019. Evaluation of Face2Gene using facial images of patients with congenital dysmorphic syndromes recruited in Japan. *Journal of human genetics*, *64*(8), pp.789–794.

NIH, 2017, *'What are common treatments for Down syndrome?'*, National Institute of Child Health and Human Development, [online] Available at: https://www.nichd.nih.gov/health/topics/down/conditioninfo/treatments [Accessed 10 Oct. 2023].

Pooch, E.H.P., Alva, T.A.P. and Becker, C.D.L., 2020. A Computational Tool for Automated Detection of Genetic Syndrome Using Facial Images. In *Intelligent Systems: 9th Brazilian Conference, BRACIS 2020, Rio Grande, Brazil, October 20–23, 2020, Proceedings, Part I 9* (pp. 361–370). Springer International Publishing.

Qin, B., Liang, L., Wu, J., Quan, Q., Wang, Z. and Li, D., 2020. Automatic identification of down syndrome using facial images with deep convolutional neural network. *Diagnostics*, *10*(7), p.487.

Ropper, A.H. and Bull, M.J., 2020. Down syndrome. *NEJM*, *382*(24), pp.2344–2352.

Setyati, E., Az, S., Hudiono, S.P. and Kurniawan, F., 2021. CNN based face recognition system for patients with Down and William syndrome. *Knowledge Engineering and Data Science (KEDS)*, *4*(2), pp.138–144.

Sideropoulos, V., Kye, H., Dukes, D., Samson, A.C., Palikara, O. and Van Herwegen, J., 2023. Anxiety and worries of individuals with down syndrome during the COVID-19 pandemic: A comparative study in the UK. *Journal of Autism and Developmental Disorders*, *53*(5), pp.2021–2036.

Thomas, M.C. and Arjunan, S.P., 2022. Deep learning measurement model to segment the nuchal translucency region for the early identification of down syndrome. *Measurement Science Review*, *22*(4), pp.187–192.

VK, V.D. and Rajesh, R., 2022. Down syndrome identification and classification using facial features with neural network. *Global Journal of Engineering and Technology Advances*, *12*(1), pp.001–011.

Wang, C., Yu, L., Su, J., Mahy, T., Selis, V., Yang, C. and Ma, F., 2023. Down Syndrome detection with Swin Transformer architecture. *Biomedical Signal Processing and Control*, *86*, p.105199.

Zhao, Q., Okada, K., Rosenbaum, K., Kehoe, L., Zand, D.J., Sze, R., Summar, M. and Linguraru, M.G., 2014. Digital facial dysmorphology for genetic screening: Hierarchical constrained local model using ICA. *Medical image analysis*, *18*(5), pp.699–710.

Automated diabetes diagnosis and risk assessment using machine learning

Deep Mathukiya & Yogesh Kumar
Department of Computer Science and Engineering, School of Technology, Pandit Deendayal Energy University, Gandhinagar, India

Apeksha Koul
Punjabi University, Patiala, Punjab, India

ABSTRACT: Currently, Diabetes is a prevalent issue worldwide, affecting millions of people. The detection machinery for this disease is only available at healthcare centers, so individuals with diabetes are only aware of their condition when they visit and undergo a checkup. Prolonged diabetes can lead to various complications like heart disorders, kidney disease, nerve damage, and diabetic retinopathy. However, the risk can be reduced if the disease at its early stage is predicted. By utilizing a deep learning approach, this model can easily determine if an individual is diabetic or not. The aim of the research is to develop machine learning model that can identify diabetic people on the basis of their clinical data. The machine learning algorithms utilized in this study include Naive Bayes (NB), k-nearest neighbor (KNN), Stochastic Gradient Descent (SGD), Decision tree (DT), Support Vector Machine (SVM), Random Forest (RF), and Logistic Regression (LR). We have implemented efficient pre-processing techniques to detect as well as handle missing/ NAN and duplicate values and later trained our models with the cleaned dataset. During execution, we found that Random Forest as well as Decision Tree performed well for both diabetic and non-diabetic class by computing 100% and 99% accuracy respectively.

1 INTRODUCTION

The transcendence of diabetes, a steady metabolic issue portrayed by raised blood glucose levels, has shown up at upsetting degrees generally, making it one of the most crushing clinical consideration troubles of the 21st 100 years. According to the World Prosperity Affiliation (WHO), a normal 463 million persons had diabetes in 2019, and if current trends continue, this figure is expected to increase to 700 million by 2045 (Saeedi *et al.* 2019). Reasonable organization and ideal intervention are essential to thwart troubles and work on the individual fulfilment for individuals living with diabetes (Parimala *et al.* 2023). Simulated intelligence (ML) strategies have emerged as essential resources for the assumption, gathering, and the chiefs of diabetes in light of their ability to research complex data models and concentrate huge encounters. The prevalence of diabetes on a global scale has reached epidemic levels, necessitating creative techniques for timely discovery and effective management. In this particular circumstance, AI (ML) algorithms have emerged as essential tools for predicting and classifying diabetes in light of a vast array of characteristics (Kaur *et al.* 2022). This study focuses on a comprehensive set of markers, including hypertension, orientation, age, a history of cardiovascular illness, a tendency to smoke, BMI, hbA1c levels, and glucose levels. Together, these characteristics present a large dataset with the potential to significantly increase the accuracy of diabetes prediction and categorization. By outfitting the force of ML calculations, medical services experts and specialists can use the complicated connections among these factors to fabricate prescient models that can recognize people in danger of diabetes or characterize them as diabetic or non-diabetic. The incorporation of different boundaries like orientation, age, and way of life propensities recognizes the intricacy of diabetes as a condition impacted by hereditary, ecological, and conduct factors. This approach adds to exact individualized expectations as well as establishes the ground work for a more nuanced comprehension of diabetes, the study of disease transmission on a worldwide scale.

2 LITERATURE WORK

Zhou et al. (2020) emphasized the global prevalence of diabetes as a pervasive and persistent ailment, posing substantial risks due to potential complications. Early detection of diabetes holds paramount importance for timely intervention and halting disease progression. Their method not only forecasts future diabetes occurrences but also identifies the specific diabetes type, crucial for tailoring individualized treatment plans. The approach transforms the challenge into a classification task, leveraging a model within the layers of a deep neural network, enhanced by dropout regularization to mitigate overfitting. With meticulous parameter tuning and binary cross-entropy loss, the proposed DLPD (Deep Learning for Anticipating Diabetes) model achieves exceptional accuracy. Notably, it attains a training accuracy of 94.02174% for the diabetes type dataset and an impressive 99.4112% for the Pima Indians diabetes dataset, affirming the efficacy of the proposed deep learning model. Fregoso-Aparicio et al. (2021) mentioned that when blood glucose levels exceed certain thresholds, the condition known as diabetes mellitus develops. In recent times, learning models have been harnessed to forecast diabetes and its associated problems. However, within the realm of constructing predictive models for type 2 diabetes, researchers and developers continue to confront two primary hurdles. The systematic review of Luis et al. endeavours to address the aforementioned challenges comprehensively. The review adheres primarily to the PRISMA methodology, augmented by the approach advocated by Keele and Durham Universities. Butt et al. (2021) presented a sizable quantity of crucial and sensitive healthcare data has been produced as a result of significant advancements in biotechnology and the growth of the public healthcare infrastructure. Intelligent data analysis methods have been used to reveal fascinating patterns that have made it easier to identify and stop the onset of a number of serious diseases. Due to its role in potentially fatal illnesses like heart, kidney, and nerve damage, diabetes mellitus has an especially dire position. Nadeem et al. (2021) mentioned about making informative driven applications and administrations for diagnosing and sorting critical ailments faces difficulties due to deficient and below average logical information for preparing and approving calculations. Internationally, the mounting costs related with overseeing diabetes, a far and wide persistent infirmity portrayed by delayed times of raised glucose levels, are applying huge tensions on medical services suppliers. The proposed arrangement holds the possibility to raise endurance rates among PwD by working with individualized treatment suggestions. Likewise, Tasin et al. (2023) discussed that globally, diabetes affects approximately 537 million people and make it the most prevalent as well as deadly non-communicable illness. Several factors, such as being overweight, having abnormal cholesterol levels, lacking any physical activity, history of any family member having diabetes, and unhealthy eating habits are responsible to increase the risk of giving rise to diabetes. A paper of them presents an automated system for predicting diabetes, where the machine learning models are trained with the private dataset collected from female patients of Bangladesh. Ahmed et al. (2021) mentioned in their paper that diabetes is a widespread disease and it can be associated with serious complications such as kidney failure and heart disease. Early detection of diabetes can lead to a longer and healthier life. Using various supervised machine learning models trained on suitable data, it is possible to diagnose diabetes in its early stages. The techniques of preprocessing which includes label encoding as well as normalization are applied to improve model accuracy. In addition, different feature selection methods are used to identify and priorities key risk factors. Rigorous experiments evaluate the performance of the model on different datasets. Birjais et al. (2019) stated that when paired with data mining techniques, machine learning offers enormous potential, particularly in the area of forecasting. Today, data generation is growing, but if it isn't turned into useful information, it's worth remains untapped. This is also true in the field of health care, where there is a wealth of information that requires its collection to improve prognosis, diagnosis, treatment, drug development and the general advancement of health care. In study, they mainly focus on the diagnosis of diabetes, a chronic disease that is rapidly spreading worldwide, as defined by the World Health Organization in 2014. Shafi & Ansari (2021) address diabetes as a persistent global health concern affecting all age groups. Leveraging technological advancements, they emphasize the potential for accurate diabetes prediction throughout a patient's life. The study proposes a paradigm utilizing machine learning techniques like DT, SVM, and NB for early diabetes identification. The research aims to maximize precision and effectiveness in diabetes detection through these techniques, contributing to the ongoing efforts to combat this global health issue. Aslan & Sabanci (2023) propose an innovative deep learning approach for early diabetes detection, addressing the numeric nature of PIMA datasets. Unable to apply traditional CNN models, the study transforms numerical data into visualizations, utilizing the significance of features. Three classification strategies are employed: feeding diabetes images into CNN models (ResNet18 or ResNet50), combining deep features with SVM, and classifying fusion features with SVM. This methodology aims to enhance the early diagnosis process by leveraging the powerful performance of CNN models on transformed diabetes image data. Awasthi *et al.* (2022) conducted assessment of multiple diabetes prediction models stemmed from the need to locate, critically evaluate, and amalgamate pertinent and high-quality individual research findings. The aim of study was the identification of optimal approaches for the selection and synthesis of high-quality studies. Medical data, which is predominantly nonlinear, characterized by intricate correlation structures, poses a significant analytical challenge. Notably, the utilization of machine learning in healthcare and medical imaging has been excluded from consideration.

The mentioned studies highlight the global prevalence of diabetes and emphasize the significance of early detection. Various approaches, including deep learning models like DLPD, machine learning techniques, and innovative methods, such as visualizing numerical data for early diagnosis, have been proposed. The studies underscore the challenges in constructing predictive models for type 2 diabetes and the potential of intelligent data analysis in healthcare. The goal is to enhance accuracy in diabetes prediction and contribute to combating this widespread health issue.

3 FRAMEWORK FOR DIABETIC DETECTION

The system design as per the Figure 1 provides a comprehensive visualization of the end-to-end process for predicting diabetes using machine learning. It starts by introducing the dataset and proceeds through crucial stages like data analysis and preprocessing, ensuring data readiness followed by data splitting. Various machine learning models have been applied and later their performances have been examined for both diabetic and non-diabetic classes.

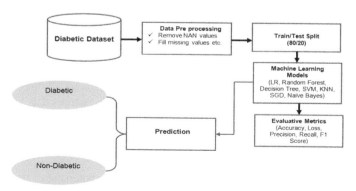

Figure 1.   System design for the prediction of diabetes.

3.1 Description of the dataset

The data for the research have been collected from Diabetes prediction dataset which encompasses of medical as well as demographic information sourced from patients, accompanied by their diabetes status (positive or negative). This dataset comprises attributes like gender, age, hypertension, body mass index (BMI), heart disease, HbA1c level, smoking history, and blood glucose level. It serves as a foundational resource for constructing machine learning models intended to forecast diabetes in patients based on their medical records and demographic particulars. This dataset holds potential utility for healthcare professionals in identifying individuals who might be prone to diabetes development, facilitating the formulation of customized treatment strategies. Table 1 presents few attributes of the dataset [https://www.kaggle.com/datasets/iammustafatz/diabetes-prediction-dataset].

Table 1. Attributes of diabetes dataset with their description.

Attributes	Description
Gender	It defines the sexuality of an individual and also impacts the suspension of having diabetes
Age	It also plays an important role because diabetes is mostly seen in adult generation.
Hypertension	It is blood pressure that consistently elevates in the arteries of the heart.
Heart Disease	This medical illness is also linked with the increasing risk of having diabetes.
Body Mass Index	It is for calculating the fat in the human body by considering their height as well as weight. In this dataset, the value of body mass index ranges from 10.16 to 71.55 where <18.5 indicates underweight and 18.5 to 24.9 classifies overweight and above 30 is obese.

3.2 Data preprocessing

Pre-processing the dataset is an important step to extract meaningful insights and construct accurate predictive models. Initially, Pandas DataFrame is used to load the data and enables efficient data manipulation and exploration. It provides robust functionality to identify missing values in any column as well as identify duplicate records and process them to ensure dataset accuracy. This thorough approach to data pre-processing lays a solid foundation for robust analysis and modeling.

3.3 Train-Test splitting

Performing a train-test split on dataset is crucial to assess the performance of machine learning model to detect diabetic and non-diabetic patients. In this research paper, we have Scikit library to split the dataset in training as well as testing in a ratio of 80:20. After the split, the training set trains the model's parameters, while the testing set assesses its predictive accuracy. This approach helps prevent overfitting, where the model performs well on the training data but fails to generalise to new data.

3.4 Applied machine learning classifiers

Logistic regression, a fundamental binary classification algorithm, is commonly employed in diabetic and non-diabetic detection due to its interpretability and efficiency. By modeling the probability of an input that belongs to a specific class, it efficiently categorizes data into two distinct groups (Awasthi et al. 2022). For binary classification, it can be computed by using Equation (1):

$$\ln\left(\frac{p}{1-p}\right) = \beta_0 + \beta_1 x_1 + \beta_2 x_2 + \ldots\ldots + \beta_3 x_3 \tag{1}$$

Here, p is the probability, ln is the natural logarithm, β_0 is the intercept of the equation and rest of them are the coefficients associated with the predicted variables x_1, x_2 ... x_n. **Naive**

Bayes, despite its simplistic assumption of independence, is surprisingly effective in tasks like diabetes prediction, especially in text-based contexts like medical records (Awasthi *et al.* 2022). It is computed by Equation (2)

$$Posterior\ Prob\ of\ class\ A = \frac{Prior\ Prob\ of\ class\ A \times Likelihood\ prob\ of\ class\ A}{Prior\ prob\ without\ any\ specific\ class} \quad (2)$$

Stochastic Gradient Descent, an optimization technique, is invaluable for training large-scale models crucial in diabetes detection. Its primary objective is to iteratively minimize the loss function by updating the model parameters. It is particularly valuable for training large models and is often used with lot of machine learning approach (Kaur *et al.* 2022). **K-Nearest Neighbors'** versatility makes it applicable for diabetic and non-diabetic classification, as it predicts based on the majority class of its nearest neighbours, a useful characteristic in handling various diabetes-related features (Thakur *et al.* 2022). It is computed by Equation (3).

$$Distance(X_{new},\ X_i) = \sqrt{\sum_{j-1}^{n} \left(X_{new.j} - X_{i,j} \right)^2} \quad (3)$$

Decision trees, with their intuitive tree-like structures, make feature-based decisions aiding in diabetes prediction scenarios. They are popular models for classification and regression tasks. They create a tree-like structure that partitions the feature space on the basis of attribute values. Every internal node and leaf node represents a feature-based decision, and a predicted class or value respectively while as **Random Forest,** an ensemble method merges predictions from multiple decision trees, enhancing accuracy and mitigating overfitting, making it pivotal in diabetes prognosis (Koul *et al.* 2022). **Support Vector Machine,** highly effective in high-dimensional data, is adept at discerning clear boundaries between diabetic and non-diabetic classes, a vital aspect in precise diabetes detection and management (Bhardwaj *et al.* 2022).

3.5   *Evaluative parameters*

Various parameters, as shown in Table 2, have been used to examine the performance of the applied learning models while being trained with the diabetic dataset (Kumar *et al.* 2022).

Table 2.   Evaluative parameters.

Parameter	Definition
Accuracy	The proportionate count of accurate predictions
Loss	The degree to which the predictions of the models are inaccurate.
Recall	The proportion of accurate predictions made for true positives.
Precision	The proportion of positive predicts that prove to be accurate.
F1 Score	A single value that integrates recall and precision as a metric.

4   RESULTS

The assessment of different machine learning models in a binary classification task involving non-diabetic and diabetic cases are presented in Table 3.

The accuracy values, along with their corresponding "loss" values, were evaluated for each model. For instance, the Logistic Regression achieved a high accuracy of 0.99 for non-diabetic cases and 0.63 for diabetic cases, corresponding to "loss" values of 0.011 and 0.37, respectively. Similarly, the Naïve Bayes model demonstrated an accuracy of 0.93 for non-diabetic cases and 0.65 for diabetic cases, with corresponding "loss" values of 0.073 and 0.35. Stochastic Gradient Descent achieved an accuracy of 0.97 for non-diabetic cases and 0.71 for diabetic cases, resulting in "loss" values of 0.025 and 0.29 for the respective classes. The K-Nearest Neighbours model attained perfect accuracy (1) for non-diabetic cases and an

Table 3. Assessing the performance of each model on diabetic and non-diabetic.

Models	Diabetic		Non-Diabetic	
	Accuracy	Loss	Accuracy	Loss
Logistic Regression	0.99	0.011	0.63	0.37
Random Forest	1	2.7e-05	0.99	0.0084
SGD	0.97	0.025	0.71	0.29
Decision Tree	1	1.4e-05	0.99	0.0084
KNN	1	0.0039	0.6	0.4
Naïve Bayes	0.93	0.073	0.65	0.35
SVM	1	1.4e-05	0.39	0.61

accuracy of 0.6 for diabetic cases, leading to "loss" values of 0.0039 and 0.4. Remarkably, both the Random Forest as well as Decision Tree models achieved perfect accuracy (1) for both non-diabetic and diabetic cases, accompanied by extremely low "loss" values of 1.4e-05 and 2.7e-05 for both classes, respectively. While the Support Vector Machine (SVM) reached perfect accuracy (1) for non-diabetic cases, its accuracy for diabetic cases was 0.39. This discrepancy resulted in "loss" values of 1.4e-05 for non-diabetic and 0.61 for diabetic cases.

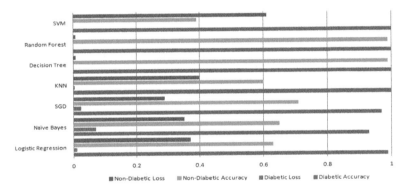

Figure 2. Graphical analysis of models.

Remark of Figure 2 depicts that the Decision Tree as well as Random Forest models demonstrated exceptional accuracy by achieving perfect scores across both classes. Other models also showed favourable performances with varying accuracy levels. Selecting the most suitable model should be based on specific task requirements and trade-offs.

The performance of various machine learning models to predict diabetic and non-diabetic cases was assessed using recall, precision, as well as F1 score metrics, as shown in Table 4.

Table 4. Evaluation parameters for diabetic and non-diabetic.

Models	Diabetic			Non-diabetic		
	Precision	Recall	F1score	Precision	Recall	F1score
SVM	1	0.39	0.56	0.95	1	0.97
Random Forest	1	1	1	1	0.99	1
Logistic Regression	0.84	0.63	0.72	0.97	0.99	0.98
SGD	0.62	0.76	0.68	0.98	0.96	0.97
Naïve Bayes	0.45	0.65	0.53	0.97	0.93	0.95
Decision Tree	1	0.99	1	1	1	1
KNN	0.94	0.60	0.73	0.96	1	0.98

Among the models evaluated, Logistic Regression demonstrated a respectable balance between precision, recall, and F1 score by computing 0.84, 0.63, and 0.72 respectively for diabetic cases. On the contrary, it has achieved the highest recall of 0.99, precision of 0.97, and F1 score of 0.98 for non-diabetic cases. Naïve Bayes, although having a lower precision (0.45) for diabetic cases, showed a relatively higher recall (0.65). However, its overall F1 score (0.53) indicated room for improvement. SGD exhibited decent performance with a precision of 0.62 and recall of 0.76 for diabetic cases, along with a respectable F1 score of 0.68. KNN showcased high precision (0.94) but a relatively lower recall (0.60) for diabetic cases, leading to an F1 score of 0.73. Decision Tree as well as Random Forest models performed exceptionally well, achieving perfect precision, recall, along with the F1 score for both diabetic and non-diabetic cases. Lastly, SVM demonstrated perfect precision (1) for diabetic cases but had a lower recall as 0.39 and F1 score as 0.56, indicating challenges in identifying all diabetic cases. In summary, Decision Tree and Random Forest models outperformed others, exhibiting flawless accuracy, while Logistic Regression also showed commendable overall performance in classifying diabetic and non-diabetic cases. The same parameters are also being graphically presented in Figure 3.

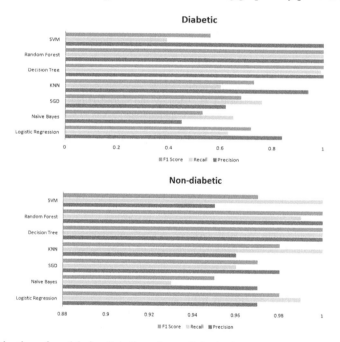

Figure 3. Evaluation of models for diabetic and non-diabetic class.

5 CONCLUSION

In this paper, a comprehensive evaluation of several machine learning models was performed, and the results are summarized in the accompanying graphs. These models which include naive Bayes, random forest, stochastic gradient, logistic regression, decision tree, K-nearest neighbours as well as support vector machine are evaluated using three basic performance metrics: precision, recall as well as F1 score. These metrics are key indicators of the models' performance in various classification tasks. In particular, Decide Tree and Random Forest achieved perfect scores for recall as well as precision, highlighting their potential suitability for applications where it is important to achieve the highest levels of precision and recall. In contrast, Naive Bayes showed a trade-off between precision and recall, making it a worthy choice when the main consideration is between false positives and false negatives.

These findings provide valuable information to select the most appropriate or best machine learning model based on specific application requirements. Nevertheless, it is crucial to recognize the constraints of our study. The success of machine learning models relies heavily on the caliber as well as volume of data. With the ongoing progress in machine learning algorithms and healthcare data collection, there is potential for future research to delve into more advanced algorithms and diverse datasets. This could lead to even more accurate and widely applicable results.

REFERENCES

Awasthi, A., Gangwal, I., and Jain, M., 2022. Diabetes prediction using machine learning: A Review. *International Journal for Research*: 1–12.

Ahmed, N., Ahammed, R., Islam, M.M., Uddin, M.A., Akhter, A., Talukder, M.A. and Paul, B.K., 2021. Machine learning based diabetes prediction and development of smart web application. *International Journal of Cognitive Computing in Engineering*, 2, pp.229–241.

Aslan, M.F. and Sabanci, K., 2023. A novel proposal for deep learning-based diabetes prediction: Converting clinical data to image data. *Diagnostics*, 13(4) : 796.

Bhardwaj, P., Kumar, S. and Kumar, Y., 2022, May. Deep learning techniques in gastric cancer prediction and diagnosis. In *2022 International Conference on Machine Learning, Big Data, Cloud and Parallel Computing (COM-IT-CON)*, 1: 843–850. IEEE.

Birjais, R., Mourya, A.K., Chauhan, R. and Kaur, H., 2019. Prediction and diagnosis of future diabetes risk: a machine learning approach. *SN Applied Sciences*, 1:1–8.

Butt, U.M., Letchmunan, S., Ali, M., Hassan, F.H., Baqir, A. and Sherazi, H.H.R., 2021. Machine learning based diabetes classification and prediction for healthcare applications. *Journal of healthcare engineering*, 2021:1–17.

Fregoso-Aparicio, L., Noguez, J., Montesinos, L. and García-García, J.A., 2021. Machine learning and deep learning predictive models for type 2 diabetes: a systematic review. *Diabetology & metabolic syndrome*, 13(1):1–22.

Kaur, K., Singh, C. and Kumar, Y., 2022, December. Artificial intelligence techniques for the detections of congenital diseases: Challenges and research perspectives. In *2022 5th International Conference on Contemporary Computing and Informatics (IC3I)* (pp. 888–893). IEEE.

Kaur, S., Bansal, K. and Kumar, Y., 2022, December. Artificial intelligence approaches for predicting hypertension diseases: Open challenges and research issues. In *2022 5th International Conference on Contemporary Computing and Informatics (IC3I)*: 338–343. IEEE

Koul, A., Kumar, Y. and Gupta, A., 2022, October. A study on bladder cancer detection using AI-based learning techniques. In *2022 2nd International Conference on Technological Advancements in Computational Sciences (ICTACS)*: 600–604. IEEE.

Kumar, Y., Patel, N.P., Koul, A. and Gupta, A., 2022, February. Early prediction of neonatal jaundice using artificial intelligence techniques. In *2022 2nd international conference on innovative practices in technology and management (ICIPTM)*, 2 : 222–226. IEEE.

Nadeem, M.W., Goh, H.G., Ponnusamy, V., Andonovic, I., Khan, M.A. and Hussain, M., 2021, October. A fusion-based machine learning approach for the prediction of the onset of diabetes. In *Healthcare*, 9(10): 1393 MDPI.

Parimala, G., Kayalvizhi, R. and Nithiya, S., 2023, January. Diabetes Prediction using Machine Learning. In *2023 International Conference on Computer Communication and Informatics (ICCCI)* (pp. 1–10). IEEE.

Saeedi, P., Petersohn, I., Salpea, P., Malanda, B., Karuranga, S., Unwin, N., Colagiuri, S., Guariguata, L., Motala, A.A., Ogurtsova, K. and Shaw, J.E., 2019. Global and regional diabetes prevalence estimates for 2019 and projections for 2030 and 2045: Results from the International Diabetes Federation Diabetes Atlas. *Diabetes research and clinical practice*, 157, p.107843.

Shafi, S. and Ansari, G.A., 2021. Early prediction of diabetes disease & classification of algorithms using machine learning approach. In *Proceedings of the International Conference on Smart Data Intelligence (ICSMDI 2021)*: 1–6.

Tasin, I., Nabil, T.U., Islam, S. and Khan, R., 2023. Diabetes prediction using machine learning and explainable AI techniques. *Healthcare Technology Letters*, 10(1–2) : 1–10.

Thakur, K., Kaur, M. and Kumar, Y., 2022, October. Artificial intelligence techniques to predict the infectious diseases: Open challenges and research issues. In *2022 2nd International Conference on Technological Advancements in Computational Sciences (ICTACS)*: 109–114. IEEE.

Zhou, H., Myrzashova, R. and Zheng, R., 2020. Diabetes prediction model based on an enhanced deep neural network. *EURASIP Journal on Wireless Communications and Networking*, 2020, pp.1–13.

Next Generation Computing and Information Systems – Gupta. (Ed.)
© 2025 The Author(s), ISBN 978-1-032-73865-9

Machine learning-based approaches for automated hypertension detection

Simranjit Kaur
Research Scholar, Desh Bhagat University, Mandi Gobindgarh, Punjab, India

Khushboo Bansal
Department of CSE, Desh Bhagat University, Mandi Gobindgarh, Punjab, India

Yogesh Kumar
Department of CSE, School of Technology, Pandit Deendayal Energy University, Gandhinagar, Gujarat, India

ABSTRACT: Early detection and timely intervention of hypertension is critical to mitigate its adverse effects as it is a leading cause of heart disease. Traditionally, to detect hypertension depends upon various clinical factors such as measuring blood pressure, assessment of medical history, etc but they also have certain limitations such as human error, time consuming, as well as chances of missed diagnosis. Hence, using AI techniques allows identifying complex relationships within the data as well as surpassing the limitations of traditional diagnostic methods. The aim of the paper is to create a model which is capable of identifying individuals with hypertension and those without by using various AI learning models. With a dataset of 14 attributes, preprocessing identifies missing data while as visualization and correlation analysis has been used for detecting outliers and feature selection respectively. Further, seven different classifiers are applied such as gradient boosting, voting classifier, light gradient boosting machine, multilayer perceptron, Catboost, eXtreme gradient boosting classifier, and Adaboost. The performance of these classifiers has been evaluated it appears that CatBoost, XGBoost, and LGBM perform with a remarkable accuracy of 100% as well as outstanding recall, precision, along with the F1 score of 1.00. It demonstrates the capability of AI-driven models to revolutionize hypertension detection.

1   INTRODUCTION

In this era of advanced technology, the fusion of AI learning models and healthcare holds great promise to revolutionize the medical sector. By coupling the power of data-driven insights, healthcare professionals have the capability to make informed decisions which improves the patient outcomes and reduces healthcare costs for a healthier population. In fact, exploring this technique for detecting diseases marks a significant step towards a future where healthcare is not just reactive but proactive, personalized, and highly effective to address any most prevalent health challenges of world (Bohr and Memarzadeh 2020).

Hypertension is a life-threatening medical condition which is caused due to elevated blood pressure levels in the arteries. It causes various cardiovascular diseases which includes heart attacks/strokes, and kidney problems. Early detection as well as management of hypertension is crucial to prevent complications and improve overall health outcomes for individuals who are affected by this (Raji *et al.* 2017). In recent years, artificial intelligence techniques are emerging as a powerful tool in healthcare and efficiently detects as well as diagnosis hypertension. AI based algorithms such as machine learning and deep learning are able to analyze large volume of data, identify the patterns in them, and make predictions or

DOI: 10.1201/9781003466383-33

classifications (Koul *et al.* 2023). On applying these learning models for detecting hypertension, they process diverse patient information, such as medical history, demographic data, lifestyle factors, and physiological measurements based on which likelihood of an individual having hypertension is predicted. In addition, this approach allows for early intervention and lifestyle adjustments for individuals who can be at risk of hypertension. It can also adapt and improve over time as more data is collected, ensuring that predictions are continuously enhanced for accuracy and reliability (Kaur *et al.* 2022).

In this context, this paper presents the development of an AI learning model which uses various algorithms such as voting classifier, Gradient Boosting, AdaBoost, XGBoost, CatBoost, Light Gradient Boosting, and Multi layer perceptron to detect and assess hypertension in individuals. These algorithms are trained using historical data of patient to learn the intricate relationships between different risk factors for finding the presence of hypertension.

2 LITERATURE SURVEY

This section informs us regarding the work conducted by researchers to detect the hypertension by using machine and deep learning models. Martinez-Ríos *et al.* (2022) proposed wavelet scattering transform to extract features from PPG data. The analysis demonstrated that on combining support vector machine with PPG features classified pre-hypertension and normotension with an accuracy and an F1-score of 71.42% and 76% respectively. Aras *et al.* (2023) assessed the impact of deep learning approach for detecting pulmonary hypertension and classifying its sub types using EEG data. The researchers performed their work by collecting the data from the University of California where patients with right heart catheterization were identified as non PH or PH retrospectively. Using 12- lead ECG voltage data, a CNN model was trained and from among 21470 patients, it computed the highest accuracy in detecting PH subtypes with an AUC-ROC of 0.79–0.91. The model also performed well when applied to ECGs obtained up to 2 years before PH diagnosis and showed an area under the curve of 0.79 or greater. Fang *et al.* (2023) mentioned that in China, the awareness, treatment, and control rate of patients who have been suffering from hypertension significantly improves to 51.6%, 45.8%, and 16.8% respectively but is still unsatisfactory. They also mentioned the opinion of the clinical studies which suggested that effective drug therapy or change in the lifestyle could significantly reduce the progression of the disease for those at risk. Keeping it in view, the researchers proposed a model that merged KNN and LightGBM for predicting the hypertension for the next five years by computing 86% accuracy and 92% recall rate. Silva *et al.* (2022) conducted a systematic review for analysing the contribution of existing techniques to predict hypertension. The authors screened as well as compared 21 articles from Jan 2018 to May 2021 using the ASRview algorithm on the basis of variable selection, data balancing, train-test split, performance metrics etc. The reviewed articles demonstrated extreme gradient boosting (XGBoost), support vector machine (SVM), and random forest emerged as the most effective algorithms in predicting hypertension risk. LaFreniere *et al.* (2016) identified various risk factors on the basis of current health conditions, demographics, and medical records of patients. These factors were used for predicting the presence of hypertension in individuals and assessing the likelihood of future hypertension development which served as an early warning system. Researchers used a neural network model which resulted in an accuracy of 82% for the dataset consisted of 185,371 patients and 193,656 controls. In addition to this, a literature review was also conducted to highlight the use of these risk factors in other research, accompanied by experimental results obtained from their developed model.

3 METHODOLOGY

In this part, the framework of the proposed system has been presented (Figure 1), which includes different phases like collection of data, preprocessing of data, exploratory data analysis, selecting features, applied classifiers, as well as performance metrics to identify and classify individuals with hypertension and without hypertension.

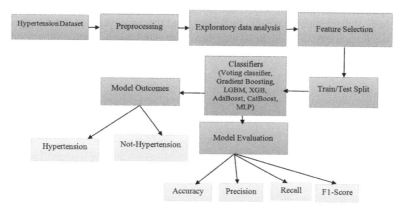

Figure 1. System design to predict hypertension and non-hypertension.

3.1 *Dataset*

The hypertension dataset has been taken in the form of a .csv file from diabetes, hypertension, and stroke prediction dataset. The dataset encompasses of 13 feature variables such as age, oldpeak, cp, chol, trestbps, fbs, restecg, ca, sex, thalach, exang, slope, and thal. Additionally, the dataset contains a binary target variable labelled 'target,' which represents two classes denoted by 0 and 1. The distribution of these classes in the target column is balanced which ensures an equal representation of both outcomes. This equilibrium in class distribution is essential for unbiased model training and accurate predictions. Figure 2 represents the distribution of binary class for each feature variable.

[https://www.kaggle.com/datasets/prosperchuks/health-dataset?select=hypertension_data.csv]

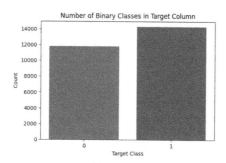

Figure 2. Distribution of target class in hypertension dataset.

3.2 *Data preprocessing*

It is an essential step as it handles missing values, converts categorical variables into numerical forms, and removes noisy data. Here in this research, initially missing values have been identified where 25 NaN values in the 'Sex' column were detected. Subsequently, the issue has been resolve by filling them with the mean of the entire column, implemented as **data['sex'].fillna(data['sex'].mean())**. This careful handling of missing data ensures the dataset's completeness and enhances the reliability of subsequent analyses and modeling efforts.

3.3 *Exploratory data analysis*

Understanding the patterns and characteristics within a dataset using visualization techniques plays an important role. Through EDA, one not only uncovers meaningful insights of the data but also detects outliers, identify trends, and understand the distribution of data. In this research, pattern of the feature variables of the dataset have been studied and presented in Figure 3.

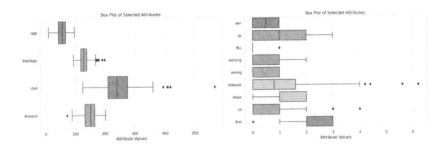

Figure 3. Bar plot of feature variables to detect outliers.

Here the values of each feature variable are shown in the form of a box plot where the rectangular box is the interquartile range (IQR) and contains the middle 50% of the data. It also contains the spans from the lower quartile (Q1) to the upper quartile (Q3). There is also a horizontal line that represents the median (Q2) which is the middle value of that particular feature as well as the whiskers which extend from the box to the minimum and maximum values from Q1 and Q3, respectively except sex. In addition to these, there are also few Individual data points that fall outside the whiskers such as in case of trestbps, chol, thalach, fbs, oldpeak, slope, ca, and thal and these points are plotted as outliers. In order to handle these outliers, Z score is used and solved via Equation (1):

$$Z = \frac{X - \mu}{\sigma} \tag{1}$$

and we can remove outliers only if $|z| > threshold \times standard\ deviation$. Here X is the data point, μ is mean, and σ is standard deviation.

3.4 *Feature selection*

In a CSV dataset, feature selection is necessary in refining the classification of AI learning models. In case of hypertension dataset, correlation coefficient (r) technique has been used to understand that how strongly each feature variable in the dataset is correlated with the target variable. In correlation coefficient, the values ranges from −1 to 1 where 1 defines the positive correlation, −1 indicates negative correlation, and 0 presents no correlation. If the correlation values are nearer to −1 and 1, it means that there is a strong relationship between feature and target which makes feature variable important for predicting the target variable (Saha *et al.* 2020). The correlation coefficient is calculated by using the Equation (2):

$$r = \frac{\sum_{i=1}^{n} \left(X_i - \overline{X} \right) \left(Y_i - \overline{Y} \right)}{\sqrt{\sum_{i=1}^{n} \left(X_i - \overline{X} \right)^2} \sqrt{\sum_{i=1}^{n} \left(Y_i - \overline{Y} \right)^2}} \tag{2}$$

Here X_i and Y_i are data points, $\overline{X}$ and $\overline{Y}$ are the mean of variables X and Y respectively.

3.5 *Applied classifiers*

Various machine as well as deep learning algorithms has been used as a classifier to detect hypertension in individuals.

3.5.1 *Voting classifier*

This is an approach to machine learning where the results of numerous independent classifiers are combined to provide a final prediction. The individual classifiers in the ensemble are trained on the same dataset. Each classifier interprets these features and delivers a binary prediction concerning the presence or absence of hypertension. The voting classifier then amalgamates these individual predictions using diverse techniques like majority voting or weighted voting and produces a final decision (Angayarkanni and Hemalatha 2023). Let LR(x) and RF(x) represent the prediction probability of logistic regression and random forest classifier respectively. The soft voting ensemble combines all these probabilities and predicts the class on the basis of highest average probability. The ensemble's prediction $\hat{y}$ for a given input x can be computed by Equation (3):

$$\hat{y} = argmax_i \left(\frac{1}{2} \sum_{j=1}^{2} p_i(x) \right) \tag{3}$$

where i denotes class labels as 0 and 1, $p_i(x)$ represent the probability estimate that the input x belongs to class i as predicted by each individual classifier, as shown in Equations (4 and 5):

$$p_0(x) = \frac{1}{1 + e^{-LR(x)}} \tag{4}$$

$$p_1(x) = RF(x) \tag{5}$$

3.5.2 *Gradient Boosting/ LGBM/XGBoost/CatBoost/AdaBoost*

In the domain of hypertension detection, these machine learning techniques did well in extracting meaningful patterns from complex medical data. *Gradient boosting* builds decision trees where each subsequent tree focuses on correcting the errors of the preceding ones and enhances the accuracy of model. *CatBoost* stands for categorical boosting and works efficiently with the medical data which is often rich in categorical variables (Chaplot *et al.* 2023). *LGBM,* optimizes the building of decision trees in a leaf-wise fashion and significantly improves training speed and makes it suitable for large healthcare dataset. *XGBoost,* an optimized version of gradient boosting, incorporates regularization techniques, prevents overfitting which thereby improves generalization on unseen data. *AdaBoost,* on the other hand, emphasizes difficult-to-classify instances by assigning higher weights to misclassified samples, and enhances the performance of the algorithm where identifying subtle patterns is crucial, such as in hypertension detection (Singh *et al.* 2023). The mathematical equation for AdaBoost's final prediction F(x) in the context of multiclass classification with α as weight and $h_m^k(x)$ as prediction of weak learners is presented in Equation (6).

$$F(x) = argmax_k \left(\sum_{m=1}^{M} \alpha_m . h_m^k(x) \right) \tag{6}$$

These algorithms collectively empower healthcare professionals by accurately recognizing intricate relationships within patient data and aid in early hypertension diagnosis and tailored treatment strategies.

3.5.3 MLP

It is a computational model inspired by the human brain which is adept at discerning intricate patterns within vast datasets. The input layer of the MLP receives essential features like age, cholesterol levels, trestbps, etc. Through hidden layers, these inputs use series of weighted computations and nonlinear transformations which allow the network to capture complex relationships between these variables. The output layer after employing a sigmoid activation function produces a prediction between 0 and 1and represents the likelihood of an individual having hypertension (Koul *et al.* 2022). Mathematically, it can be computed as shown in Equation (7)

$$y = f(WxT + b) \tag{7}$$

where activation function is f, W defines the set of parameters, x in the input vector and b is the bias vector.

3.6 *Evaluative metrics*

Precision, accuracy, recall, as well as F1 score are fundamental metrics used for assessing the performance of classifiers (Kumar *et al.* 2021) (Table 1).

Table 1. Formulae of evaluative parameters.

Metrics	Formulae
Accuracy	$\dfrac{True\ Positive + True\ Negative}{True\ Positive + True\ Negative + False\ Positve + False\ Negative}$
Precision	$\dfrac{True\ Positive}{True\ Positive + False\ Positive}$
Recall	$\dfrac{True\ Positive}{True\ Positive + False\ Negative}$
F1 score	$2\dfrac{Precision * Recall}{Recall + Precision}$

4 RESULTS

In this section, the performance of the applied learning approaches has been assessed using the parameters listed in sec 3.6. Figure 4 shows a confusion matrix of size 2x2 that has been generated to compute the predicted and actual value of the binary class of hypertension dataset for different methods.

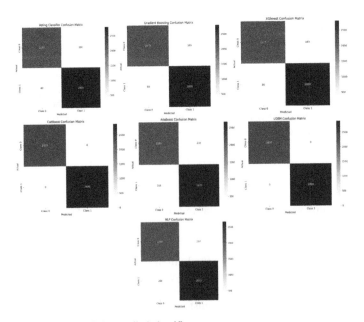

Figure 4. Confusion matrix of the applied classifiers.

At first, the values of the parameters such as accuracy, recall, precision, as well as F1 score for the entire dataset have been generated based on the confusion matrix values to evaluate the performance of the employed algorithms (Table 2).

Table 2.   Analysis of classifiers using hypertension dataset.

Classifiers	Accuracy	Precision	Recall	F1Score
Voting Classifier	95.39	0.96	0.95	0.95
Gradient Boosting	98.43	0.99	0.98	0.98
Adaboost	91.56	0.92	0.91	0.91
XGBoost	100	1.00	1.00	1.00
CatBoost	100	1.00	1.00	1.00
LGBM	100	1.00	1.00	1.00
MLP	89.86	0.92	0.89	0.89

The Voting Classifier exhibits an accuracy of 95.39%, with balanced precision, recall, and F1 score values at 0.96, 0.95, and 0.95, respectively. Gradient Boosting surpasses others with an impressive accuracy of 98.43% and outstanding precision at 0.99, recall at 0.98, and F1 score at 0.98. AdaBoost, whereas having a relatively lower accuracy at 91.56%, still maintains good precision of 0.92, F1 score and recall of 0.91. CatBoost, XGBoost, and LGBM demonstrate flawless performance with a perfect accuracy of 100% and impeccable precision, recall, and F1 score all at 1.00. Lastly, the MLP model lags behind others with an accuracy of 89.86% and slightly lower precision, recall, and F1 score at 0.92, 0.89, and 0.89. In summary, Gradient Boosting, XGBoost, CatBoost, and LGBM stand out as top performers by achieving a perfect accuracy and optimal precision, recall, and F1 score. Besides, the execution of the algorithms have been also evaluated for different two classes of the dataset i.e. hypertensive (class 1) and non-hypertensive (class 0) whose results are mentioned in Table 3.

Table 3.   Examining the performance of classifiers for different classes of hypertension dataset.

Classifiers	Hypertensive			Non-hypertensive		
	Precision	Recall	F1Score	Precision	Recall	F1Score
Voting Classifier	0.96	0.93	0.94	0.94	0.97	0.95
Gradient Boosting	1.00	0.96	0.98	0.97	1.00	0.99
Adaboost	0.91	0.90	0.91	0.92	0.93	0.92
XGBoost	1.00	1.00	1.00	1.00	1.00	1.00
CatBoost	1.00	1.00	1.00	1.00	1.00	1.00
LGBM	1.00	1.00	1.00	1.00	1.00	1.00
MLP	0.99	0.78	0.87	0.85	0.99	0.92

Voting classifier performed well with hypertensive and non-hypertensive class of the dataset by computing 0.96 and 0.97 as a precision, 0.93 and 0.97 as recall , which results in 0.94 and 0.95 F1 score respectively Likewise, Gradient Boosting achieved a perfect precision (1.00) for hypertensive case and high recall (0.96), leading to a strong F1 score of 0.98. For non-hypertensive cases, it exhibits perfect recall and slightly lower precision, resulting in a F1 score of 0.99. Adaboost performed reasonably well with balanced precision and recall for both classes, leading to an F1 score of 0.91 for hypertensive and 0.92 for non-hypertensive cases. XGBoost, CatBoost, and LGBM outperform with perfect precision, recall, and F1 score for both classes which indicates their flawless performance in classifying both hypertensive and non-hypertensive cases. MLP computed high precision (0.99) for hypertensive cases but relatively lower recall (0.78) that results in an F1 score of 0.87. For non-hypertensive cases, the model performs better with a precision of 0.85, recall of 0.99, and an F1 score of 0.92. In summary, XGBoost, CatBoost, and LGBM performed extremely well for both the classes while as MLP performs decently and it indicates room for improvement in classifying the hypertensive and non-hypertensive class. In addition, a comparison table (Table 4) has been created to assess the accuracy of the existing techniques with the techniques discussed in this paper for hypertension detection and classification.

Table 4. Assessing the performance of models.

Author's Name	Dataset	Techniques	Accuracy
Martinez-Ríos *et al.* (2022)	PPG data	Support Vector Machine	71.42%
Fang *et al.* (2023)	Data of 30k records	KNN+LGBM	86%
LaFreniere *et al.* (2016)	Data of 185,371 patients	Neural Network Model	82%
Soh *et al.* (2020)	Raw ECG data	CNN	99.99%
Our paper	*Hypertension dataset*	*XGBoost, CatBoost, LGBM*	*100%*

5 CONCLUSION

In the field of hypertension detection, the use of AI learning techniques has shown significant advancements and promise. It is worth mentioning that classifiers like XGBoost, CatBoost, and LGBM have shown exceptional precision, recall, F1 score, and accuracy in distinguishing between hypertensive and non-hypertensive cases. In spite of these achievements, there have been some challenges, especially with the Multi-Layer Perceptron (MLP) model. It has faced difficulties in attaining high accuracy. This emphasizes the need for ongoing optimization and refinement of neural network models to enhance their performance in detecting comprehensive hypertension.

The challenges encountered involve complex data patterns and the need to continuously research and develop model architecture. Understanding and navigating through these complexities is essential to furthering the integration of machine learning and deep learning techniques, which holds great potential for detecting hypertension. Future work should prioritize a deep exploration of data patterns, fine-tuning model architectures, and investigating innovative optimization strategies. This iterative process will help advance the development of more dependable, accurate, and easily accessible diagnostic tools for detecting hypertension, further strengthening the collaboration between computer science and healthcare.

REFERENCES

Angayarkanni, G. and Hemalatha, S., 2023, March. Evaluating the performance of supervised machine learning algorithms for predicting multiple diseases: A comparative study. In *2023 9th International Conference on Advanced Computing and Communication Systems (ICACCS)* (Vol. 1, pp. 1–7). IEEE.

Aras, M.A., Abreu, S., Mills, H., Radhakrishnan, L., Klein, L., Mantri, N., Rubin, B., Barrios, J., Chehoud, C., Kogan, E. and Gitton, X., 2023. Electrocardiogram detection of pulmonary hypertension using deep learning. *Journal of Cardiac Failure.*

Bohr, A. and Memarzadeh, K., 2020. The rise of artificial intelligence in healthcare applications. In *Artificial Intelligence in healthcare* (pp. 25–60). Academic Press.

Chaplot, N., Pandey, D., Kumar, Y. and Sisodia, P.S., 2023. A comprehensive analysis of artificial intelligence techniques for the prediction and prognosis of genetic disorders using various gene disorders. *Archives of Computational Methods in Engineering*, 30(5), pp.3301–3323.

Fang, M., Chen, Y., Xue, R., Wang, H., Chakraborty, N., Su, T. and Dai, Y., 2023. A hybrid machine learning approach for hypertension risk prediction. *Neural Computing and Applications*, 35(20), pp.14487–14497.

Kaur, S., Bansal, K. and Kumar, Y., 2022, December. Artificial intelligence approaches for predicting hypertension diseases: Open challenges and research issues. In *2022 5th International Conference on Contemporary Computing and Informatics (IC3I)* (pp. 338–343). IEEE.

Koul, A., Bawa, R.K. and Kumar, Y., 2022. Artificial intelligence in medical image processing for airway diseases. In *Connected e-Health: Integrated IoT and Cloud Computing* (pp. 217–254). Cham: Springer International Publishing.

Koul, A., Bawa, R.K. and Kumar, Y., 2023. Artificial intelligence techniques to predict the airway disorders illness: a systematic review. *Archives of Computational Methods in Engineering*, 30(2), pp.831–864.

Kumar, Y., Gupta, S., Singla, R. and Hu, Y.C., 2022. A systematic review of artificial intelligence techniques in cancer prediction and diagnosis. *Archives of Computational Methods in Engineering*, 29(4), pp.2043–2070.

LaFreniere, D., Zulkernine, F., Barber, D. and Martin, K., 2016, December. Using machine learning to predict hypertension from a clinical dataset. In *2016 IEEE symposium series on computational intelligence (SSCI)* (pp. 1–7). IEEE.

Martinez-Ríos, E., Montesinos, L. and Alfaro-Ponce, M., 2022. A machine learning approach for hypertension detection based on photoplethysmography and clinical data. *Computers in Biology and Medicine*, 145, p.105479.

Raji, Y.R., Abiona, T. and Gureje, O., 2017. Awareness of hypertension and its impact on blood pressure control among elderly nigerians: report from the Ibadan study of aging. *The Pan African Medical Journal*, 27.

Saha, P., Patikar, S. and Neogy, S., 2020, October. A correlation-sequential forward selection based feature selection method for healthcare data analysis. In *2020 IEEE International Conference on Computing, Power and Communication Technologies (GUCON)* (pp. 69–72). IEEE.

Silva, G.F., Fagundes, T.P., Teixeira, B.C. and Chiavegatto Filho, A.D., 2022. Machine learning for hypertension prediction: a systematic review. *Current Hypertension Reports*, 24(11), pp.523–533.

Singh, J., Sandhu, J.K. and Kumar, Y., 2023. An analysis of detection and diagnosis of different classes of skin diseases using artificial intelligence-based learning approaches with hyper parameters. *Archives of Computational Methods in Engineering*, pp.1–28.

Soh, D.C.K., Ng, E.Y.K., Jahmunah, V., Oh, S.L., San Tan, R. and Acharya, U.R., 2020. Automated diagnostic tool for hypertension using convolutional neural network. *Computers in Biology and Medicine*, 126, p.103999.

Deep learning based multiple regression analysis approach for designing medical image recognition system

Shweta Singh
Electronics and Communication Department, IES College of Technology, Bhopal, India

A.N. Jagadish
Department of Computer Science, Bapuji Institute of Engineering and Technology, Davanagere Karnataka, India

Bhuman Vyas
Credit Acceptance, Canton, USA

Mohammed Wasim Bhatt
Model Institute of Engineering and Technology, Jammu, J&K, India

Aws Zuhair Sameen
College of Medical Techniques, Al-Farahidi University, Baghdad, Iraq

Sheshang Degadwala
Department of Computer Engineering, Sigma University, Vadodara, Gujarat, India

ABSTRACT: Machine learning models are optimization models that make it possible to gather data, evaluate it, and provide the experts and management with the reports they need to make the best decisions. This study uses quantitative research, as it seeks to gather information and analyse it with statistical models such as regression analysis and analysis of variance. Researchers want to use statistical techniques such as descriptive analysis, regression analysis, and ANOVA (analysis of variance) to analyse the hypothesis. The purposive sample technique is used by the authors to choose respondents who work in the healthcare sector. A non-probability sampling technique called purpose sampling is used. The R squared value is 0.744, or 74.4 percent, according to the analysis. According to the investigation, there is a strong correlation between improved picture quality and ML-based techniques in digital image recognition.

1 INTRODUCTION

The emergence of new technologies like artificial intelligence, machine learning and deep learning are transforming different areas of business and related processes. The application of machine learning n the area of image processing, mainly in the domain related to massive image analysis and providing the clear output, technologies related to machine learning has been highly supportive as it can process complex information and offer better output. Machine learning enables in supporting the users to separate the images so as to recognise the image in a clearer form, it should be noted that the machine learning based image processing technique has been increasing in the modern world as the users' needs better classification, segmentation, and overall recognition of the image (Kim *et al.* 2017). For example, in the health care industry medical practitioners and nurses are looking to bisect the images so as to identify the ailments and other issues, so that appropriate medical treatment can be

DOI: 10.1201/9781003466383-34

offered to the patients. Machine learning algorithms support in processing the signal to engineering and focus in enhancing the video image processing so that it can be applied in various fields from engineering, medical, prevention of road accidents etc. Figure 1 representing the digital image processing.

1.1 Block diagram

The following is the basic block diagram of digital image processing through machine learning

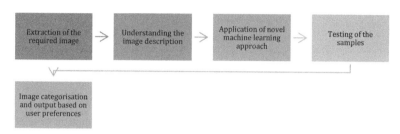

Figure 1. Block diagram representing the digital image processing.

The initial step usually requires pre-processing, such as scaling, called image acquisition. Image enhancement, one of the easiest and most visually attractive aspects of digital image processing, comes next in the process. The purpose of enhancement techniques is to give details that are hidden or to simply emphasize certain appealing aspects of an image. A common example of enhancing is improving an image's appearance by increasing contrast (Bishop 2015). In representation and description, an image is represented using several techniques, such as boundary representation, which emphasizes outward shape elements like corners and inflections, and regional representation, which emphasizes interior features like texture and skeletal structure (Adadi and Berrada 2018).

1.2 A brief history on the available tools and techniques and approaches

The oldest surviving photograph on the real-world aspects was created in 1827, however the Lumiere brothers created the cinematographic film in the late 1900s and was available for the screening to the public. However, with more than 20 years related to the overall development, whereas the most recent enhancement has impacted from the later part of 20th century, this has enabled in the development of new technologies in an efficient manner, this has paved way for the digital revolution (Jain et al. 2019). The emergence of critical imaging has enabled in surpassing the usage of film-based cameras and the digital revolution n the imaging industry has begun in 1990. In the early 1992, the JPEG format (Joint Photographic Experts Group) was created and it was created as a standard for the still images, also MPEG (Moving picture experts' group) was created for coding the moving pictures and related audio. The main purpose of the study is to analyse the major factors influencing the medical practitioners and others in using the KNN for better image processing mainly for X Rays in health care industry.

1.3 Proposed model

The proposed model is to use the novel machine learning algorithms like Hybrid feature extraction, Convolutional neural network, K Nearest neighbours and support vector mechanism Figure 2 present the proposed model. The application of machine learning for analysing the X ray in the health care industry is gaging in the current scenario as the medical practitioners need to focus in analysing the digital images effectively, estimate the pattern and provide better medical support to the patients.

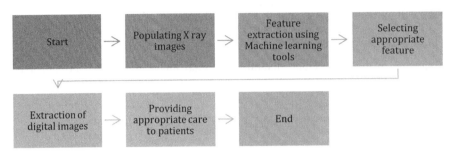

Figure 2. Proposed model.

2 LITERATURE REVIEW

Text recognition, digital editing, picture recognition, and target recognition were all stages in the evolution of image recognition. In general, if an industry has a need that cannot be met by domestic technology, new technology is developed to meet that need. The same may be said for image recognition software (Jain and Pandey 2019). The development of this technology is designed to allow computers to generate more factual data on an individual basis and to solve the problem of unknown or incomplete data. Computer images are a variant of body image is the source. Structural analysis is essential in any image analysis. is a source. Preliminary processing often refers to the steps of drying, bathing and rotating the image processing, which improves the basic characteristics of the image. The application of different dense network is involved in more manageable. These aspects tend to consist of introducing the overall information so that the model can forecast the output effectively, support in understanding the patterns and offer better clarity and information. Literature shows in Table 1.

Table 1. Showing the literature review related to machine learning in digital imaging (Geirhos *et al.* 2019; Jain and Pandey 2019; Mahony *et al.* 2020; Panwar *et al.* 2021; Yang *et al.* 2017).

Hybrid feature extraction	Various data are extracted from the information available which includes alphabets, symbols, characteristics etc. Neural networks are applied in order to classify the information and appropriate patterns are extracted. The individuals and users are trained in order to apply the appropriate tool so that clear images can be obtained and zoning of the images can be ascertained for better clarity and analysis
Neural network	This method is highly useful in the analysing the mathematical model, support in analysing the image more clearly for effective diagnosis. This model applied two different layers which support in parameter sharing, enhanced interaction and better representation.
KNN	It is based on the nearest distance of the classes, KNN enable in computing the best weight for analysing the correct point so that the image quality can be enhanced. KNN image classifier supports the user in providing better image when there is suspicious in the on the image generated through the conventional sources. Through application of normalisation the user can increase the pixel aspects for decomposed image and enhance digital image can be generated in a better manner.
Support vector mechanism	SVM is applied in order to classify the data through the regression problems, SVM also enables in recognising the pattern effectively, recognise the data and enhance feature extraction for better quality
Light fields and Volumetric imaging	In this method, 7D method is applied which supports in light field magnification as it enables in collating the radiance from different directions. Hence, it possesses better picture clarity than the traditional model of 2D imaging.

3 METHODOLOGY

The purpose of the study is to focus in analysing the major factors influencing the medical practitioners and others in using the KNN for better image processing mainly for X Rays in health care industry. The implementation of machine learning in the diagnostics process like X ray and other aspects has been transforming the medical industry as the practitioners can quick map the image in an easier manner, support in making quick analysis on the reports and offer better guidance to the patients for speedy recovery (Ke *et al.* 2019). Hence, this study uses quantitative research study as it intends to collect the information and analyse them using statistical models like regression analysis, Analysis of variance (Tang and Shabaz 2021). The application of machine learning in the digital imaging mainly in healthcare industry as supported in enhancing the image quality, help in forming the patterns, supportive in making appropriate diagnosis and address the patient requirements in an effective manner. The researchers intend to apply the statistical techniques like descriptive analyses, regression analysis and ANOVA (Analysis of variance) to analyse the hypothesis (Sharma *et al.* 2022).

4 RESULTS AND DISCUSSION

This section deals in presenting detailed data analysis based on the data collected from the respondents. The analysis includes descriptive analysis, regression analysis and chi quare test.

Table 2. Descriptive analysis.

Descriptive Stats	Gender	Age	Qualification	Current Designation	Experience
Mean	1.46	2.24	1.90	1.93	2.67
Std. Deviation	0.50	0.90	0.67	0.75	1.47
Skewness	0.16	0.41	0.11	0.12	0.40
Kurtosis	−2.00	−0.53	−0.74	−1.19	−1.23

From Table 2, of the descriptive statistics it is noted that the demographic variables of the samples population are in the better range, the standard deviation of the demographic variables are less showing that the data lies near to the mean value, furthermore the skewness is nearly 0 for most of the demographic variables and hence can be considered as normally distributed (Shabaz and Soni 2023).

Furthermore, 54.9% of the respondents have completed professional medical degree, 27.5% of them have completed their post-graduation course in nursing, and 17.6% have completed radiology certification and related courses. 44% of them are currently working as doctors and medical practitioner, 31.6% of them were working as nurses and 24.4% of them were radiologist in the medical industry (Al-Kahtani *et al.* 2022; Aski *et al.* 2023; Parashar *et al.* 2023). 28% of the respondents poses less than 5 years of experience, 25.9% of them poses experience between 5 – 10 years, 18.7% possess experience of more than 20 years, 15.5% possess experience between 10 – 15 years and the remaining possess experience between 15 – 20 years (Farooqi *et al.* 2019).

Table 3. Role of machine learning in X-Ray imaging.

ML in Imaging	Values in %
Respondents Disagreeing most	7.3
Respondents Disagreeing	9.3
Respondents neutral	16.6
Respondents Agreeing	37.3
Respondents Agreeing most	29.5

Based on the Table 3 analysis, it is noted that 29.5% of the respondents have stated that they strongly agree to the statement that the machine learning plays a vital role in X ray imaging in the health care sector, furthermore, 37.3% of the respondents have agreed to the statement (Gupta *et al.* 2023; Mazhar *et al.* 2023). Figure 3 representing the Role of machine learning in x ray imaging. Hence nearly 66.3% are highly positive that machine learning plays a crucial role in X ray imaging, however 16.6% of the respondents have been neutral to the statement, 9.3% have disagreed to the statement and 7.3% of the respondents have strongly disagreed.

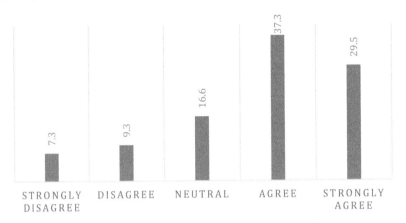

Figure 3. Chart representing the Role of machine learning in X Ray imaging.

3.1 *Regression analysis*

Table 4. Regression weights.

Regression	B	Std. Error	T	Sig.
(Constant)	0.078	0.17	0.47	0.64
Better image quality	0.319	0.10	3.18	0.001
Identification of pattern	0.329	0.09	3.53	0.001
Addressing patient requirements effectively	0.016	0.08	0.19	0.85
Supportive in making proper diagnosis	0.247	0.09	2.73	0.01
R Squared	0.744			
F data	140.480			
Signi	0.000			

The generic regression model is:

Y = Constant + a x 1st variable + b x 2nd variable + c x 3rd variable + d x 4th variable
Based on analyses from Table 4, the regression weights are being stated as

Better image quality is 0.319; Identification of pattern is 0.329; Addressing patient requirements effectively is 0.016 and Supportive in making proper diagnosis is 0.247. Hence, the regression equation is stated as follows:

Y (ML in Digital Image recognition) = 0.078 + 0.319 x better image quality + 0.329 x Identification of pattern + 0.016 x Addressing patient requirements effectively + 0.247 x Supportive in making proper diagnosis. The analysis further states the R squared value is 0.744 or 74.4% hence the significance value of all the variables except Addressing patient requirements effectively is less than 0.05, hence they are statistically significant (Sharma *et al.* 2019).

Test of hypothesis using Analysis of variance test

Null: No major connection between the better image quality and application of machine learning in digital image recognition. ANOVA shows in Table 5 and Table 6.

From analysis it is stated there is a significant association between the better image quality and ML based approaches in digital image recognition

Table 5. ANOVA between better image quality and application of machine learning in digital image recognition.

ANOVA	Data value	Overall Sig
Among the groups	288.01	0.001
F Data	151.84	

2nd test

Null: No major connection between the identification of patterns and application of machine learning in digital image recognition

Table 6. ANOVA between the identification of patterns and application of machine learning in digital image recognition.

ANOVA	Data value	Overall Sig
Among the groups	288.01	0.001
F Data	150.45	

From analysis it is stated there is a major connection between the identification of patterns and application of machine learning in digital image recognition.

3rd test

Null: No major connection between supporting in proper diagnosis and application of machine learning in digital image recognition (Soni *et al.* 2022). From analysis it is stated that there is a major connection between supporting in proper diagnosis and application of machine learning in digital image recognition *4th test*

Null: No major connection between supporting in Addressing patient requirements effectively and application of machine learning in digital image recognition. ANOVA between supporting in Addressing patient requirements effectively and application of machine learning in digital image recognition from analysis it is stated that there is a significant association between supporting in Addressing patient requirements effectively and application of machine learning in digital image recognition. The results of the data analysis have been collated in an extensive manner in this section, the regression equation has mentioned that major variables like better image quality; identification of pattern and supportive in making proper diagnosis are highly influencing the application of machine learning in enhancing the digital image processing in health care sector (Soni and Jain 2018; Yao *et al.* 2021)

5 CONCLUSION

In healthcare, doctors and nurses share images to identify diseases and other problems so that patients can receive appropriate medical treatment. On the other hand, due to the increasing complexity of image distribution, various applications are implemented to help users get the clarity of the necessary image and make the right decision. Whereas, ANOVA is a statistical method used to examine if there is a common mean among unrelated groups. The results of data analysis are collected in detail in this section, using the regression equation as a reference for the main variables, as the best image quality. Sample identification and support for correct diagnoses strongly influence the use of machine learning to improve digital image processing in healthcare. Doctors said that ML had a great deal of influence in determining the X-ray pattern, which enabled a clear diagnosis.

REFERENCES

Adadi, A. and Berrada, M., 2018. Peeking inside the black-box: a survey on explainable artificial intelligence (XAI). *IEEE access*, *6*: 52138–52160.

Al-Kahtani, M.S., Khan, F. and Taekeun, W., 2022. Application of internet of things and sensors in healthcare. *Sensors*, *22*(15): 5738.

Aski, V.J., Dhaka, V.S., Parashar, A. and Rida, I., 2023. Internet of Things in healthcare: A survey on protocol standards, enabling technologies, WBAN architectures and open issues. *Physical Communication*: 102103.

Farooqi, M.M., Shah, M.A., Wahid, A., Akhunzada, A., Khan, F., ul Amin, N. and Ali, I., 2019. Big data in healthcare: A survey. *Applications of Intelligent Technologies In Healthcare*: 143–152.

Geirhos, R., Rubisch, P., Michaelis, C., Bethge, M., Wichmann, F.A. and Brendel, W., 2018. ImageNet-trained CNNs are biased towards texture; increasing shape bias improves accuracy and robustness. *arXiv preprint arXiv:1811.12231*.

Gupta, S., Shabaz, M., Gupta, A., Alqahtani, A., Alsubai, S. and Ofori, I., 2023. Personal HealthCare of Things: A novel paradigm and futuristic approach. *CAAI Transactions on Intelligence Technology*.

Jain, A. and Pandey, A.K., 2017. Multiple quality optimizations in electrical discharge drilling of mild steel sheet. *Materials Today: Proceedings*, *4*(8): 7252–7261.

Jain, A. and Pandey, A.K., 2019. Modeling and optimizing of different quality characteristics in electrical discharge drilling of titanium alloy (Grade-5) sheet. *Materials Today: Proceedings*, *18*: 182–191.

Jain, A., Yadav, A.K. and Shrivastava, Y., 2020. Modelling and optimization of different quality characteristics in electric discharge drilling of titanium alloy sheet. *Materials Today: Proceedings*, *21*: 1680–1684.

Ke, Q., Zhang, J., Wei, W., Połap, D., Woźniak, M., Kośmider, L. and Damaševičius, R., 2019. A neuro-heuristic approach for recognition of lung diseases from X-ray images. *Expert Systems With Applications*, *126*: 218–232.

Kim, H., Sefcik, J.S. and Bradway, C., 2017. Characteristics of qualitative descriptive studies: A systematic review. *Research in Nursing & Health*, *40*(1): 23–42.

Mazhar, T., Irfan, H.M., Haq, I., Ullah, I., Ashraf, M., Shloul, T.A., Ghadi, Y.Y., Imran and Elkamchouchi, D.H., 2023. Analysis of Challenges and Solutions of IoT in Smart Grids Using AI and Machine Learning Techniques: A Review. *Electronics*, *12*(1): 242.

Mazhar, T., Nasir, Q., Haq, I., Kamal, M.M., Ullah, I., Kim, T., Mohamed, H.G. and Alwadai, N., 2022. A Novel Expert System for the Diagnosis and Treatment of Heart Disease. *Electronics*, *11*(23): 3989.

O'Mahony, N., Campbell, S., Carvalho, A., Harapanahalli, S., Hernandez, G.V., Krpalkova, L., Riordan, D. and Walsh, J., 2020. Deep learning vs. traditional computer vision. In *Advances in Computer Vision: Proceedings of the 2019 Computer Vision Conference (CVC)*, Volume 1 1 (pp. 128–144). Springer International Publishing.

Panwar, V., Sharma, D.K., Kumar, K.P., Jain, A. and Thakar, C., 2021. Experimental investigations and optimization of surface roughness in turning of en 36 alloy steel using response surface methodology and genetic algorithm. *Materials Today: Proceedings*, *46*: 6474–6481.

Parashar, A., Rishi, R., Parashar, A. and Rida, I., 2023. Medical imaging in rheumatoid arthritis: A review on deep learning approach. *Open Life Sciences*, *18*(1): 20220611.

Shabaz, M. and Soni, M., 2023. Cognitive digital modelling for hyperspectral image classification using transfer learning model. *Turkish Journal of Electrical Engineering and Computer Sciences*, *31*(6): 1039–1060.

Sharma, M., Singh, J. and Gupta, A., 2019, August. Intelligent resource discovery in inter-cloud using blockchain. In *2019 IEEE SmartWorld, Ubiquitous Intelligence & Computing, Advanced & Trusted Computing, Scalable Computing & Communications, Cloud & Big Data Computing, Internet of People and Smart City Innovation (SmartWorld/SCALCOM/UIC/ATC/CBDCom/IOP/SCI)* (pp. 1333–1338). IEEE.

Sharma, S., Rattan, P., Sharma, A. and Shabaz, M., 2022. Voice activity detection using optimal window overlapping especially over health-care infrastructure. *World Journal of Engineering*, *19*(1): 118–123.

Soni, M. and Jain, A., 2018, February. Secure communication and implementation technique for sybil attack in vehicular Ad-Hoc networks. In *2018 Second International Conference on Computing Methodologies and Communication (ICCMC)* (pp. 539–543). IEEE.

Soni, M., Khan, I.R., Babu, K.S., Nasrullah, S., Madduri, A. and Rahin, S.A., 2022. Light weighted healthcare CNN model to detect prostate cancer on multiparametric MRI. *Computational Intelligence and Neuroscience, 2022*.

Tang, S. and Shabaz, M., 2021. A new face image recognition algorithm based on cerebellum-basal ganglia mechanism. *Journal of Healthcare Engineering, 2021*.

Yang, L., Zhang, Y., Chen, J., Zhang, S. and Chen, D.Z., 2017. Suggestive annotation: A deep active learning framework for biomedical image segmentation. In *Medical Image Computing and Computer Assisted Intervention– MICCAI 2017: 20th International Conference, Quebec City, QC, Canada, September 11–13, 2017, Proceedings, Part III 20* (pp. 399–407). Springer International Publishing.

Yao, Q., Shabaz, M., Lohani, T.K., Wasim Bhatt, M., Panesar, G.S. and Singh, R.K., 2021. 3D modelling and visualization for vision-based vibration signal processing and measurement. *Journal of Intelligent Systems*, *30*(1): 541–553.

Yardley, L. and Bishop, F.L., 2015. Using mixed methods in health research: Benefits and challenges. *British Journal Of Health Psychology*, *20*(1): 1–4.

Next Generation Computing and Information Systems – Gupta. (Ed.)
© 2025 The Author(s), ISBN 978-1-032-73865-9

Survey on detection of bone cancer using deep learning-based techniques

Harshit Vora & Seema Mahajan
Indus University, Ahmedabad, India

Yogesh Kumar
Pandit Deendayal Energy University, Gandhinagar, India

ABSTRACT: The early detection and precise diagnosis of bone cancer are essential for efficient treatment planning because it is a serious and potentially fatal condition. Among the various types of diseases, cancer may be one of the most dangerous. The manual interpretation of medical imaging scans used in traditional methods for bone cancer detection can be time-consuming and prone to human error. Deep learning techniques have recently shown promise as tools for automatically detecting bone cancer. Among all cancers, bone cancer may be the leading cause of death. Based on this survey researchers are using some different techniques for the prediction of bone cancer. The first test performed when there is a possible bone tumor is suspected is a bone X-ray. The best method for determining osteosarcoma from the bones is to use imaging tests and X-rays. The suggested procedure for obtaining a conclusive diagnosis is a biopsy. This is a difficult and laborious technique that can be done automatically. We present many supervised deep-learning methods and select the optimal model. These decisions are made by analyzing user data to detect bone cancer. We show that the selected model obtains the maximum accuracy of 90.36% and the highest precision of 89.51% in the prediction tests using the residual neural network (ResNet101) algorithm.

1 INTRODUCTION

One of the conventional types of bone cancer is osteosarcoma, also known as osteogenic sarcoma. There are three classifications for this cancer critical, medium, and basic. The grade indicates cancerous cells spread quickly to other bodily parts. Osteosarcoma typically develops between the ages of 10 and 25, particularly in adolescence, and recurs in later life around age 60. This tumor damages good bone tissues and results in malignancy because the malformed bone cells lead to bone fractures, wound infections, and sluggish recovery. One method for detection is to look for osteosarcoma (Shukla and Patel 2020). Approximately 20 out of every 100 patients have an infection when they are initially diagnosed with bone cancer. Numerous researchers have attempted to employ Deep-Learning techniques to identify bone cancer in people and have found them to be rather effective in doing so. We have developed a model in this study to identify bone cancer in its very early stages. As such, bone cancer needs to be detected automatically. Predicting cancer, preventing it, and controlling it are major difficulties in the healthcare sector. This paper proposes to automatically forecast the likelihood of sickness by using supervised deep-learning algorithms. Four methods were used in the construction of this conceptual framework: Residual Network 101 (ResNet101), Dense Convolutional Network 201 (DenseNet201), Visual Geometry Group 19 (VGG19), and Visual Geometry Group 16 (Shukla and Patel 2020).

DOI: 10.1201/9781003466383-35

1.1 *Motivation*

The main driving force behind this research is the large annual number of cancer cases and deaths. The projected number of new cases of male cancer in 2020 is shown in Figure 1 (Sharma *et al.* 2021). Deep learning should be utilized to diagnose bone cancer for a number of reasons, including the necessity of early detection for successful treatment and better patient outcomes. Conventional techniques for detecting bone cancer often depend on radiologists evaluating medical pictures by hand, a laborious and human error-prone process. By automatically evaluating and extracting relevant features from massive volumes of medical imaging data, deep learning systems can automate this process. Personalized treatment is made possible by deep learning algorithms in the area of bone cancer screening. Deep learning systems can identify deep learning algorithms that can learn to recognize biomarkers by training models on a variety of patient populations and incorporating clinical data, genetic information, and treatment outcomes (Sharma *et al.* 2021).

1.2 *Related work*

1.2.1 *VGG16*
Visual Geometry Group-16's Convolution Neural Network (CNN) architecture. while maintaining the same filling and maxpool layer of 2*2 filter with stride 2. VGG16 is an architecture for a 16 different shades of CNN. (Sharma *et al.* 2021).

1.2.2 *VGG19*
Convolutional neural networks, such as the VGG19 architecture, are extensively employed in image classification. The architecture is straightforward but effective, and many creative models have been built around it (Shukla and Patel 2020).

1.2.3 *DenseNet201*
Building upon the DenseNet foundation, DenseNet201 is a Part of convolutional neural network(CNN) architecture. In 2017, Gao Huang and associates created it. Feed-forward linking, or connecting each layer to every other layer, is the main idea behind DenseNet. This leads to the creation of a dense connection network that improves feature reuse and mitigates fading gradient problems. DenseNet201 is a version of the DenseNet architecture with 201 layers. Applications in computer vision, such as object detection and image classification, commonly use it (Gawade *et al.* 2023).

1.2.4 *Resnet101*
The architecture of the convolutional neural System ResNet101 in 2015. A variation of this paradigm known as the "Residual Network" model is the ResNet model. Consequently, it is now possible to train very deep System 152 different shades or more deprived of running into problems with diminishing gradients. ResNet101 is one version of ResNet that has 101 layers. It is widely used for object recognition and image arrangement applications and has fragmented previous records for image arrangement performance in datasets (Gawade *et al.* 2023).

1.3 *Bone cancer types and challenges*

There are several types of bone cancer, each originating from different cells within the bone. The basic differences of Bone Tumors are.

1.3.1 *Osteosarcoma*
The most prevalent kind of bone cancer, osteosarcoma, usually affects young adults and children. It grows in the cells that give rise to new bone. It could spread to other body parts and has a tendency to grow quickly.

1.3.2 *Ewing sarcoma*

Ewing sarcoma primarily affects children and young adults. It arises from primitive nerve tissue in the bone and most commonly occurs in the pelvis, thigh bone, or shinbone. Ewing sarcoma can spread to other bones or distant organs, and it is known for its aggressive nature.

1.3.3 *Chondrosarcoma*

Chondrosarcoma develops in cartilage cells, which are found at the ends of bones or between joints. It can occur in individuals of any age but is more commonly seen in adults. Chondrosarcoma tends to grow slowly and has a higher likelihood of occurring in the pelvis, thigh bone, or shoulder.

1.3.4 *Giant cell bone cancer*

Giant cell cancer is a relatively not common bone cancer that mostly affects adults between the ages of 20 and 40. It typically occurs near the ends of long bones, such as the knee or wrist. While most giant cell tumors are benign, some can be locally aggressive or recur after treatment.

1.3.5 *Chordoma*

Chordoma is a rare type of bone cancer that arises from remnants of the notochord, a structure present during fetal development. It commonly occurs in the base of the skull or the bones of the spine. Chordomas grow slowly and tend to be locally invasive, making complete surgical removal challenging.

1.3.6 *Fibrosarcoma*

Fibrosarcoma is a rare malignant bone tumor that arises from fibrous tissue. It can occur in any bone but is more commonly found in the legs, arms, or jaw. Fibrosarcoma tends to grow rapidly and has the potential to metastasize to other parts of the body.

2 EXECUTION SEQUENCE OF BONE CANCER IDENTIFICATION

The execution of bone tumor identification using deep learning typically involves several key steps. The working methodology of bone cancer detection is discussed in Figure 2.

2.1 *Input data collection*

The researcher collects the X-ray Photos from the given database and then starts the Pre-processing of the Given input X-ray photos.

2.2 *Preprocessing*

In medical imaging data the Images may need to be resized, intensities may need to be normalized, and image enhancement techniques may be used.

2.3 *Data augmentation*

Amplify the dataset by subjecting the images to various transformations, such as rotation, scaling, flipping, and noise addition. The model's generalization and variability are increased with the aid of data augmentation.

2.4 *Design of the model architecture*

Select the best deep-learning architecture for detecting bone cancer. The use of Convolutional Neural Networks (CNNs) is widespread because the noise and clatter are removed from the image by the object processing system.

Evaluate the trained model on a separate validation set to assess its performance. Accuracy, precision, recall, F1-score, and area under the receiver operating characteristic curve (AUC-ROC) are examples of common evaluation measures.

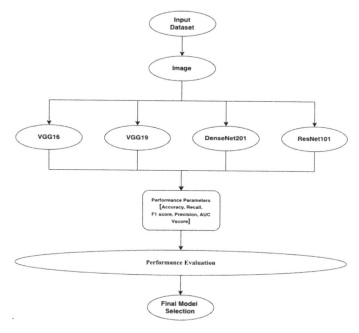

Figure 1.   System design for bone cancer detection (Gawade *et al.* 2023).

3   RESULTS & LITERATURE REVIEW

The suggested architecture is designed to employ supervised deep-learning techniques for model construction. Its architecture was more appropriate for osteosarcoma detection. The DenseNet201 model has an F1-score of 85.5% and an accuracy of 87.15%. v score, an accuracy of 84.38%, recall of 87.15%, and an AUC of 0.9093 of 1.907, meaning that the metrics were the lowest of all the tested models. are comparable to VGG16, however, the time factor causes the v-score to be low. With an F1-score of 89.35% and an accuracy of 90.36%, the ResNet101 model v score, an accuracy of 89.51%, recall of 89.59%, and an AUC of 0.9461 of 2.72, the greatest value about other models, the residual In comparison with different replicas, the neural System (ResNet101) performed excellent, taking the shortest amount of time to train and having the highest accuracy (Gawade *et al.* 2023).

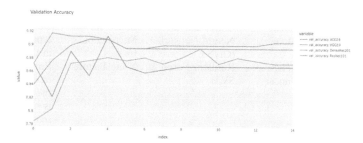

Figure 2.   Validation of all models with accuracy (Gawade *et al.* 2023).

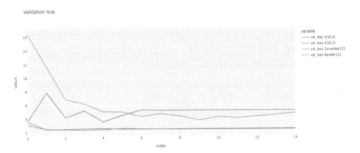

Figure 3.   Validation loss of all models (Gawade *et al.* 2023).

Table 1.   Algorithms' performance using the dataset (Gawade *et al.* 2023).

Parameters	VGG16	VGG19	DenseNet201	ResNet101
Accuracy Results	86.6%	89.4%	87.2%	90.6%
F1 Score Result	85.5%	88.3%	85.5%	89.3%
Precision Results	84.3%	87.7%	84.3%	89.5%
Recall score	87.2%	88.9%	87.2%	89.5%
AUC Results	0.90	0.932	0.0	0.94

Table 2.   Literature review of bone cancer.

Sr No.	Authors	Disease Types	Techniques used	Data set	Result
1	Gawade, S.Bhansali, A. Patil, k. (2023)	Bone Cancer	CNN((ResNet101, VGG16, VGG19)	X-Ray images	ResNet101 Highest Accuracy 90.36%).
2	Anisuzzaman, D. Hosein, B. Ling, T. Jake, L. (2021)	Osteosarco-ma	CNN (Con-Volutional Neural Networks) VGG-19, Inception V3.	Histological Images(Osteo-sarcoma data-set)	With accuracy of 93.91% and 95.65% in multi-class and binary classifications, respectively, the VGG19 model performed best.
3	Vandana, B. Sathyavathi, A. (2021)	Bone Cancer	Convolution Neural Network (CNN)	Images from histopathology.	Threatening, harmless, and ordinary and the achievement rate is 92%.
4	Madhuri, A. Narasimha, P. (2014)	Bone Cancer	K Means Clustering	MRI Images	Bone cancer detection with a 95 percent accuracy rate.
5	Nikita, J. Dee-pak, K. Ayush, K. (2020)	White Blood	Convolutional Neural Network (CNN)	Bone Marrow-Microscopic Images.	Determine the various types of cancer found in the bone marrow.
6	Sharma, A. Yadav, D. Garg, H. (2021)	Bone Cancer	Random Forest & SVM	X-ray Pictures.	For the categorization of malignant bones, the F1-score of 0.94 is superior to the F1-score of 0.88 in the work. RF 89% & SVM 92%.

(continued)

Table 2. Continued

Sr No.	Authors	Disease Types	Techniques used	Data set	Result
7	Abhilash, S. Atul, P. (2020)	Bone Cancer	Image Segmentation	X-Ray & MRI.	Detects Osteosarcoma cancer present in bone.
8	Satheetsh, B. Sathiyaprasad N. (2021)	Bone Cancer	KNN classifier.	X-rays and MRI.	Identifying the present or absent of a tumor in the input MRI and whether it has been diagnosed as benign or malignant after this
9	Jabber, J. Shankar, M. Rao, V. (2020)	Benign and malignant.	SVM & PNN	MRI Images	SVM with 92% accuracy is achieved. PNN accuracy 84%.

So, we are currently conducting a inclusive study on the prediction of bone cancer utilizing Machine Learning and Deep Learning techniques. covers nine distinct publications on different types of bone cancer, including primary and secondary bone cancer. There are several methods for making predictions.

4 DATA SET

One kind of malignant tumor that develops in a bone is called osteosarcoma. Although any bone can experience it, long bones are more likely to do so. Osteosarcoma is most common in people between the ages of 10 and 25, particularly in adolescence, and it recurs in older adults over 60. This dataset contains images of the micro anatomy of osteosarcoma stained with hematoxylin and eosin (H and E). A team of clinical professionals from the University of Texas, Dallas oversaw the data collection process. Fifty patients receiving treatment at the Medical Center of Children in Dallas served as an acceptable sample for the creation of this dataset. The medical examiner selected 4 out of 50 patients based on the tumor samples' dissimilarities following clinical surgery. The images were labeled with the three most common kinds of cancer: necrosis, viable tumor, and non-tumor. The clarification was done by two clinical scientists. This dataset consists of 1144 1024 x 1024 images with ten times the resolution.

5 CONCLUSION & FUTURE SCOPE

According to the study's findings, Cancerous Tumor was identified by consuming deep learning-based replicas. Using transfer learning, Groups (DenseNet201), and (ResNet101) were linked with modified ANN layers on top of the (VGG16). The (ResNet101) model performed better in this study than the previously used image net models, with an accuracy rate of 90.36%. Future cancer identification may make use of additional replicas such as different available deep learning Models as they might be able to identify cancer more accurately.

REFERENCES

A, A. *et al.*, 2018. Feature extraction to detect Bone Cancer Using Image Processing. *Research Journal of Pharmaceutical, Biological and Chemical*, 8(3), pp. 434–442.

Avula, M., Lakkakula, N. P. & Raja, M. P., 2014. *Bone Cancer Detection from MRI Scan Imagery Using Mean Pixel Intensity*. Taipei, Taiwan, IEEE.

Bhukya, J. *et al.*, 2020. *SVM Model-based Computerized Bone Cancer Detection*. Coimbatore, India, IEEE.

Gawade, S., Bhansali, A., Patil, K. & Shaikh, D., 2023. Application of the convolutional neural networks and supervised. *elsevier*, 2023(3), pp. 2772–4425.

Hela, B., Hela, M. & Kamel, H., 2018. *Bone Cancer Diagnosis Using GGD Analysis*. Yasmine Hammamet, Tunisia, IEEE.

K, S. *et al.*, 2018. *Screening and Identify the Bone Cancer/Tumor using Image Processing*. Coimbatore, India, IEEE.

Kumar, S. & B, S., 2021. Bone cancer detection using feature extraction with classification using k-nearest neighbor and decision tree algorithm. *Smart Intelligent Computing and Communication Technology*, 4(10), pp. 347–353.

Papandrianos, N., Papageorgiou , E. & Anagnostis, A., 2020. Efficient bone metastasis diagnosis in bone scintigraphy using a fast convolutional neural network architecture. *Multidisciplinary Digital Publishing Institute*, 10(8), p. 532.

S, P. *et al.*, 2021. An approach to detect and classify bone tumour using fast and robust. *Annals of the Romanian Society for Cell Biology*, 25(6), p. 13736–13741.

Sharma, A. *et al.*, 2021. Bone cancer detection using feature extraction based machine. *Computational and Mathematical Methods in Medicine*, Volume 2021.

Shrivastava, D., Sanyal, S., Maji, A. K. & Kandar, D., 2020. Bone cancer detection using machine learning techniques. In: S. P. a. D. Bhatia, ed. *Smart Healthcare for Disease Diagnosis and Prevention*. India: Academic Press, pp. 175–183.

Shukla, A. and Patel, D., March 30, 2020. Bone cancer detection from x-ray and mri images through image segmentation techniques. *International Journal of Recent Technology and Engineering (IJRTE)*, 8 (6), pp. 273–278.

Next Generation Computing and Information Systems – Gupta. (Ed.)
© 2025 The Author(s), ISBN 978-1-032-73865-9

Optimized deep learning models based-sentiment analysis approach for identifying healthcare rumors over social media

Pramod Kumar
Computer Science and Technology, Ganga Institute of Technology and Management, Kablana, Jhajjar, Haryana, India

Ela Vashishtha
Lead Analytics, Texas Health Resources, Texas, USA

Faisal Yousef Alghayadh
Computer Science and Information Systems Department, College of Applied Sciences, AlMaarefa University, Riyadh, Saudi Arabia

Mohammed Wasim Bhatt
Model Institute of Engineering and Technology, Jammu, J&K, India

Suchitra Bala
ICT & Cognitive Systems, Sri Krishna Arts and Science College, India

Sreejitha T.S.
St. Marys College Thrissur, India

ABSTRACT: Research the mechanism of fake news refuting rumors in the real social network environment. Propose methods to evaluate the effect of rumor refuting and explore the factors that affect the impact of rumor refuting. Based on the current research results and assumptions, eight factors that affect the effect of rumor refuting are summed up, such as the proportion of the original fake news content. Experiments were conducted based on the rumor-refuting Weibo data. The experimental analysis showed that the proportion of the original fake news in the rumor-refuting information was negatively correlated with the rumor-refuting effect, and the explanation was positively correlated with the rumor-refuting effect suggestions for refuting rumors, including explaining why the original fake news is wrong to guiding social network fake news denying rumors.

1 INTRODUCTION

Necessary convenience, high dissemination and low social network threshold facilitated the widespread dissemination of fake news (Shrivastava *et al.* 2020). Consistent with trustworthy information, Fake news has the characteristics of fast dissemination, vast influence, and dissemination. The network topology and other features have caused significant social problems—negative impacts. Since the end of January 2020, fake news of pneumonia has spread widely on social networks for epidemic prevention workers. The orderly development has brought significant obstacles. Social network fake news (Matsumoto *et al.* 2021). Build a dissemination model for disinformation based on accurate data, but Most of these studies ignore the mechanism of refuting rumors and the mechanism of audience cognition. In psychology and sociology, the research on the tool of rumor refutation is more in-depth. (Yang 2021) analysis revealed the perceived reasons why fake news is difficult to refute. One of the factors is motivated reasoning; (Alkawaz *et al.* 2021) found a Targeted refutation of rumors that will negatively impact people; this effect Is called the backfire effect of refuting rumors. In addition,

DOI: 10.1201/9781003466383-36

(Ramezani *et al.* 2019), based on psychological theory, put forward an effective form of refuting stories and false news. The challenges are the lack of complete and usable research on the impact of rumors factor datasets. Most of the existing datasets are used for fake news detection. However, the measured data set does not contain fake news and its effects. Therefore, there is currently no effectiveness in dispelling rumors on social networks quantitative evaluation method of Reliable and practical advice based on analyzing large amounts of data.

2 RELATED WORK

The dissemination of fake news is mainly divided into the elimination of counterfeit news nodes, Competitive Cascading and Network Monitoring. Fake news node elimination refers to contacting contacts through rumor-refuting methods. Strategies for eliminating the adverse effects of fake news, such as Based on the greedy approach proposed by the Author. Method. In the simulated network model, the greedy algorithm chooses to refute rumors. The dissemination of trustworthy news and the cascading competition of false information reduce the Strategies for the negative impact of small fake news. (Syrovatkova and Pavlicek 2021) calculate the optimal strategy for the objective function that affects the blocking maximization, enhancing the Dissemination of rumors. (Assaf and Saheb 2021) proposed the first combination of robust A multi-stage intervention framework for chemical learning and point-process network activity models, Modeling the dissemination of fake news and external intervention behaviors, intervention strategies Slightly adjusted to the dynamics of artificial news dissemination, the Intervention activity effect as a reward function. Internet monitoring refers to using semi-automated social accounts or people as an intervention strategy to block fake news from suspicious sources. Net Network monitoring locations have the highest probability of cross-site transmission and are used at questionable sources is a site or partition with at most k users nearby (Garg and Jeevaraj 2021); another the solution for finding network monitoring locations is based on attackers and defenders Stackelberg game between (Ren *et al.* 2020). Combining manual and machine inspection methods can improve the robustness of fake news detection. When the machine misses some fake news, the human can detect and block its phone information; when human detection fails, machines can make up for these mistakes.

3 PROPOSED MODEL FOR DATA ACQUISITION

3.1 *Research question definition*

Fake news usually refers to false or inaccurate information, along with the development of social networks, fake news has gradually emerged in various forms. For example, rumors, fake news, headline parties, etc., as shown in Table 1.

3.2 *Social network data acquisition*

1) Collection of important fake news events. Due to the large number of Low-confidence information spread on social networks, from February 2020 from the 7th, Weibo launched the Weibo Little Secretary "Weibo Rumor Information Exchange". Total" function, which summarizes recent occurrences with significant impact, Powerful disinformation and debunking.
2) Weibo and user comment collection. Use the collected fake keyword of the news event matches the related Weibo and its comments. The blog data and comments formats are shown in Tables 2.
3) Publisher data collection. Subsequent data used to evaluate the influence of rumor-refuting Weibo publishers mainly include the number of followers, the number of followers, The number of mutual relations, the number of Weibo yesterday, the number of readings yesterday and the number of interactions yesterday etc.

Table 1. Data format of post.

Data Name	Describe	Type	Data Example
Weibo ID	Weibo's Unique & Unique Identifier	String	IvZBqpH55
Announcer	Username for Posting Weibo	Text	Weibo to Refuting Rumors
Weibo Text	Weibo Text Comment	Text	Slightly
Release Time	Weibo Release Time	Time	20221/8/30 20:09
Number of Like	Number of Like in Weibo	Integer	4325
Number Of Re-Tweets	Number of Tweets being Forwarded	Integer	1190
Number of Comments	Number of Weibo Comments	Integer	1667

Table 2. Data format of comment.

Data Name	Description	Type	Data Example
Userid	The user who posted the comment and unique identifier	string of numbers	2643842897
Release time	Comment posting Time	Time	20221/8/30 21:09
Comment	The Content of Comment	Text	Hunan Daily reported that have not
Likes	Comments Liked	Integers	3

4 EVALUATION OF THE EFFECT OF REFUTING RUMORS

Problem Definition Using social context to evaluate fake news Rumor-refuting effect, where social context includes likes, reposts, User's commenting behavior, etc. (Kaliyar *et al.* 2020) found that information Acceptance of likes, retweets, and positive comments on social media. Behavior has a significant positive effect, i.e., the user believes a piece of information is more likely to have social interaction behaviors. Use social the following can partly reflect people's trust in information degree, according to the number of likes, reposts, positive comments, invalid comments the number of users who have views to evaluate the rumor-refuting effect of a Weibo. The evaluation of the rumor-refuting effect is defined as the number of likes, retweets, and a weighting of the number of active, positive reviews.

$$E_i = \alpha L_i + \beta T_i + \gamma M_i P_i \tag{1}$$

According to the correlation between the degree of information acceptance and the behavior of liking, reposting, and expressing supportive opinions in (Alkawaz and Khan 2020), set $\alpha = 0.5$, $\beta = 0.4$, $\gamma = 0.2$.

Comment sentiment analysis the evaluation of the effect of refuting rumors needs to use the sentiment analysis results of the comments of the refuting posts. SnowNLP is used to perform sentiment analysis on the comments. The distribution of the training set used is as follows. The positive: negative is 5 497:4 503, the sentiment of the test set the distribution is 7:3. The corpus data format. The data needs to be preprocessed before model training. The preprocessing process is as follows. 1) Data cleaning: remove impurity information such as English, numbers, expressions, URLs, special characters, etc., but do not delete topic tags because it has relevant Semantic information that can assist in sentiment analysis. 2) Word segmentation: Decompose Chinese text into the smallest unit with semantic information – words; this article uses Jieba word segmentation, which has the advantages of high performance, high accuracy and scalability 3) Word vector model training: use the training set and test set to train the word vector model through python's genism package. Co-occurrence features and context features between the first two steps of data preprocessing. The next step

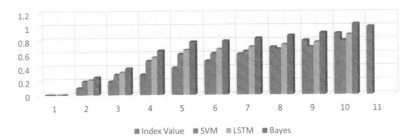

Figure 1. Comparison of model indicator.

Table 3. Examples of unrelated comments.

Comment User ID	Comments	Describe
6094950087	What do these slanderers think?	Not expressing views on fake news, It's just a critique of those who spread rumors
3896912225	Israel may already be on the way	Mentioned not related to current fake news
5133431256	An Iraqi in China is talking about IraqLack	Mentions, not current fake cancellations news, but other fake news

is model training. Based on the corpus and the pre-trained word2vec model, SnowNLP Bayesian classifier, SVM and LSTM are used to train the sentiment binary classification model. The index comparison of the training model is shown in Figure 1.

Before the official analysis of Weibo comments, since Weibo comments contain content that is not very relevant to the original Weibo content, it is necessary to clean up the comments that are not related to Weibo content and then conduct sentiment analysis (Yanagi *et al.* 2020). One thousand medical teams were dispatched to some irrelevant comments under Weibo of "Wuhan" Examples of unrelated comments are shown in Table 3.

Use word2vec to generate word vectors because compared with the classic vector space model, word2vec considers the semantic information of the context and maps high-dimensional information to low-dimensional details to solve the dimensional disaster problem (Kumar, Jagota and Shabaz 2021). The steps are as follows. 1) Corpus processing and word2vec model training: Perform data preprocessing on the obtained epidemic comment data set; 2) Text vector construction: Use the open-source python library Jieba to segment the text and weight the word vector to obtain the text vector v(d).

$$v(d) = \frac{v(t_1)w_1 + v(t_2)w_2 + v(t_3)w_3 + \cdots\cdots + v(t_n)w_n}{w_1 + w_2 + w_3 + \cdots\cdots\cdots + w_n} \qquad (2)$$

In the formula: t_i represents the word, w_i is the corresponding weight of the word, $v(t_i)$ is the word vector of t_i.

3) Similarity calculation: use cosine distance to calculate between vectors' similarity of:

$$sim(d_i, d_j) = \frac{\sum_{k=1}^{n} x_{ki} x_{kj}}{\sqrt{\sum_{k=1}^{n} x_{ki}^2} \sqrt{\sum_{k=1}^{n} x_{kj}^2}} \qquad (3)$$

Where: x_{ki} is a text vector $v(d_i)$ Bk dimension corresponding to the value. After cleaning irrelevant comments, use Bayesian SnowNLP for sentiment classification; the flowchart. Detailed steps are as follows. 1) Comment on impurity cleaning: remove English, numbers, website, and Impurity information such as special characters. 2) Remove irrelevant comments: select threshold, If the value is 0.15, the comments with a less than 0.15 will be deleted. 3) Tokenization & Remove stop words: Segment the reserved comments and use the stop word table with SnowNLP to remove stop words. 4) Bayesian sentiment score class: output sentiment score, score $\geq$ 0.6 is positive; otherwise, it is negative. The comment sentiment analysis is completed, and the rumor is completed according to formula (1). Effect evaluation.

5 EXPERIMENTAL SET UP AND RESULT ANALYSIS

5.1 *Experimental method setup*

To study the factors and refutation of presuppositions that may affect the effect of refuting rumors, SPSS software was used for correlation analysis and difference analysis. The data classification method of SPSS is as follows. Quantitative data: the size of the numbers has comparative significance, such as questionnaires the options of the survey are "dislike" = 0, "general" = 1, "like" = 2. According to the SPSS data classification method, rumors can be refuted. The factors and the effect of refuting rumors determine the data classification, as shown in Table 4. The normality test does not conform to a normal distribution (significance < 0.05); only nonparametric tests can be selected for all methods.

5.2 *Experimental results*

When the significance level is less than 0.05, and the different analysis results are shown in Table 4, it is considered that this factor has a different effect on the rumor refuting effect. Text warnings and picture warnings have no difference in refuting rumors." Whether it contains the accurate picture" and "whether the user is authenticated" are effective in dispelling rumors (Chaudhury *et al.* 2021; Chaitra *et al.* 2023). Only "whether to explain the reason" can refute rumors' differential impact if there is no impact. The results of the Spearman correlation analysis are shown in Table 5. When the correlation is less than 0.001, the correlation is very substantial, and the correlation coefficient indicates whether the correlation is Strong or weak; the closer the absolute value of the correlation coefficient is to 1, the stronger the correlation.

Table 4.   Results of difference analysis.

Factor	salience
Whether it contains rumor text warning	0.678
Whether to include rumor image warning	0.61
Whether to include the truth picture	0.58
Whether to explain the reason	0.02
Is it an authenticated user	0.425

Table 5.   Results of spearman correlation analysis.

Factors	Correlation coefficient	salience
The proportion of original fake news content	−0.321	0.002
Post word count	0.035	0.498
Source Influence	0.275	0

5.3 *Experimental discussion*

Analyze words of text warnings, picture warnings, and the frequency distribution of alerts containing pictures or text is shown in Table 6.

From Table 7, it can be found that 74% of the rumor-refuting Weibo contains text Word warnings or picture warnings, and at least a rumor-refuting microblog containing text warnings Bo accounted for 72%, indicating that the media attaches great importance to warnings through text to attract people's attention, but this has no significant effect on rumors enhancement (Mazhar *et al.* 2023; Tufail *et al.* 2022).

The influencing factor of "whether it contains the true picture" is because Dispelling rumors can leave people's previous mental models vacant. Adequate explanation fills the gap, and the graphical description is more than the text. The literal interpretation is valid. They are all screenshots of other rumor-refuting Weibo, blurred webpages. Pictures or screenshots of chats without a convincing explanation of the truth will significantly reduce rumors (Sharma *et al.* 2019). It is hoped that the relevant media will Add concise and clear facts to social network fake news. . Comparison of effectives between posts explaining or not is shown in Table 8.

The setting of the influencing factor "whether the user is authenticated" assumes that the credibility of the source of information will have an impact on the effect of rumor dispelling; however, the experiment in this paper denies this hypothesis, which may be due to the uneven distribution of the data. The irrelevance caused by the balance, that is, most Weibo publishers refute rumors Authenticate users. The distribution of "whether the user is authenticated" is shown in Figure 2. Experiments show that "whether to explain the reason" significantly affects rumors, as shown in Figure 3 for posts explaining why vs. The difference is in dispelling rumors of posts that do not explain the reason. E is the rumor refuting impact in the figure after taking the logarithm (Gupta *et al.* 2023). The set of rumor-refuting effects for unexplained reasons In [0, 1.4], the product is poor, and the rumors on Weibo explain the cause. The impact of refuting rumors is mainly concentrated in higher, explaining why it promotes the effects of refuting rumors, and more efforts are needed to deny rumors Domain, under the same proportion of original fake news content, the influence of the source.

Table 6. Frequency distributions of posts whether contains warnings of false information in text graphic or either format respectively.

variable name	Yes	No
Whether it contains rumor text warning	0.75	0.3
Whether to include rumor picture warning	0.32	0.72
Whether to include rumor warning	0.75	0.25

Table 7. Distribution of verified users.

Frequency	Yes	No
0	0	0
0.2	0.01	0.2
0.4	0.02	0.4
0.6	0.03	0.6
0.8	0.04	0.8
1	0.05	0.9

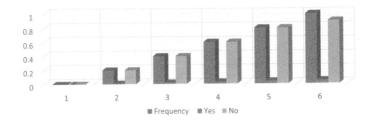

Figure 2. Distribution of verified users.

Table 8. Comparison of effectives between posts explaining or not.

distribution density	Reason not explained	explain the reason
0	0	0
0.1	1	1.2
0.2	0.45	0.54
0.3	0.2	0.24
0.4	0.1	0.12
0.5	0.04	0.048
0.6	0	0

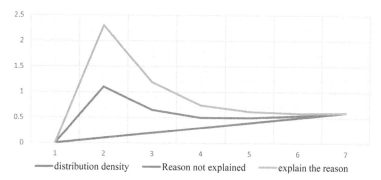

Figure 3. Comparison of effectives between posts explaining or not.

To explore the effect of timeliness on the development of refuting rumors, choose two things. The first fake news incident was "Academician Zhong Nanshan" It is recommended to gargle with salt water to prevent viruses", refuting rumors on Weibo about the incident. Next, the effect of denying rumors, the number of refuting rumors, and the influence of sources over time Changes are visualized, Finally, explain "why fake news is wrong", or fake news has spread the reason, instead of stating that the message is wrong (Dhiman *et al.* 2021; Soni and Kumar 2020).

6 CONCLUSION

This paper is based on a dataset of epidemic-related fake news on Weibo. Evaluate the effectiveness of rumor refutation and explore the factors that affect the point. Via Experimental Analysis Summarizes 6 Guidance for Social Network Disinformation Debunking Sexual suggestions, referring to the effective governance of fake news. The work of this paper has many shortcomings: the preset rumor-refuting influence factors can be further extended based on other psychological theories; also, more fine-grained research on the effect of rumor refuting in different types of fake elimination. The audience's previous perceptions will refute rumors Acceptance has a strong impact but assessing the audience's prior perceptions small challenge, so further research on the audience's previous perceptions to refute rumors, The impact of acceptance and methods to address the eradication of stubborn cognitive errors.

REFERENCES

Alkawaz, M.H. and Khan, S.A., 2020, February. Use of fake news and social media by main stream news channels of India. In *2020 16th IEEE International Colloquium on Signal Processing & Its Applications (CSPA)* (pp. 93–97). IEEE.

Alkawaz, M.H., Khan, S.A. and Abdullah, M.I., 2021, April. Plight of social media users: The problem of fake news on social media. In *2021 IEEE 11th IEEE Symposium on Computer Applications & Industrial Electronics (ISCAIE)* (pp. 289–293). IEEE.

Assaf, R. and Saheb, M., 2021, October. Dataset for arabic fake news. In *2021 IEEE 15th International Conference on Application of Information and Communication Technologies (AICT)* (pp. 1–4). IEEE.

Chaitra, H.V., Manjula, G., Shabaz, M., Martinez-Valencia, A.B., Vikhyath, K.B., Verma, S. and Arias-Gonzáles, J.L., 2023. Delay optimization and energy balancing algorithm for improving network lifetime in fixed wireless sensor networks. *Physical Communication*, 58: 102038.

Chaudhury, S., Shelke, N., Sau, K., Prasanalakshmi, B. and Shabaz, M., 2021. A novel approach to classifying breast cancer histopathology biopsy images using bilateral knowledge distillation and label smoothing regularization. *Computational and Mathematical Methods in Medicine*, 2021: 1–11.

Dhiman, G., Soni, M., Pandey, H.M., Slowik, A. and Kaur, H., 2021. A novel hybrid hypervolume indicator and reference vector adaptation strategies based evolutionary algorithm for many-objective optimization. *Engineering with Computers*, 37: 3017–3035.

Garg, R. and Jeevaraj, S., 2021, November. Effective fake news classifier and its applications to COVID-19. In *2021 IEEE Bombay Section Signature Conference (IBSSC)* (pp. 1–6). IEEE.

Gupta, S., Shabaz, M., Gupta, A., Alqahtani, A., Alsubai, S. and Ofori, I., 2023. Personal HealthCare of Things: A novel paradigm and futuristic approach. *CAAI Transactions on Intelligence Technology*.

Kaliyar, R.K., Kumar, P., Kumar, M., Narkhede, M., Namboodiri, S. and Mishra, S., 2020, October. DeepNet: an efficient neural network for fake news detection using news-user engagements. In *2020 5th International Conference on Computing, Communication and Security (ICCCS)* (pp. 1–6). IEEE.

Kumar, M.N., Jagota, V. and Shabaz, M., 2021. Retrospection of the optimization model for designing the power train of a formula student race car. *Scientific Programming*, 2021: 1–9.

Matsumoto, H., Yoshida, S. and Muneyasu, M., 2021, October. Propagation-based fake news detection using graph neural networks with transformer. In *2021 IEEE 10th Global Conference on Consumer Electronics (GCCE)* (pp. 19–20). IEEE.

Mazhar, T., Irfan, H.M., Khan, S., Haq, I., Ullah, I., Iqbal, M. and Hamam, H., 2023. Analysis of Cyber Security Attacks and Its Solutions for the Smart Grid Using Machine Learning and Blockchain Methods. *Future Internet*, 15(2): 83.

Ramezani, M., Rafiei, M., Omranpour, S. and Rabiee, H.R., 2019, August. News labeling as early as possible: Real or Fake?. In *Proceedings of the 2019 IEEE/ACM International Conference on Advances in Social Networks Analysis and Mining* (pp. 536–537).

Ren, Y., Wang, B., Zhang, J. and Chang, Y., 2020, November. Adversarial active learning based heterogeneous graph neural network for fake news detection. In *2020 IEEE International Conference on Data Mining (ICDM)* (pp. 452–461). IEEE.

Sharma, M., Singh, J. and Gupta, A., 2019, August. Intelligent resource discovery in inter-cloud using blockchain. In *2019 IEEE SmartWorld, Ubiquitous Intelligence & Computing, Advanced & Trusted Computing, Scalable Computing & Communications, Cloud & Big Data Computing, Internet of People and Smart City Innovation (SmartWorld/SCALCOM/UIC/ATC/CBDCom/IOP/SCI)* (pp. 1333–1338). IEEE.

Shrivastava, G., Kumar, P., Ojha, R.P., Srivastava, P.K., Mohan, S. and Srivastava, G., 2020. Defensive modeling of fake news through online social networks. *IEEE Transactions on Computational Social Systems*, 7(5): 1159–1167.

Soni, M. and Kumar, D., 2020, September. Wavelet based digital watermarking scheme for medical images. In *2020 12th international conference on computational intelligence and communication networks (CICN)* (pp. 403–407). IEEE.

Syrovatkova, J. and Pavlicek, A., 2021, October. Comparison of student news sharing in the Czech Republic and South Africa. In *2021 International Conference on Engineering Management of Communication and Technology (EMCTECH)* (pp. 1–4). IEEE.

Tufail, A.B., Ullah, I., Rehman, A.U., Khan, R.A., Khan, M.A., Ma, Y.K., Hussain Khokhar, N., Sadiq, M. T., Khan, R., Shafiq, M. and Eldin, E.T., 2022. On disharmony in batch normalization and dropout methods for early categorization of Alzheimer's disease. *Sustainability*, 14(22): 14695.

Yanagi, Y., Orihara, R., Sei, Y., Tahara, Y. and Ohsuga, A., 2020, July. Fake news detection with generated comments for news articles. In *2020 IEEE 24th International Conference on Intelligent Engineering Systems (INES)* (pp. 85–90). IEEE.

Yang, Y., 2021, December. COVID-19 fake news detection via graph neural networks in social media. In *2021 IEEE International Conference on Bioinformatics and Biomedicine (BIBM)* (pp. 3178–3180). IEEE.

Author index

Printed and bound by CPI Group (UK) Ltd, Croydon, CR0 4YY

17/10/2024

01775659-0007